Numerical Grid Generation

Foundations and Applications

Numerical Grid Generation

Foundations and Applications

Joe F. Thompson
Department of Aerospace Engineering, Mississippi State University

Z.U.A. Warsi
Department of Aerospace Engineering, Mississippi State University

C. Wayne Mastin
Department of Mathematics, Mississippi State University

North-Holland
New York • Amsterdam • Oxford

Elsevier Science Publishing Co., Inc.
52 Vanderbilt Avenue, New York, New York 10017

Sole distributors outside the United States and Canada:

Elsevier Science Publishers B.V.
P.O. Box 211, 1000 AE Amsterdam, The Netherlands

Library of Congress Cataloging in Publication Data

Thompson, Joe F.
Numerical grid generation.

Includes bibliographical references and index.
1. Numerical grid generation (Numerical analysis)
2. Differential equations, Partial—Numerical solutions.
I. Warsi, Z.U.A. II. Mastin, C. Wayne. III. Title.
QA377.T524 1985 519.4 85-4543
ISBN 0-444-00985-X

Manufactured in the United States of America

to a sense of place, Mississippi

PREFACE

Numerical grid generation has now become a fairly common tool for use in the numerical solution of partial differential equations on arbitrarily shaped regions. This is especially true in computational fluid dynamics, from whence has come much of the impetus for the development of this technique, but the procedures are equally applicable to all physical problems that involve field solutions. Numerically generated grids have provided the key to removing the problem of boundary shape from finite difference methods, and these grids also can serve for the construction of finite element meshes. With such grids all numerical algorithms, finite difference or finite element, are implemented on a square grid in a rectangular computational region regardless of the shape and configuration of the physical region. (Finite volume methods are effectively a type of conservative finite difference method on these grids.)

In this text, grid generation and the use thereof in numerical solutions of partial equations are both discussed. The intent was to provide the necessary basic information, from both the standpoint of mathematical background and from that of coding implementation, for numerical solutions of partial differential equations to be constructed on general regions. Since these numerical solutions are ultimately constructed on a square grid in a rectangular computational region, any solution algorithm that can treat equations with variable coefficients is basically applicable, and therefore discussion of specific algorithms is left to classical texts on the numerical solution of partial differential equations.

The area of numerical grid generation is relatively young in practice, although its roots in mathematics are old. This somewhat eclectic area involves the engineer's feel for physical behavior, the mathematician's understanding of functional behavior, and a lot of imagination, with perhaps a little help from Urania. The physics of the problem at hand must ultimately direct the grid points to congregate so that a functional relationship on these points can represent the physical solution with sufficient accuracy. The mathematics controls the points by sensing the gradients in the evolving physical solution, evaluating the accuracy of the discrete representation of that solution, communicating the needs of the physics to the points, and by providing mutual communication among the points as they respond to the physics.

Numerical grid generation can be thought of as a procedure for the orderly distribution of observers, or sampling stations, over a physical field in such a way that efficient communication among the observers is possible and that all physical phenomena on the entire continuous field may be represented with sufficient accuracy by this finite collection of observations. The structure of an intersecting net of families of coordinate lines allows the observers to be readily identified in relation to each other, and results in much more simple coding than would the use of a triangular structure or a random distribution of points. The grid generation system provides some influence of each observer on the others, so that if one moves to get into a better position for observation of the solution, its neighbors will follow to some extent in order to maintain smooth coverage of the field.

Another way to think of the grid is as the structure on which the numerical solution is built. As the design of the lightest structure requires consideration of the load distribution, so the most economical distribution of grid points requires that the grid be influenced by both the geometric configuration and by the physical solution being done thereon. In any case, since resources are limited in any numerical solution, it is the function of the numerical grid generation to make the best use of the number of points that are available, and thus to make the grid points an active part of the numerical solution.

This is a rapidly developing area, being now only about ten years old, and thus is still in search of new ideas. Therefore no book on the subject at this time could possibly be considered to be definitive. However, enough material has now accumulated in the literature, and enough basic concepts have emerged, that a fundamental text is now needed to meet the needs of the rapidly expanding circle of interest in the area. It is with the knowledge of both these needs and these limitations that this text has been written. Some of the techniques discussed will undoubtedly be superceded by better ideas, but the fundamental concepts should serve for understanding, and hopefully also for some inspiration, of new directions. The only background assumed of the student is a senior-level understanding of numerical analysis and partial differential equations. Concepts from differential geometry and tensor analysis are introduced and explained as needed.

Numerical grid generation draws on various areas of mathematics, and emphasis throughout is placed on the development of the relations involved, as well as on the techniques of application. This text is intended to provide the student with the understanding of both the mathematical background and the application techniques necessary to generate grids and to develop codes based on numerically generated grids for the numerical solution of partial differential equations on regions of arbitrary shape.

The writing of this text has been a cooperative effort over the last two years, spurred on by the institution of a graduate course in numerical grid generation, as well as an annual short course, at Mississippi State. The students in both of these courses have contributed significantly in revising the text as it evolved. The last appendix is the result of a class assignment prepared by Col. Hyun Jin Kim, graduate student in the computational fluid dynamics program, who also compiled the index. Our colleage, Dr. Helen V. McConnaughey of Mathematics contributed significantly through continual discussions and wrote most of Chapter IV.

We are indebted to a large number of former students and fellow researchers around the world for the development of the ideas that have crystallized into numerical grid generation. The complete debt can be acknowledged only through mention of the bibliographies contained in the several surveys cited herein. A list here would either be too long to note the strongest influences or too short to acknowledge all the significant ones. We must, however, acknowledge the many long and fruitful discussions with Peter Eiseman of Columbia University.

Of vital importance is the support that has been provided for the research from which the developments discussed in this book have emerged, including NASA; the research offices of the Air Force, Army, and Navy; the National Science Foundation, and various industrial concerns. The interest and contributions of a number of contract monitors has been essential over the years. We are especially appreciative of Bud Bobbitt and Jerry South of NASA Langley Research Center, who provided the initial support for an unknown with an idea.

Particular debts are owed to W. H. Chu for an idea in the Journal of Computational Physics in 1971, and to Frank Thames who put the idea into a dissertation.

In the preparation of the text we had the conscientious and untiring efforts of two most able secretaries, Rita Curry and Susan Triplett, who typed on in good spirits through a year of numerous revisions and frustrations as the text evolved.

Finally, we were particularly fortunate to have the services of Yeon Seok Chae, graduate student in the computational fluid dynamics program and illustrator par excellence, who did all the figures with understanding of the intended meaning as well as artistic competence. His meticulous efforts were extensions of our thoughts.

Joe F. Thompson
Z. U. A. Warsi
C. Wayne Mastin

Mississippi State, Mississippi
January 1985

TABLE OF CONTENTS

CHAPTER

I. INTRODUCTION

The numerical solution of partial differential equations requires some discretization of the field into a collection of points or elemental volumes (cells). The differential equations are approximated by a set of algebraic equations on this collection, and this system of algebraic equations is then solved to produce a set of discrete values which approximates the solution of the partial differential system over the field. The discretization of the field requires some organization for the solution thereon to be efficient, i.e., it must be possible to readily identify the points or cells neighboring the computation site. Furthermore, the discretization must conform to the boundaries of the region in such a way that boundary conditions can be accurately represented. This organization is provided by a coordinate system, and the need for alignment with the boundary is reflected in the routine choice of cartesian coordinates for rectagular regions, cylindrical coordinates for circular regions, etc., to the extent of the handbook's resources.

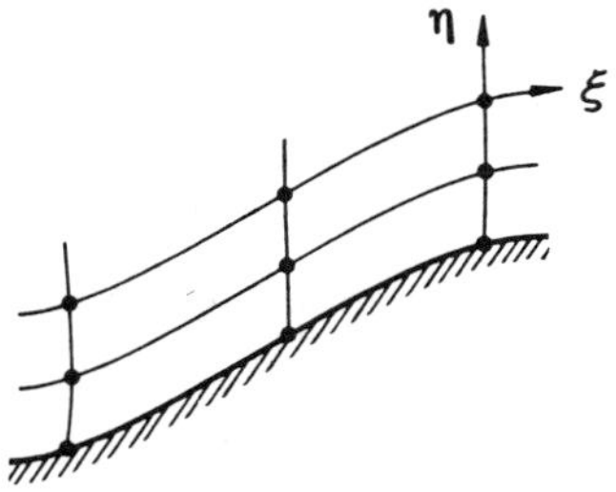

The current interest in numerically-generated, boundary-conforming coordinate systems arises from this need for organization of the discretization of the field for general regions, i.e., to provide computationally for arbitrary regions what is available in the handbook for simple regions. The curvilinear coordinate system covers the field and has coordinate lines (surfaces) coincident with all boundaries. The distribution of lines should be smooth, with concentration in regions of strong solution variation, and the system should ultimately be capable of sensing these variations and dynamically adjusting itself to resolve them.

A numerically-generated grid is understood here to be the organized set of points formed by the intersections of the lines of a boundary-conforming curvilinear coordinate system. The cardinal feature of such a system is that some coordinate line (surface in 3D) is coincident with each segment of the boundary of the physical region. The use of coordinate line intersections to define the grid points provides an organizational structure which allows all computation to be done on a fixed square grid when the partial differential equations of interest have been transformed so that the curvilinear coordinates replace the cartesian coordinates as the independent variables.

This grid frees the computational simulation from restriction to certain boundary shapes and allows general codes to be written in which the boundary shape is specified simply by input. The boundaries may also be in motion, either as specified externally or in response to the developing physical solution. Similarly, the coordinate system may adjust to follow variations developing in the evolving physical solution. In any case, the numerically-generated grid allows all computation to be done on a fixed square grid in the computational field which is always rectangular by construction.

In the sections which follow, various configurations

for the curvilinear coordinate system are discussed in Chapter II. In general, the computational field will be made rectangular, or composed of rectangular sub-regions, and a wide variety of configurations is possible. Coordinate systems may also be generated separately for sub-regions in the physical plane and patched together to form a complete system for complex configurations. The basic transformation relations applicable to the use of general curvilinear coordinate systems are developed in Chapter III; the construction of numerical solutions of partial differential equations on those systems is discussed in Chapter IV; and consideration is given in Chapter V to the evaluation and control of truncation error in the numerical representations.

Basically, the procedures for the generation of curvilinear coordinate systems are of two general types: (1) numerical solution of partial differential equations and (2) construction by algebraic interpolation. In the former, the partial differential system may be elliptic (Chapter VI), parabolic or hyperbolic (Chapter VII). Included in the elliptic systems are both the conformal (Chapter X), and the quasi-conformal mappings, the former being orthogonal. Orthogonal systems (Chapter IX) do not have to be conformal, and may be generated from hyperbolic systems as well as from elliptic systems. Some procedures designed to produce coordinates that are nearly orthogonal are also discussed. The algebraic procedures, discussed in Chapter VIII, include simple normalization of boundary curves, transfinite interpolation from boundary surfaces, the use of intermediate interpolating surfaces, and various other related techniques.

Coordinate systems that are orthogonal, or at least nearly orthogonal near the boundary, make the application of boundary conditions more straightforward. Although strict orthogonality is not necessary, and conditions involving normal derivatives can certainly be represented by difference expressions that combine one-sided differences along the line emerging from the boundary with central expressions along the boundary, the accuracy deteriorates if the departure from orthogonality is too large. It may also be more desirable in some cases not to involve adjacent boundary points strongly in the representation, e.g., on extrapolation boundaries. The implementation of algebraic turbulence models is more reliable with near-orthogonality at the boundary, since information on local boundary normals is usually required in such models. The formulation of boundary-layer equations is also much more straightforward and unambiguous in such systems. Similarly, algorithms based on the parabolic Navier-Stokes equations require that coordinate lines approximate the flow streamlines, and the lines normal thereto, especially near solid boundaries. It is thus better in general, other considerations being equal, for coordinate lines to be nearly normal to boundaries.

Finally, dynamically-adaptive grids are discussed in Chapter XI. These grids continually adapt during the course of the solution in order to follow developing gradients in the physical solution. This topic is at the frontier of numerical grid generation and may well prove to be one of its most important aspects.

The emphasis throughout is on grids formed by the intersections of coordinate lines of a curvilinear coordinate system, as opposed to the covering of a field with triangular elements or a random distribution of points. Neither of these latter collections of points is suitable for really efficient numerical solutions (although numerical representations can be constructed on each, of course) because of the cumbersome process of identification of neighbors of a point and the lack of banded structure in the matrices. Thus the subject of triangular mesh generators, per se, is not addressed here. (Obviously a triangular mesh can be produced by construction rectangular mesh diagonals.)

Considerable progress is being made toward the development of the techniques of numerical grid generation and toward casting them in forms that can be readily applied. A comprehensive survey of numerical grid generation procedures and applications thereof through 1981 was given by Thompson, Warsi, and Mastin in Ref. [1], and the conference proceedings published as Ref. [2] contains a number of expository papers on the area, as well as current results. Other collections of papers on the area have also appeared (Ref. [3] and [4]), and a later review through 1983 has been given by Thompson in Ref. [5]. Some other earlier surveys are noted in Ref. [1]. A later survey by Eiseman is given in Ref. [37]. The present text is meant to be a developmental treatment of the techniques of grid generation and its applications, not a survey of results, and therefore no attempt is made here to cite all related references, rather only those needed to illustrate particular points are noted. The surveys mentioned above should be consulted directly for references to examples of various applications and related contributions. (Ref [1] gives a short historical development of the ideas of grid generation.) Other surveys of particular areas of grid generation are cited later as topics are introduced.

Finally, in regard to implementation, a configuration for the transformed (computational) field is first established as discussed in Chapter II. The grid is generated from a generation system constructed as discussed in Chapters VI - X. (If the grid is to be adaptive, i.e., coupled with the physical solution done thereon, then the grid must be continually updated as discussed in Chapter XI.) In the construction of the grid, due account must be taken of the truncation error induced by the grid discussed in Chapter V. The partial differential equations of the physical problem of interest are transformed according to the relations given in Chapter III. These transformed equations are then discretized, cf. Chapter IV, and the resulting set of algebraic equations is solved on the fixed square grid in the rectangular transformed field.

II. BOUNDARY-CONFORMING COORDINATE SYSTEMS

1. Basic Concepts

To provide a familiar ground from which to view the general development to follow, consider first a two-dimensional cylindrical coordinate system covering the annular region between two concentric circles:

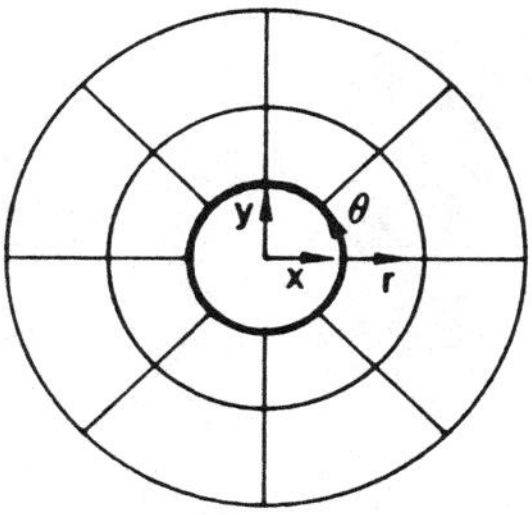

Here the curvilinear coordinates (r,θ) vary on the intervals $[r_1,r_2]$ and $[0,2\pi]$, respectively. These curvilinear coordinates are related to the cartesian coordinates (x,y) by the transformation equations

$$x(r,\theta) = r\cos\theta$$
$$y(r,\theta) = r\sin\theta \qquad (1)$$

The inverse transformation is given by

$$r(x,y) = \sqrt{x^2+y^2}$$
$$\theta(x,y) = \tan^{-1}\frac{y}{x} \qquad (2)$$

Note that one of the curvilinear coordinates, r, is constant on each of the physical boundaries, while the other coordinate, θ, varies monotonically over the same range around each of the boundaries. Note also that the system can be represented as a rectangle on which the two physical boundaries correspond to the top and bottom sides:

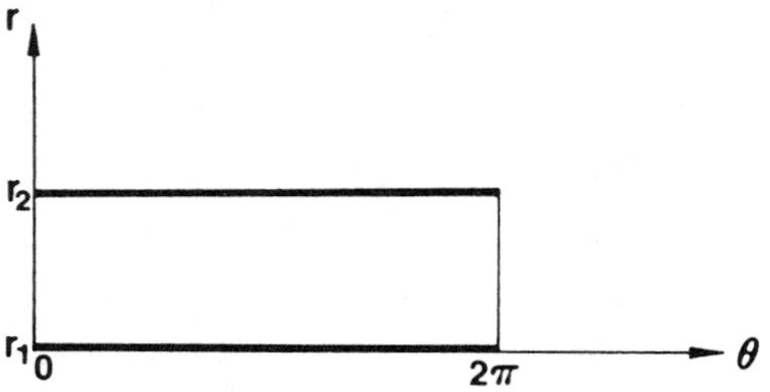

The transformed region, i.e., where the curvilinear coordinates, r and θ, are the independent variables, thus can be thought of as being rectangular, and can be treated as such from a coding standpoint. These points will be central to what follows.

The curvilinear coordinates (r,θ) can be normalized to the interval [0,1] by introducing the new curvilinear coordinates (ξ,η), where

$$\xi = \theta/2\pi, \qquad \eta = \frac{r - r_1}{r_2 - r_1} \tag{3}$$

or,

$$\theta(\xi) = 2\pi\xi, \; r(\eta) = r_1 + (r_2 - r_1)\eta \tag{4}$$

The transformation then may be written

$$x(\xi,\eta) = [r_1 + (r_2 - r_1)\eta]\cos(2\pi\xi) \tag{5a}$$

$$y(\xi,\eta) = [r_1 + (r_2 - r_1)\eta]\sin(2\pi\xi) \tag{5b}$$

where now ξ and η both vary on the interval [0,1]. This is thus a mapping of the annular region between the two circles in the physical space onto the unit square in the transformed space, i.e., each point (x,y) on the annulus corresponds to one, and only one, point (ξ,η) on the unit square:

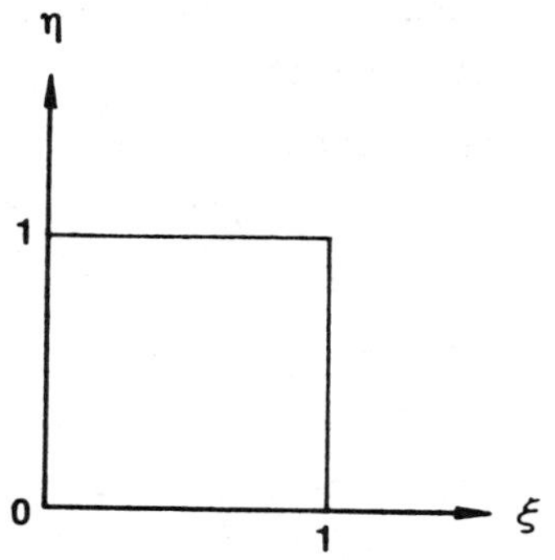

The bottom ($\eta = 0$) and top ($\eta = 1$) of the square correspond, respectively, to the inner and outer circles, $r = r_1$, and $r = r_2$. The sides of the square, $\xi = 0$ and $\xi = 1$, correspond to $\theta = 0$ and $\theta = 2\pi$, respectively, and hence to the two coincident sides of a branch cut in the physical space. Therefore, boundary conditions are not to be specified on these sides of the unit square in the transformed space. Rather these sides are to be considered re-entrant on each other with points adjacent to one, outside the square, being equivalent to points adjacent to the other, inside the square.

Conceptually, the physical region can be considered to have been opened at the cut($\theta = 0$ and 2π) and then deformed into a rectangle to form the transformed region:

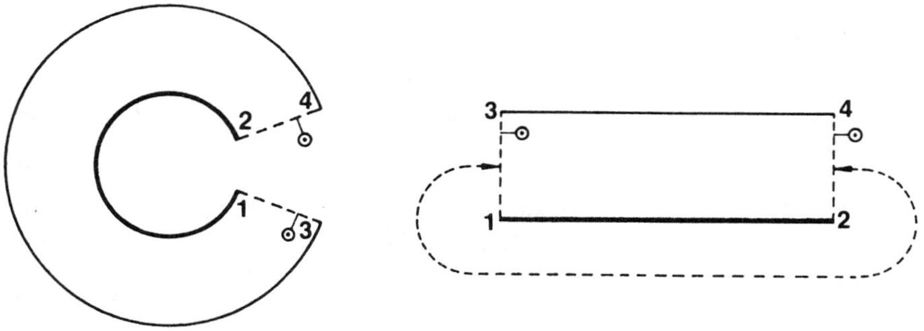

Here point correspondence across the re-entrant boundaries (indicated by the dashed connecting line) in the transformed region is illustrated by the coincidence of the pair of circled points. This conceptual device and mode of illustration for the point correspondence across re-entrant boundaries will serve later for more general configurations.

These simple concepts extend to more complicated two-dimensional configurations, the central feature being that one of the curvilinear coordinates is made to be constant on a boundary curve (as was r above), while the other varies monotonically along that boundary curve (as does θ). The transformation to the rectangle is achieved by making the range and direction of variation of the varying coordinate the same on each of two opposing boundaries (as θ varies from 0 to 2π on each circle above).

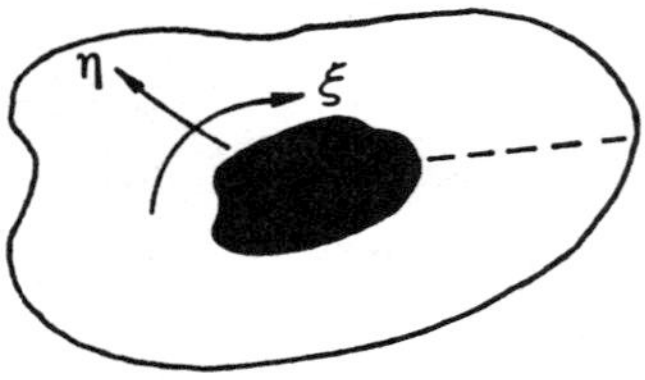

The physical space thus transforms to the rectangle shown above regardless of the shape of the physical region. (It is not necessary to normalize the curvilinear coordinates to the interval [0,1], and in fact, any normalization can be used. In computational applications the normalization is more conveniently done to different intervals for each coordinate. The field in the transformed space is then rectangular, rather than square.) Familiar examples of this are elliptical coordinates for the region between two confocal ellipses, spherical coordinates for two spheres, parabolic coordinates for two parabolas, etc.

These same concepts will be extended later to completely general configurations involving any number of boundary curves and branch cuts. The extension to three dimensions follows directly, using boundary surfaces instead of curves, i.e., one curvilinear coordinate will be made constant on a boundary surface, with the other two forming a two-dimensional coordinate system on the surface.

Returning to the concentric circles, if the functional dependence of θ on ξ, and/or that of r on η, had been made more general than the simple linear normalizations given by Eq. (4), the corresponding coordinate lines would have become unequally spaced in the physical space, while remaining as radial lines and concentric circles:

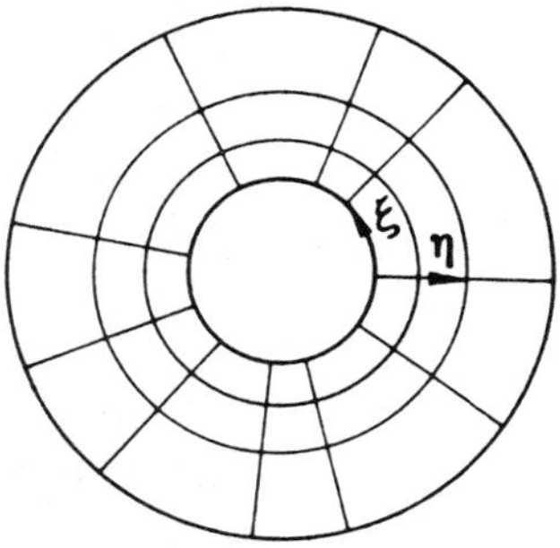

The transformation, from Eq. (1), is now given by

$$x(\xi,\eta) = r(\eta)\cos\theta(\xi) \tag{6a}$$

$$y(\xi,\eta) = r(\eta)\sin\theta(\xi) \tag{6b}$$

In this case the points on the inner and outer circular boundaries are not equally spaced around the circles in the physical space for equal increments of ξ, although they remain equally spaced on the top and bottom of the unit square in the transformed space by construction. The spacing around these circles is determined by the functional dependence of θ on ξ, and, since the points are located at equal increments of ξ by construction, this functional relationship is defined by the placement of these points around the circles. This point, that the coordinate system in the field is determined from the boundary point distribution, will be central to the discussion of grid generation to follow. The distribution of circumferential lines is controlled here by the functional relationship between r and η, which is not related to any boundary point distribution. Thus factors other than the boundary point distribution may be expected to be involved in grid generation, as well. That the point distribution on the boundaries may be controlled by direct placement of the points, while the coordinate line distribution in the field must be controlled by other means will also continue to appear in the developments to follow.

The one-dimensional functional relationship between θ and ξ in Eq. (6) requires that the relative distributions of boundary points around the inner and outer circles be the same. This restriction can be removed by making θ a function of η, as well as of ξ, while retaining the periodic nature of the dependence on ξ. In this case the coordinate lines of constant ξ will no longer be straight radial lines, although they will continue to connect corresponding points on the inner and outer circular boundaries. Similarly the circumferential coordinate lines (lines of constant η here) can be made to depart from circles by making r dependent on both ξ and η, but with the restriction that the dependence vanishes on the inner and outer circular boundaries (where $\eta = 0$ and $\eta = 1$, respectively, here).

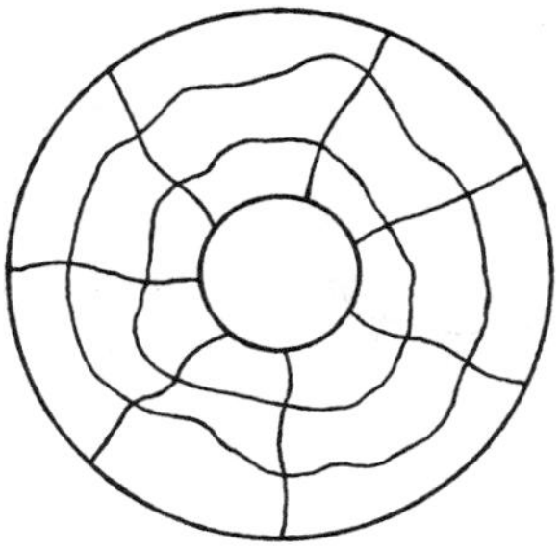

Obviously certain constraints will have to be placed on the functions $\theta(\xi,\eta)$ and $r(\xi,\eta)$ to keep the mapping one-to-one. All of these considerations will reappear in the general developments that follow.

Finally, it should be realized that the intermediate use here of the cylindrical coordinates (r,θ) in defining the transformation between the curvilinear coordinates (ξ,η) and the cartesian coordinates (x,y) has been only in deference to the familiarity of the cylindrical coordinates, and such intermediary coordinates will not appear in general. The generalized statement for the simple configuration under consideration here is as follows: Find $\xi(x,y)$ and $\eta(x,y)$ in the annular region bounded by the curves $x^2 + y^2 = r_1^2$ and $x^2 + y^2 = r_2^2$, subject to the boundary conditions

$$\eta = 0 \text{ on } x^2 + y^2 = r_1^2$$

$$\eta = 1 \text{ on } x^2 + y^2 = r_2^2$$

Specified monotonic variation of ξ over [0,1] on $x^2 + y^2 = r_1^2$ and on $x^2 + y^2 = r_2^2$ with same sense of direction on each of these two curves.

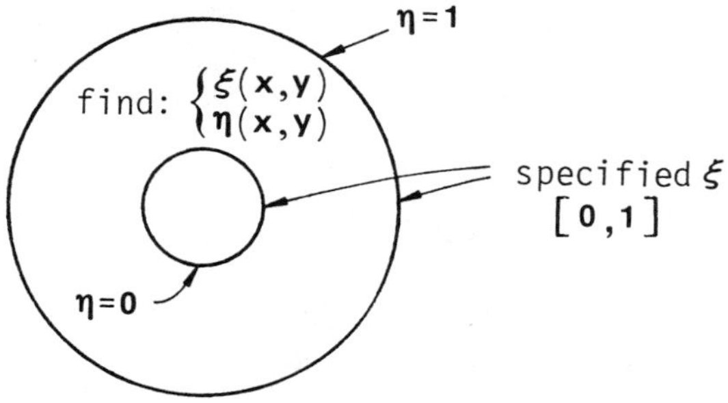

It is the inverse problem that will be treated in fact, however, i.e., find $x(\xi,\eta)$ and $y(\xi,\eta)$ on the unit square in the transformed space ($0 \leq \xi \leq 1$, $0 \leq \eta \leq 1$), subject to the boundary conditions

$$x(\xi,0) \text{ and } y(\xi,0) \text{ specified on } \eta = 0$$
$$\text{such that } x^2(\xi,0) + y^2(\xi,0) = r_1^2$$

$$x(\xi,1) \text{ and } y(\xi,1) \text{ specified on } \eta = 1$$
$$\text{such that } x^2(\xi,1) + y^2(\xi,1) = r_2^2$$

$$\text{Periodicity in } \xi\text{: } x(1 + \xi,\eta) = x(\xi,\eta)$$
$$y(1 + \xi,\eta) = y(\xi,\eta)$$

The simple form for the transformation given by Eq. (6) is made possible by choosing the same functional dependence of x and y on ξ on the boundaries, $\eta = 0$ and $\eta = 1$. The familiar cylindrical coordinate system is thus a special case of the general grid generation problem for this simple configuration applicable to the region between two concentric circles, as is the elliptical coordinate system for two ellipses, etc.

2. Generalization

Generalizing from the above consideration of cylindrical coordinates, the basic idea of a boundary-conforming curvilinear coordinate system is to have some coordinate line (in 2D, surface in 3D) coincident with each boundary segment, analogous to the way in which lines of constant radial coordinate coincide with circles in the cylindrical coordinate system. The other curvilinear coordinate, analogous to the angular coordinate in the cylindrical system, will vary along the boundary segment and clearly must do so monotonically, else the same pair of values of the curvilinear coordinates will occur at two different physical points. (It should be clear that the curvilinear coordinate that varies along a boundary segment must have the same direction and range of variation over some opposing segment, e.g., as the angular variable varies from 0 to 2π over both of two concentric circles in cylindrical coordinates).

With the values of the curvilinear coordinates thus specified on the boundary, it then remains to generate values of these coordinates in the field from these boundary values. There must, or course, be a unique correspondence between the cartesian (or other basis system) and the curvilinear coordinates, i.e., the mapping of the physical region onto the transformed region must be one-to-one, so that every point in the physical field corresponds to one, and only one, point in the transformed field, and vice versa. Coordinate lines of the same family must not cross, and lines of different families must not cross more than once.

In this chapter a two-dimensional region will be considered in most of the discussions in the interest of economy of presentation. Generalization to three dimensions will be evident in most cases and will be mentioned specifically only when necessary. As noted above, the curvilinear coordinates may be normalized to any intervals, just as the radial and angular coordinates of the cylindrical coordinate system can be expressed in many different units. Since the interest of the present discussion is numerical application, it will be generally convenient to define the increments of all the curvilinear coordinates to be uniformly unity, and then to normalize these coordinates to the interval $[1,N^{(i)}]$, where $N^{(i)}$ is the total number of grid points to be used in the ξ^i direction. (The three curvilinear coordinates will be indicated as ξ^i, $i = 1,2,3$, in general. In two dimensions, however, the notation ξ, η will often be used for the two coordinates ξ^1 and ξ^2.) The computational field, i.e., the field in the transformed space, thus will have rectangular boundaries and will be covered by a square grid. (It will become clear later that the actual values of the increments in the curvilinear coordinates are immaterial since they do not appear in the final numerical expressions. Therefore no generality is lost in making the grid square and of unit increment in the transformed field.)

A. Boundary-value Problem - Physical Region

The generation of the curvilinear coordinate system may be treated as follows: with the curvilinear coordinates specified on the boundaries, e.g., $\xi(x,y)$ and $\eta(x,y)$ on a boundary curve Γ (this specification amounting to a constant value for either ξ or η on each segment of Γ, with a specified monotonic variation of the other over the segment), generate the values, $\xi(x,y)$ and $\eta(x,y)$, in the field bounded by Γ. This is thus a boundary value problem on the physical field with the curvilinear coordinates (ξ,η) as the dependent variables and the cartesian coordinates (x,y) as the independent variables, with boundary conditions specified on curved boundaries:

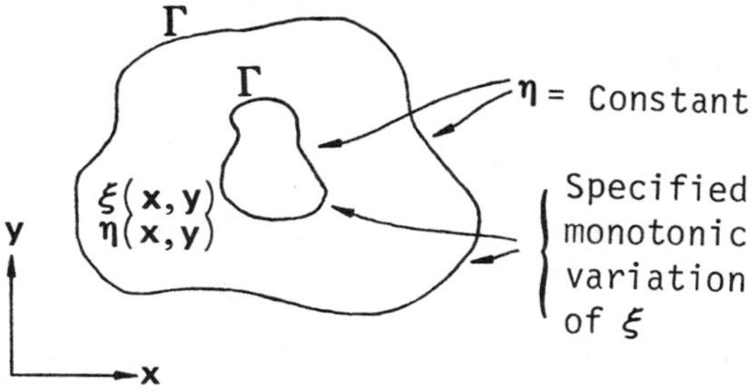

(In these discussions, the transformation is assumed to be from cartesian coordinates in the physical space. The transformation can, however, be from any system of coordinates in the physical space.)

B. Boundary value Problem - Transformed Region

The problem may be simplified for computation, however, by first transforming so that the physical cartesian coordinates (x,y) become the dependent variables, with the curvilinear coordinates (ξ,η) as the independent variables. Since a constant value of one curvilinear coordinate, with monotonic variation of the other, has been specified on each boundary segment, it follows that these boundary segments in the physical field will correspond to vertical or horizontal lines in the transformed field. Also, since the range of variation of the curvilinear coordinate varying along a boundary segment has been made the same over opposing segments, it follows that the transformed field will be composed of rectangular blocks.

The boundary value problem in the transformed field then involves generating the values of the physical cartesian coordinates, $x(\xi,\eta)$ and $y(\xi,\eta)$, in the transformed field from the specified boundary values of $x(\xi,\eta)$ and $y(\xi,\eta)$ on the rectangular boundary of the transformed field, the boundary being formed of segments of constant ξ or η, i.e., vertical or horizontal lines. With η = constant on a boundary segment, and the increments in ξ taken to be uniformly unity as discussed above, this boundary value specification is implemented numerically by distributing the points as desired along the boundary segment and then assigning the values of the cartesian coordinates of each successive point as boundary values at the equally spaced boundary points on the bottom (or top) of the transformed field in the following figure.

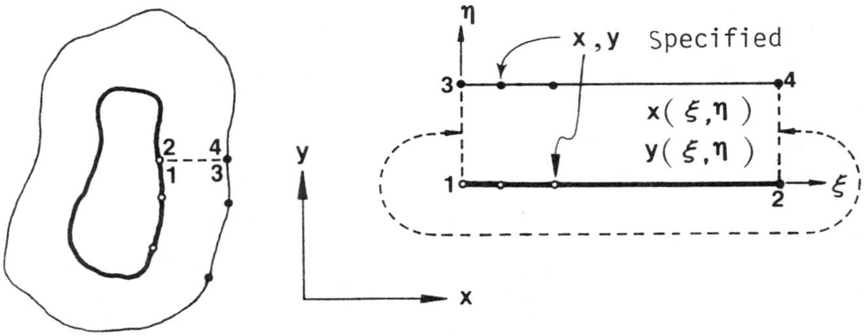

Boundary values are not specified on the left and right sides of the transformed field since these boundaries are re-entrant on each other (analogous to the 0 and 2π lines in the cylindrical system), as discussed above, and as indicated by the connecting dotted line on the figure. Points outside one of these re-entrant boundaries are coincident with points at the same distance inside the other. The problem is thus much more simple in the transformed field, since the boundaries there are all rectangular, and the computation in the transformed field thus is on a square grid regardless of the shape of the physical boundaries.

With values of the cartesian coordinates known in the field as functions of the curvilinear coordinates, the network of intersecting lines formed by contours (surfaces in 3D) on which a curvilinear coordinate is constant, i.e., the curvilinear coordinate system, provides the needed organization of the discretization with conformation to the physical boundary. It is also possible to specify intersection angles for the coordinate lines at the boundaries as well as the point locations.

3. Transformed Region Configurations

As noted above, the generation of the curvilinear coordinate system is done by devising a scheme for determination of the field values of the cartesian coordinates from specified values of these coordinates (and/or curvilinear coordinate line intersection angles) on portions of the boundary of the transformed region. Since the boundary of the transformed region is comprised of horizontal and vertical line segments, portions of which correspond to segments of the physical boundary on which a curvilinear coordinate is specified to be constant, it should be evident that the configuration of the resulting coordinate system depends on how the boundary correspondence is made, i.e., how the transformed region is configured.

Some examples of different configurations are given below, from which more complex configurations can be inferred. In these examples only a minimum number of coordinate lines are shown in the interest of clarity of presentation. In all of these examples, boundary values of the physical cartesian coordinates (and/or curvilinear coordinate line intersection angles) are understood to be specified on all boundaries, both external and internal, of the transformed region except for segments indicated by dotted lines. These latter segments correspond to branch cuts in the physical space, as is explained in the examples in which they appear. Such re-entrant boundary segments always occur in pairs, the members of which are indicated by the dashed connecting lines on each of the configurations shown. Points outside the field across one segment of such a pair are coincident with points inside the field across the other member of the pair. The conceptual device of opening the physical field at the cuts is used here to help clarify the correspondence between the physical and transformed fields. In many cases an example of an actual coordinate system is given as well. References to the use of various configurations may be found in the surveys given by Ref. [1] and [5], and a number of examples appear in Ref. [2].

A. Simply-connected Regions

It is natural to define the same curvilinear coordinate to be constant on each member of a pair of generally opposing boundary segments in the physical plane. Thus, a simply-connected region formed by four curves is logically treated by transforming to an empty rectangle:

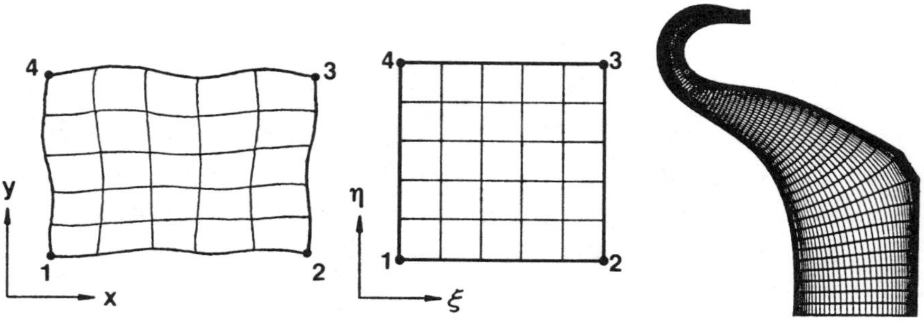

Similarly, an L-shaped region could remain L-shaped in the transformed region:

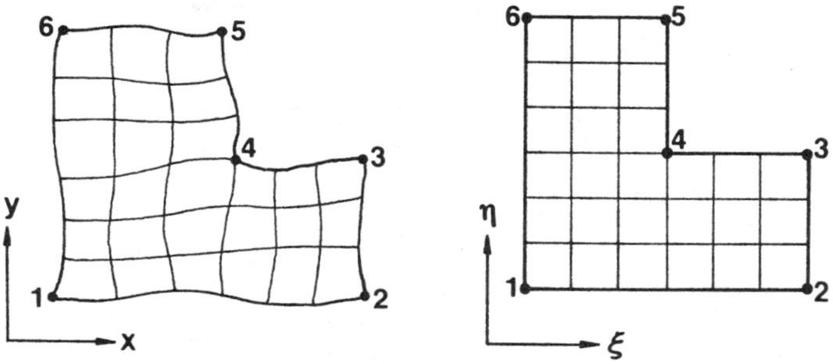

Here, for instance, the cartesian coordinates of the desired points on the physical boundary segment 4-5 are specified as boundary conditions on the vertical line 4-5, in corresponding order, which forms a portion of the boundary of the transformed region.

The generalization of these ideas to more complicated regions is obvious, the transformed region being composed of contiguous rectangular blocks. An example follows:

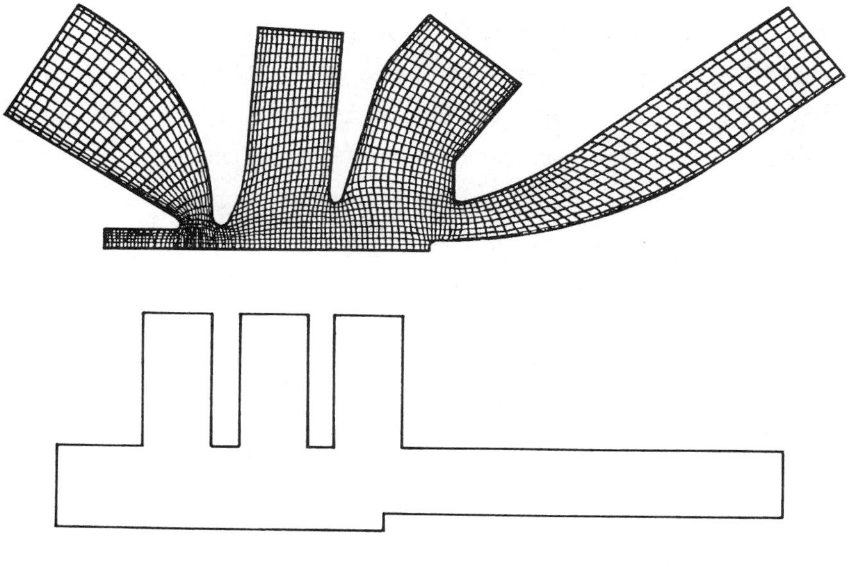

The physical boundary segment on which a single curvilinear coordinate is constant can have slope discontinuities, however, so that the L-shaped region above could have been considered to be composed of four segments instead of six, so that the transformed region becomes a simple rectangle:

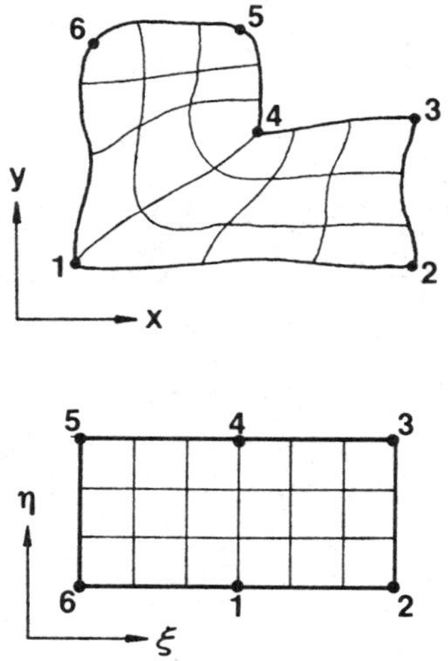

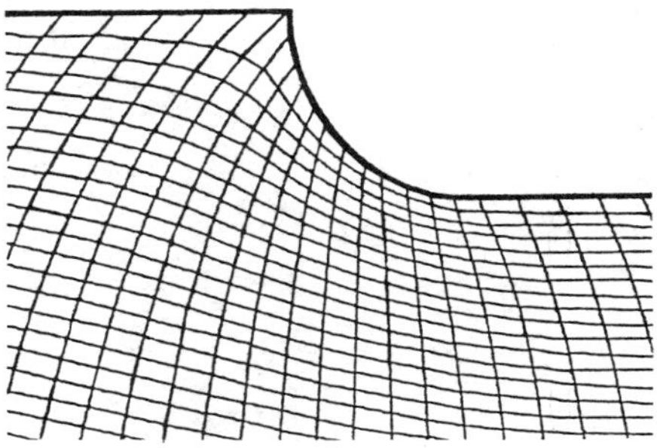

Here the cartesian coordinates of the desired points on the physical boundary 5-4-3 are the specified boundary values from left to right across the top of the transformed region. Whether or not the boundary slope discontinuity propagates into the field, so that the coordinate lines in the field exhibit a slope discontinuity as well, depends on how the coordinate system in the field is generated, as will be discussed later.

It is not necessary that corners on the boundary of the transformed region correspond to boundary slope discontinuities on the physical boundary and a counter-example follows next:

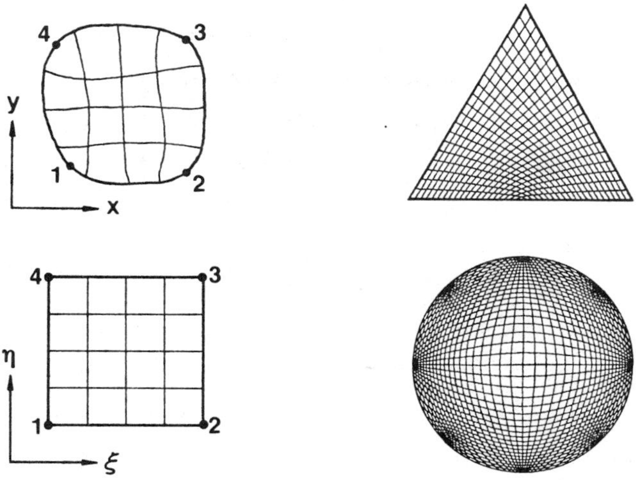

In this case, the segment 1-2 on the physical boundary is a line of constant η, while the segment 1-4 is a line of constant ξ. Thus at point 1 we have the following coordinate line configuration:

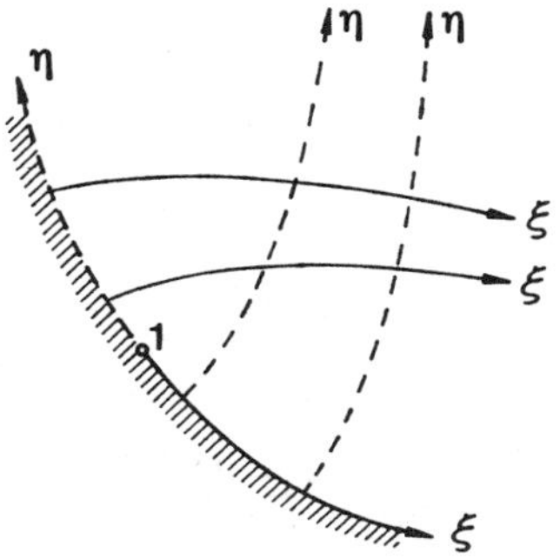

The lines through point 1 are as follows:

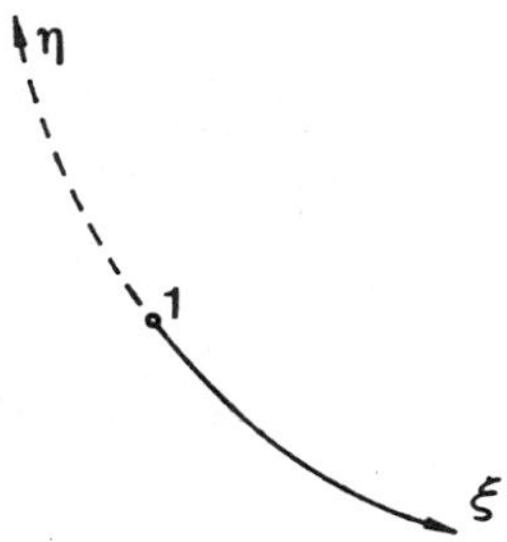

so that the angle between the two coordinate lines is π at point 1, and consequently the Jacobian of the transformation (the cell area, cf. Chapter III) will vanish at this point. The coordinate species thus changes on the physical boundary at point 1. (Difference representations at such special points as this, and others to appear in the following examples, are discussed in Chapter IV.) Since the species of curvilinear coordinate necessarily changes at a corner on the transformed region boundary, the identification of a concave corner on the transformed region boundary with a point on a smooth physical boundary will always result in a special point of the type illustrated here. (A point of slope discontinuity on the physical boundary also requires special treatment in difference solutions, since no normal can be defined thereon. This, however, is inherent in the nature of the physical boundary and is not related to the construction of the transformed configuration.)

Some slightly more complicated examples of the alternatives introduced above now follow:

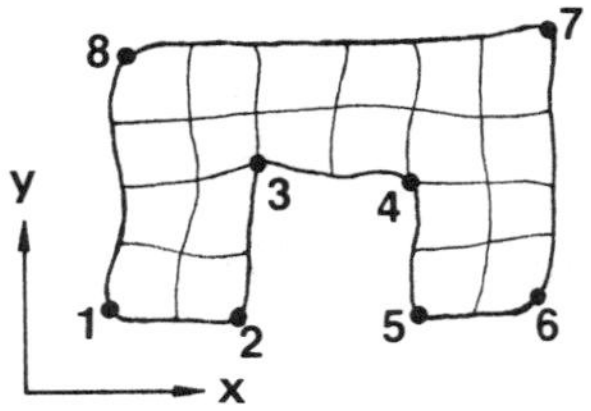

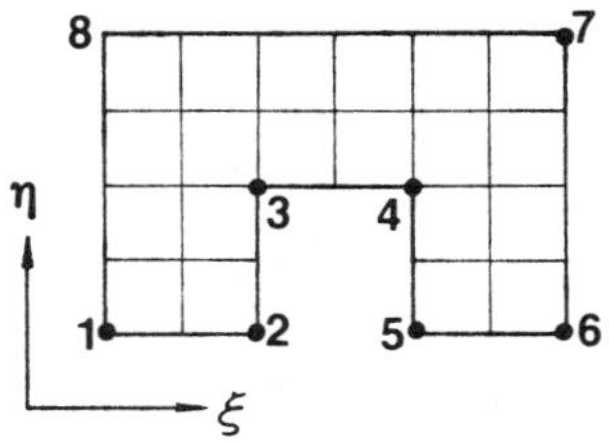

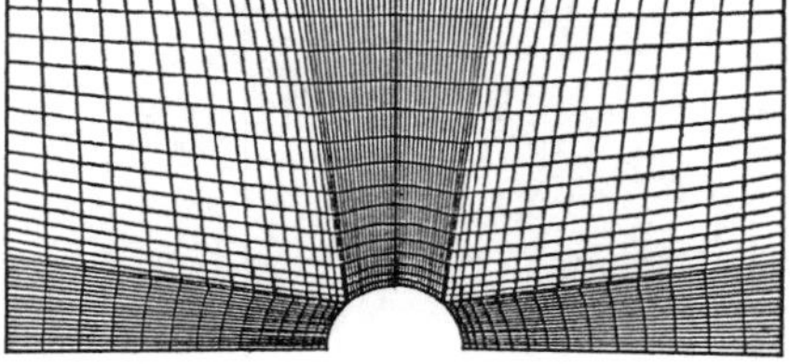

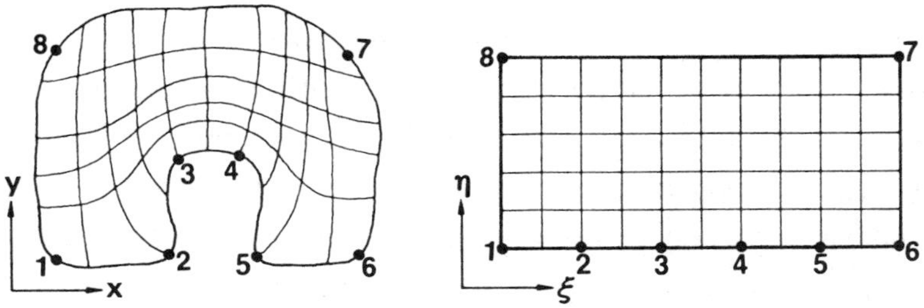

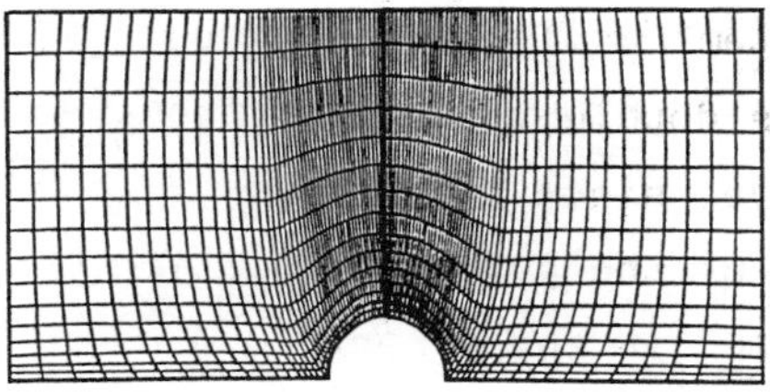

Still another alternative in this case would be to collapse the intrusion 2-3-4-5 to a slit in the transformed region:

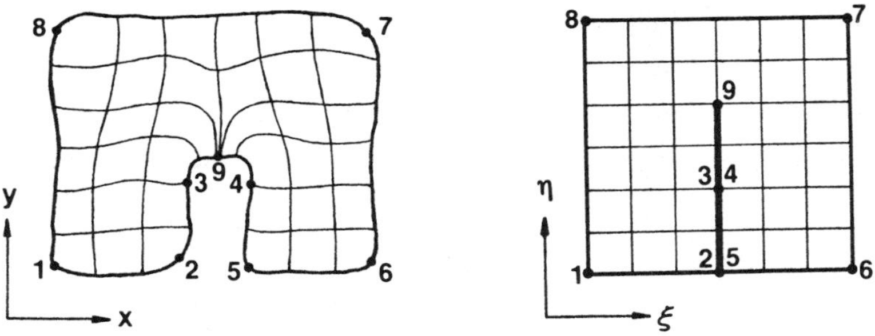

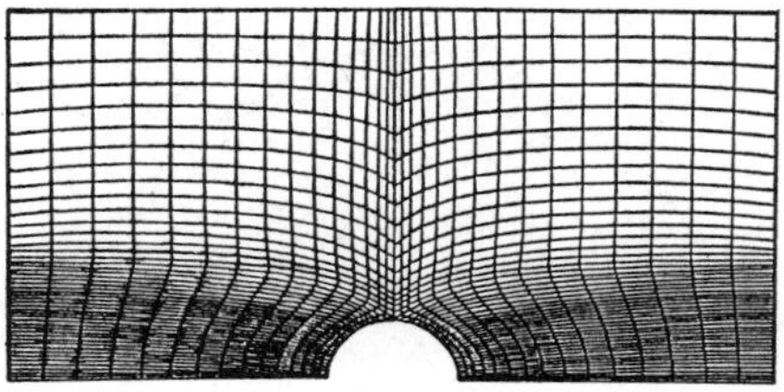

Here the physical cartesian coordinates are specified and are double-valued on the vertical slit, 2-9-5, in the transformed region. The cartesian coordinates of the desired points on the physical boundary 2-9 are to be used on the slit in the generation of the grid to the left of the slit in the transformed region, while those on the physical boundary 5-9 are used for generation to the right of the slit. Solution values in a numerical solution on such a coordinate system would also be double-valued on the slit, of course. This double-valuedness requires extra bookkeeping in the code, since two values of each of the cartesian coordinates and of the physical solution must be available at the same point in the transformed region so that difference representations to the left of the slit use the slit values appropriate to the left side, etc. Difference representations near slits are discussed in Chapter IV. With the composite grid structure discussed in Section 4, however, this need for double-valuedness, and the concomitant coding complexity, with the slit configuration can be avoided.

The point 9 here requires special treatment, since the coordinate line configuration there is as follows:

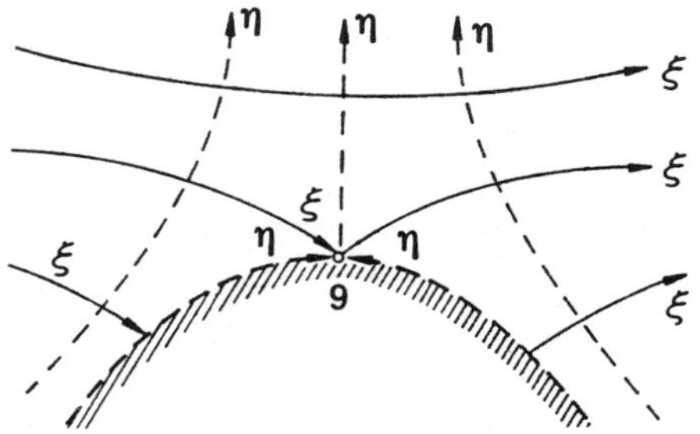

The coordinate lines through point 9 are as follows:

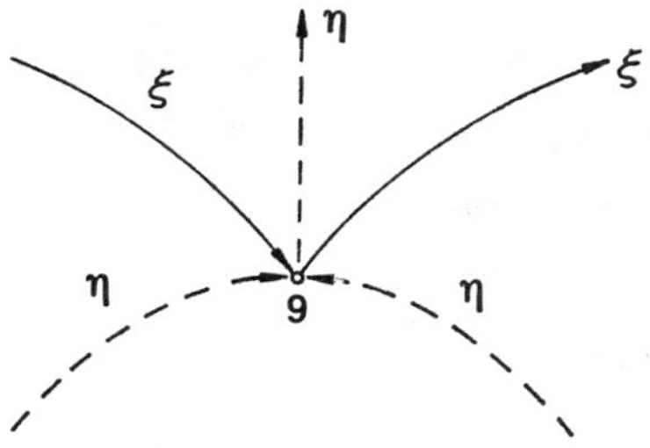

Here the slope of the coordinate line on which ξ varies is discontinuous at point 9, and the line on which η varies splits at this point. Such a special point will always occur at the slit ends with the slit configuration.

B. Multiply-connected Regions

With obstacles in the interior of the field, i.e., with interior boundaries, there are still more alternative configurations of the transformed region. One possibility is to maintain the connectivity of the transformed region the same as that of the physical region, as in the following examples showing two variations of this approach using interior slabs and slits, respectively, in the transformed region. The slab configuration is as follows:

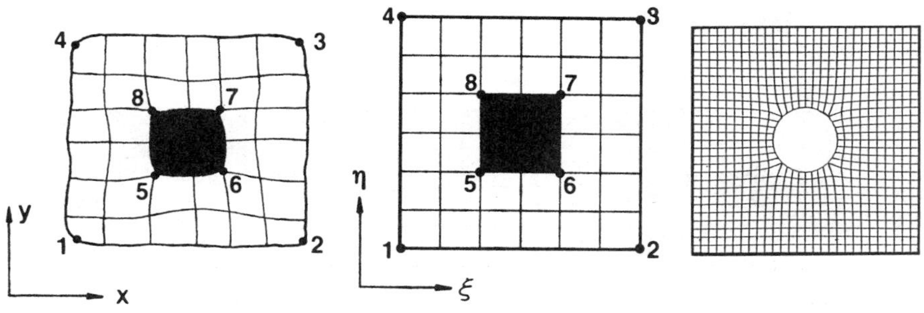

In coding, points inside the slab in the transformed region are simply skipped in all computations.

This configuration introduces a special point of the following form at each of the points corresponding to the slab corners in the transformed field:

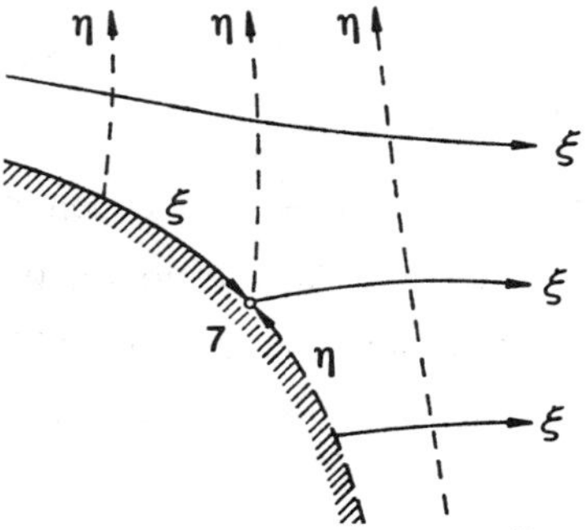

The coordinate lines through point 7 are shown below:

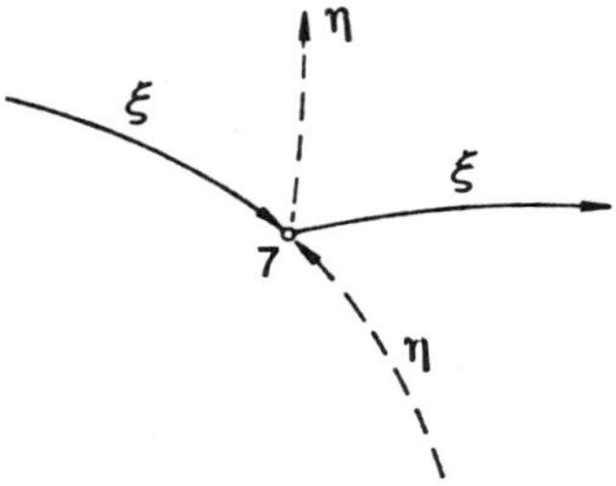

This type of special point, where the coordinate species changes on a smooth line, occurs when a convex corner in the transformed field is identified with a point on a smooth contour in the physical field. Both coordinate lines experience slope discontinuities at this point.

The slit configuration is as shown below:

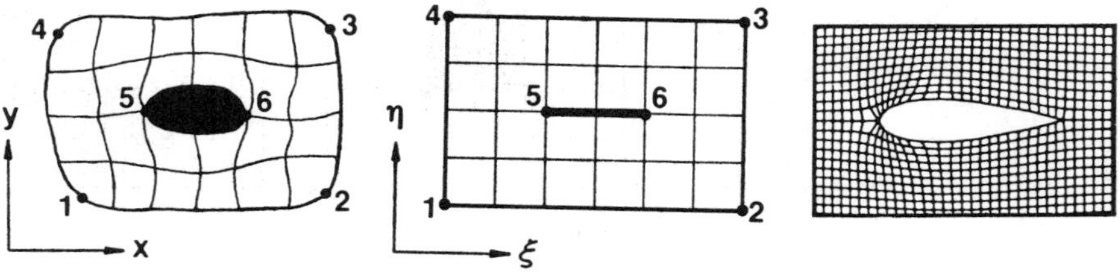

(An obvious varition would be to have the slit vertical.) In this slit configuration, the point 5 and 6 are special points of the form shown on p. 26 characteristic of the slit configuration, and will require special treatment in difference solutions.

The transformed region could, however, be made simply-connected by introducing a branch cut in the physical region as illustrated below:

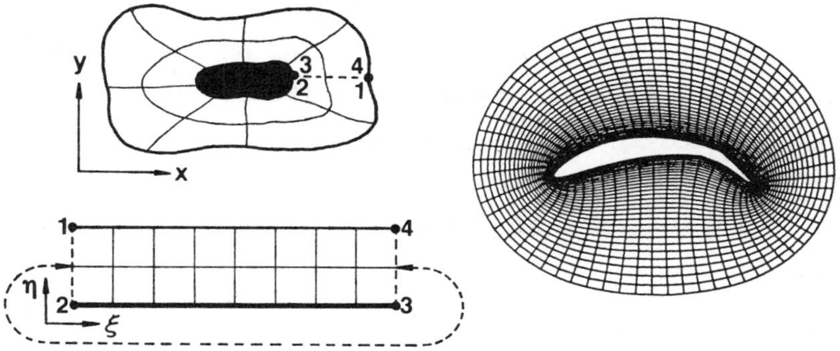

Conceptually this can be viewed as an opening of the field at the cut and then a deformation into a rectangle:

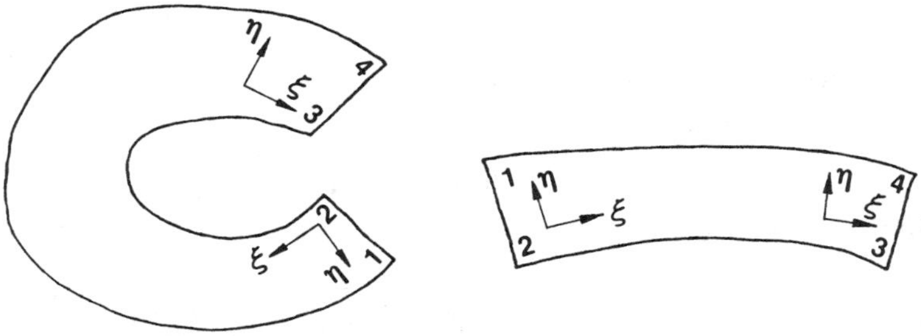

Here the coincident coordinate lines 1-2 and 4-3 form a branch cut, which becomes re-entrant boundaries on the left and right sides of the transformed region. All derivatives are continuous across this cut, and points at a horizontal distance outside the right-side boundary in the transformed region are the same as corresponding points at the same horizontal distance on the same horizontal line inside the left-side boundary, and vice versa. (In all discussions of point correspondence across cuts, "distance" means distance in the transformed region). In coding, the use of a layer of points outside each member of a pair of re-entrant boundaries in the transformed region holding values corresponding to the appropriate points inside the other boundary of the pair avoids the need for conditional choices in difference representations, as discussed in Section 6 of this chapter.

Boundary values are not specified on the cut. (This cut is, of course, analogous to the coincident 0 and 2π lines in the cylindrical coordinate system discussed above.) At the cut we have the following coordinate line configuration, as may be seen from the conceptional deformation to a rectangle:

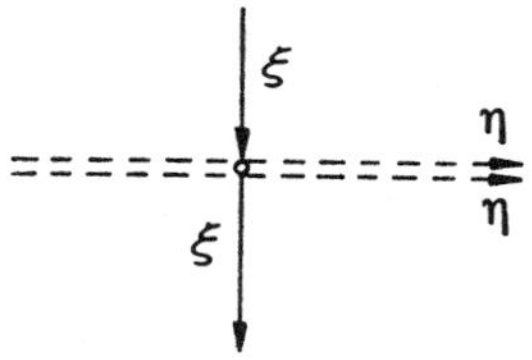

so that the coordinate species and directions are both continuous across the cut.

This type of configuration is often called an O-type. Another possible configuration is as shown below, often called a C-type:

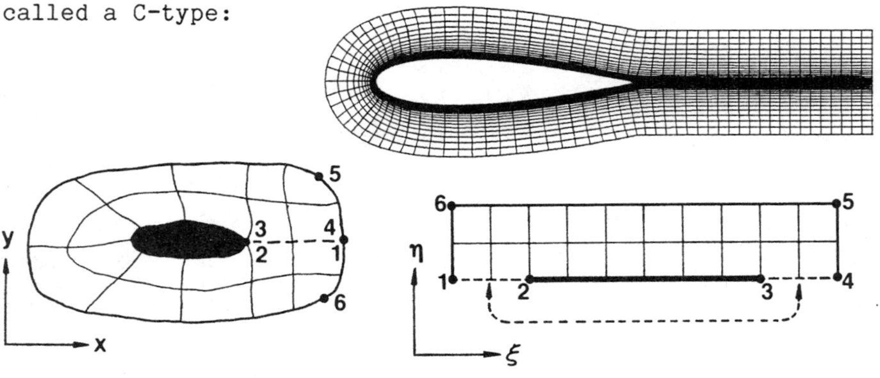

Opening the field at the cut we have, conceptually,

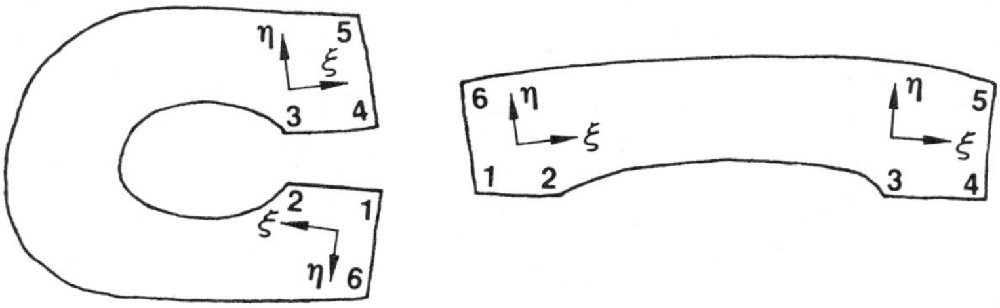

with 1-2-3-4 to flatten to the bottom of the rectangle. Here the two members of the pair of segments forming the branch cut are both on the same side of the transformed region, and consequently points located at a vertical distance below the segment 1-2, at a horizontal distance to the left of point 2, coincide with points at the same vertical distance above the segment 4-3, at the same horizontal distance to the right of point 3. The point 2(3) is a special point of the type shown on p. 26 for slit configurations.

The coordinate line configuration at the cut in this configuration is as follows:

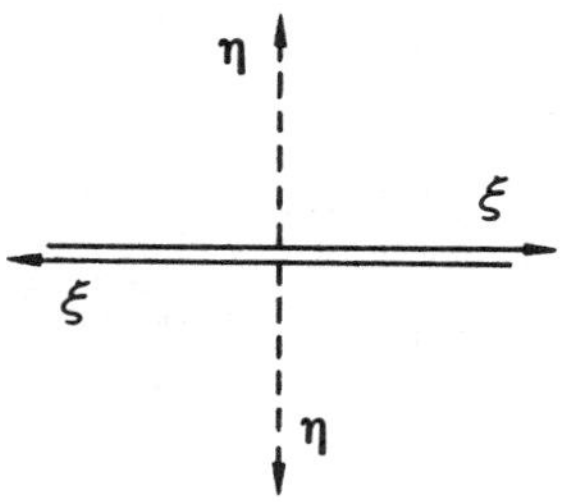

where it is indicated that ξ varies to the right on the upper side of the cut, but to the left on the lower side. The direction of variation of η also reverses at the cut, so that although the species and slope of both lines are continuous across the cut, the direction of variation reverses there.

It is possible to pass onto a different sheet across a branch cut, and discontinuities in coordinate line species and/or direction occur only when passage is made onto a different sheet. It is also possible, however, to remain on the same (overlapping) sheet as the cut is crossed, in which case the species and direction are continuous, and this must be the interpretation when derivatives are evaluated across the cut, as is discussed in Section 5 to follow. These concepts are illustrated in the following figure, corresponding to the C-type configuration given on p. 30:

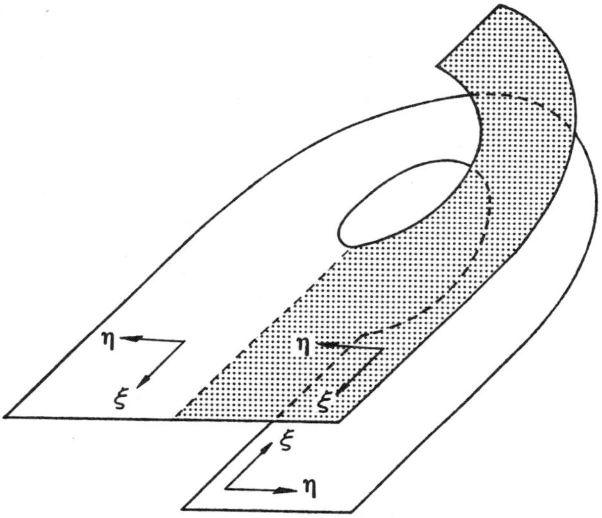

In the present discussion of configurations, the behavior of the coordinate lines across the cut will always be described in regard to the passage onto a different sheet, since this is in fact the case in codes. It is to be understood that complete continuity can always be maintained by conceptually remaining on the same sheet as the cut is crossed. Much of this complexity can, however, be avoided with the use of an extra layer of points surrounding the transformed region as will be discussed in Section 6.

Although in principle any region can be transformed into an empty rectangular block through the use of branch cuts, the resulting grid point distribution may not necessarily be reasonable in all of the region. Furthermore, an unreasonable amount of effort may be required to properly segment the boundary surfaces and to devise an appropriate point distribution thereon for such a transformation. Some configurations are better treated with a computational field that has slits or rectangular slabs in it.

Regions of higher connectivity than those shown above are treated in a similar manner. The level of connectivity may be maintained as in the following illustration:

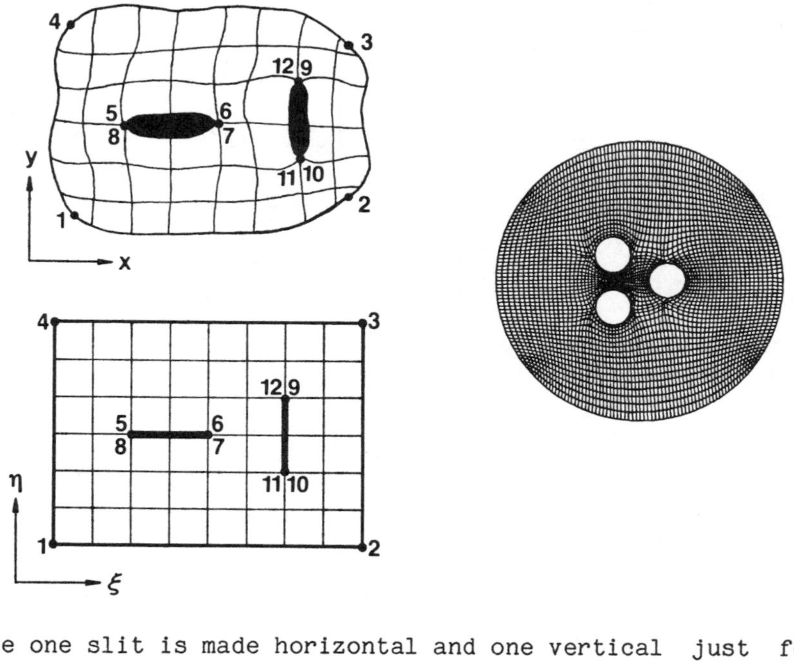

Here one slit is made horizontal and one vertical just for generality of illustration. Both could, of course, be of the same orientation. Slabs, rather than slits, could also have been used. The example has three bodies.

With the transformed region made simply-connected we have, using two branch cuts, a configuration related to the O-type shown above for one internal boundary:

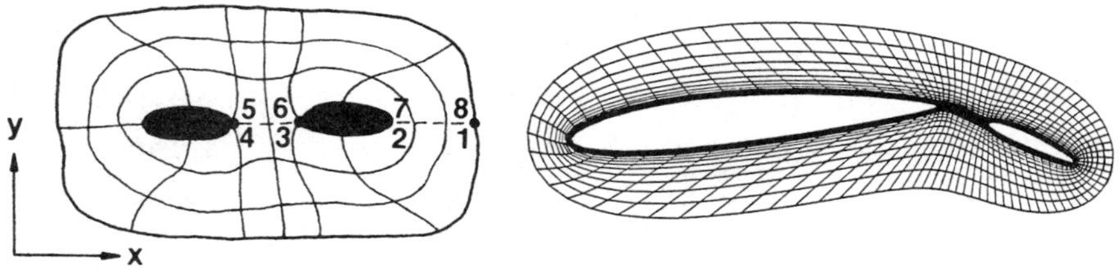

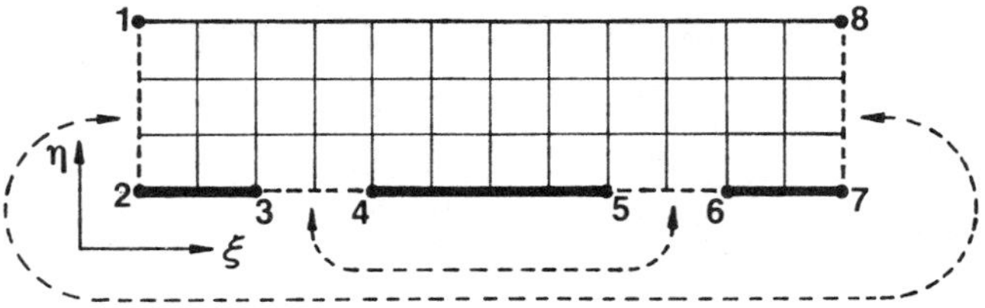

The conceptual opening here is as follows:

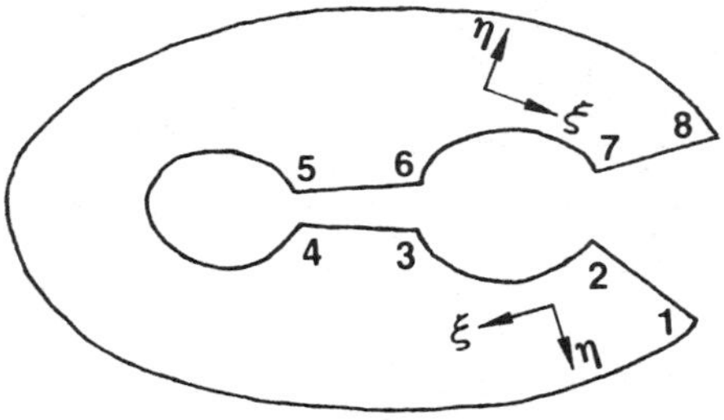

with segment 2-3-4-5-6-7 opening to the bottom. Here the pairs of segments (1-2,8-7) and (3-4,6-5) are the branch cuts, which form re-entrant boundaries in the transformed region as shown. In this case, points outside the right side of the transformed region coincide with points inside the left side, and vice versa. This cut is of the form described on p. 30, where both the coordinate species and direction are continuous across the cut. Points below the bottom segment 3-4, to the left of point 4, coincide with points above the bottom segment 6-5 to the right of point 5. This cut is of the form discussed on p. 31, for which the coordinate species is continuous across the cut but the direction changes there. There are a number of other possibilities for placement of the two cuts on the boundary of the transformed region, of course, some examples of which follow.

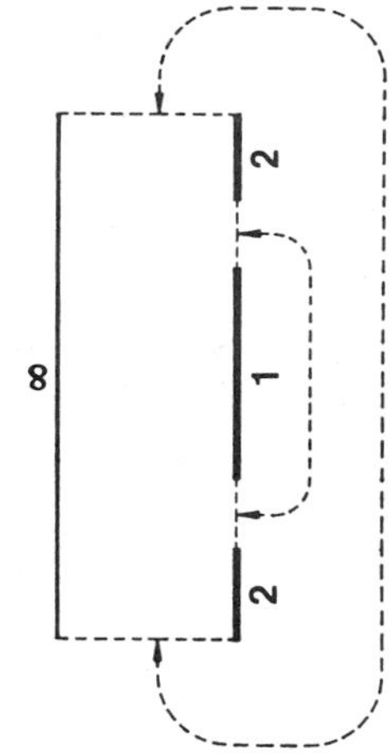

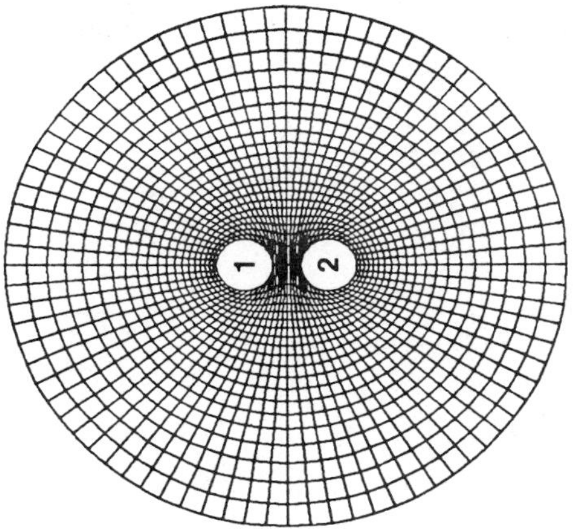

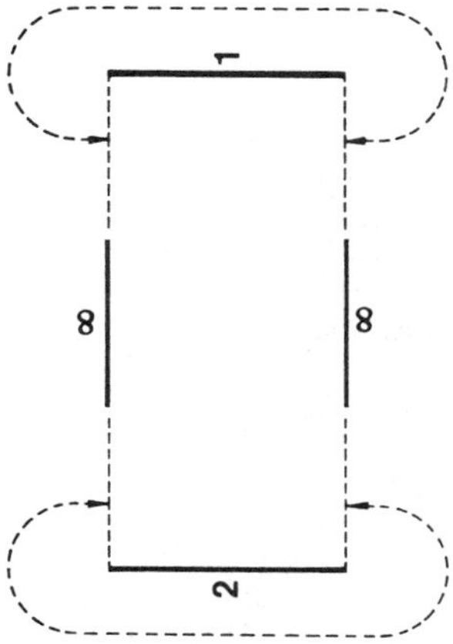

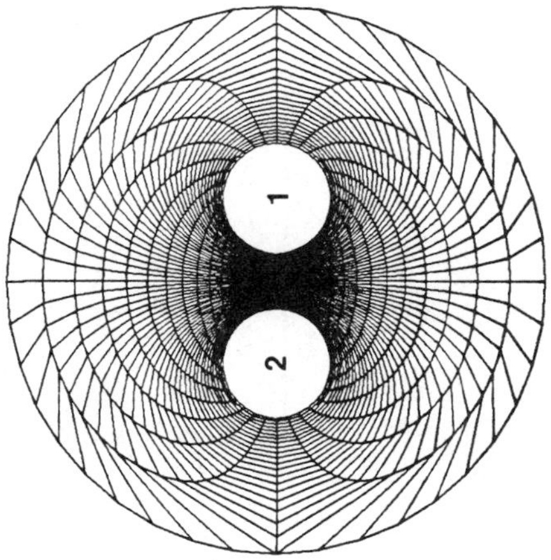

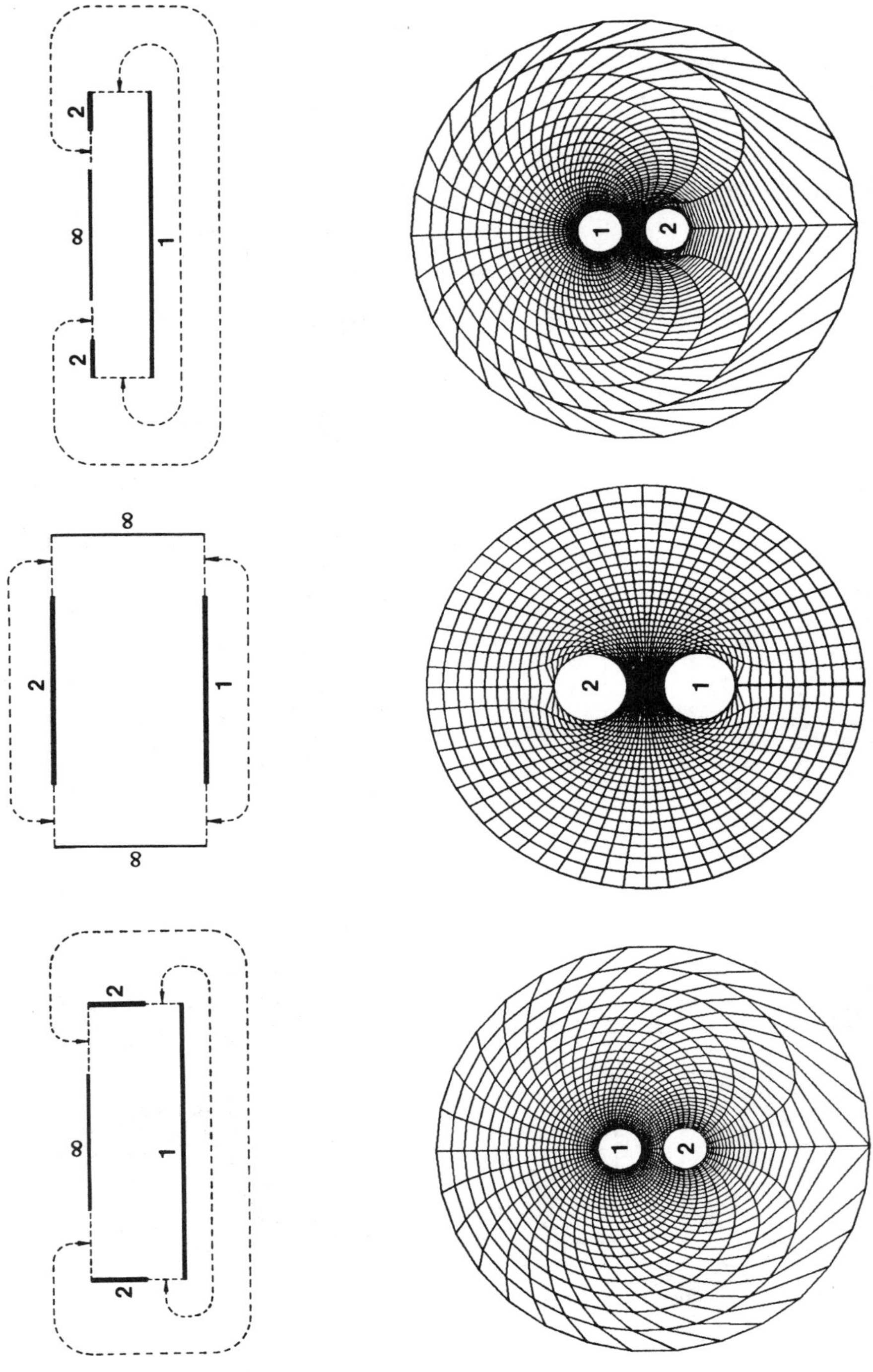
2
∞
2
1
∞
2
1
∞
2
∞
2
1
1
2
2
1
1
2

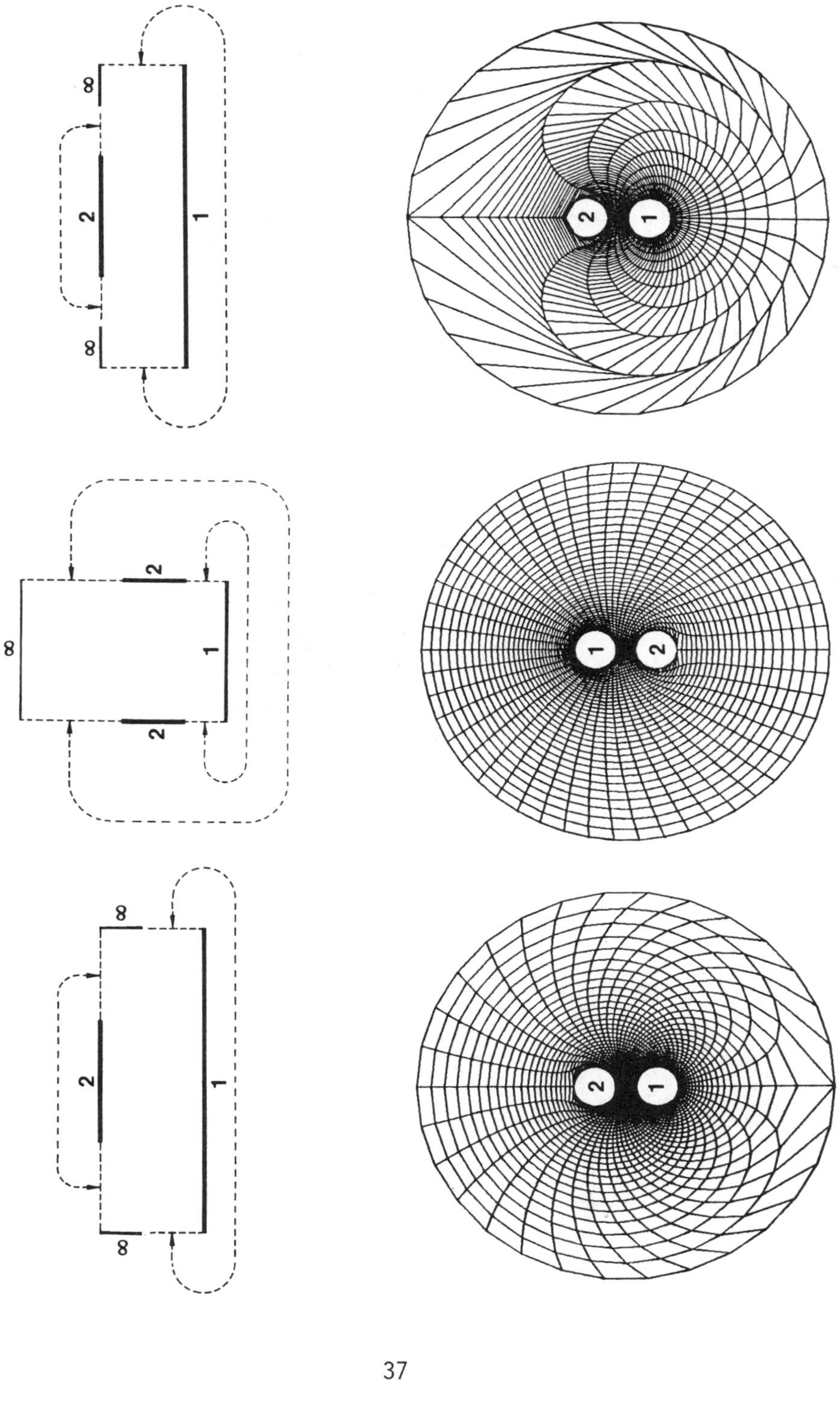
∞
2
1

It is not necessary to reduce the connectivity of the region completely; rather, a slit or slab can be used for some of the interior boundaries, while others are placed on the exterior boundary of the transformed region.

One other possibility in two dimensions is the use of a preliminary analytical transformation of infinity to a point inside some interior boundary, with the coodinates resulting therefrom replacing the cartesian coordinates in the physical region. The grid generation then operates from these transformed coordinates rather than from the cartesian coordinates. This typically gives a fine grid near the bodies, but may give excessively large spacing away from the body.

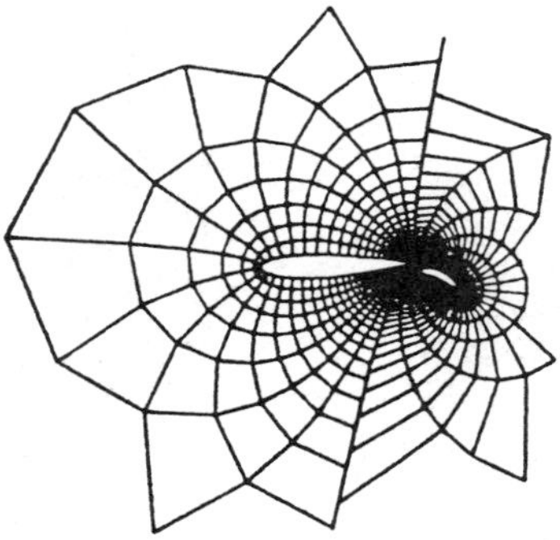

Thus, for example, if points on the two physical boundaries shown below

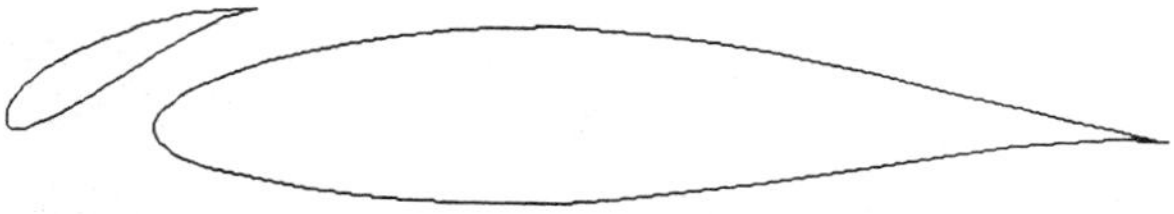

are transformed according to the complex transformation

$$z' = 1/z$$

where z = x+iy and z' = x'+iy', infinity in the x,y system will transform to the origin in the x', y' system, as shown below.

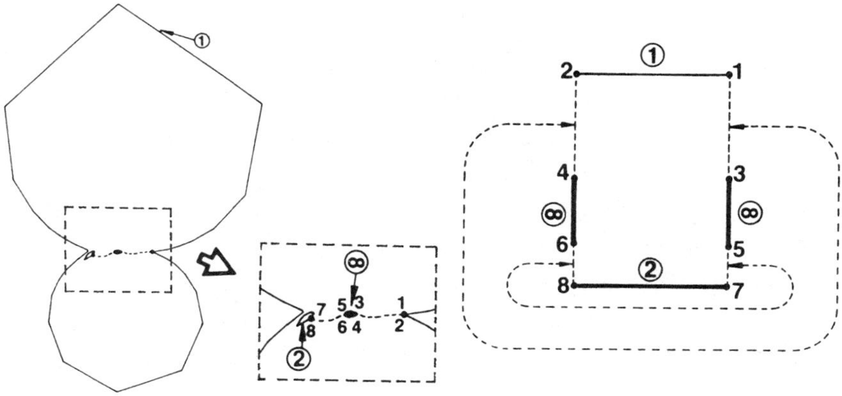

Then with the grid generated numerically from the x', y' system the following configuration results:

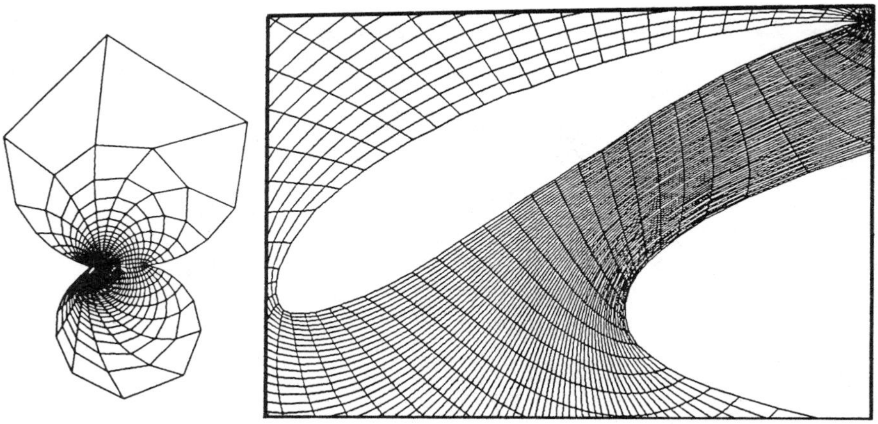

References to the use of this approach are made in the survey of Ref. [1]. Somewhat related to this are various two-dimensional configurations which arise directly from conformal mapping, cf. Ref. [6] and the survey of Ives on this subject, Ref. [7]. (Conformal mapping is discussed in Chapter X.)

C. Embedded Regions

In more complicated configurations, one type of coordinate system can be embedded in another. A simple example of this is shown below, where an O-type system surrounding an internal boundary is embedded in a system of a more rectangular form, using what amounts to a slit configuration.

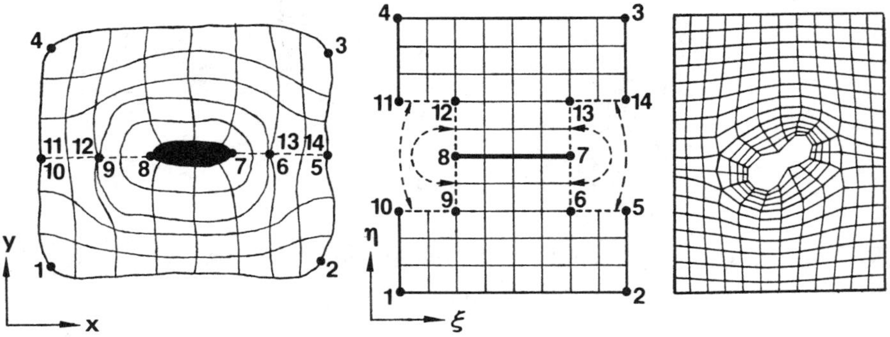

The conceptual opening of this system is best understood in stages: First considering only the embedded O-type system surrounding the interior boundary, we have the region inside the contour 12-13-6-9 opening as follows:

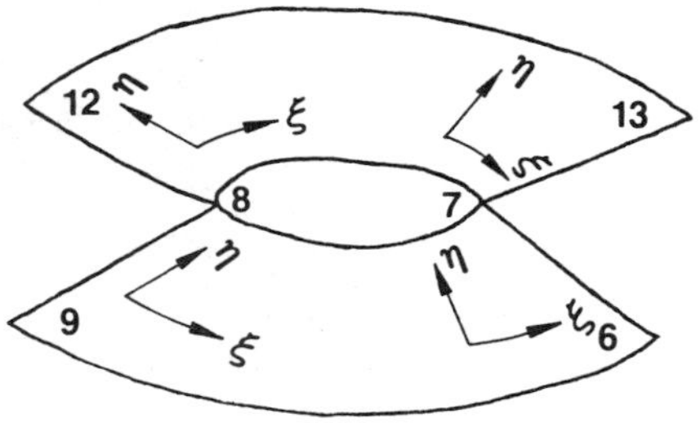

This then opens to the rectangular central portion of the transformed region shown above, with the inner boundary contour 8-7-8 collapsing to a slit. The rest of the physical region then opens as shown below:

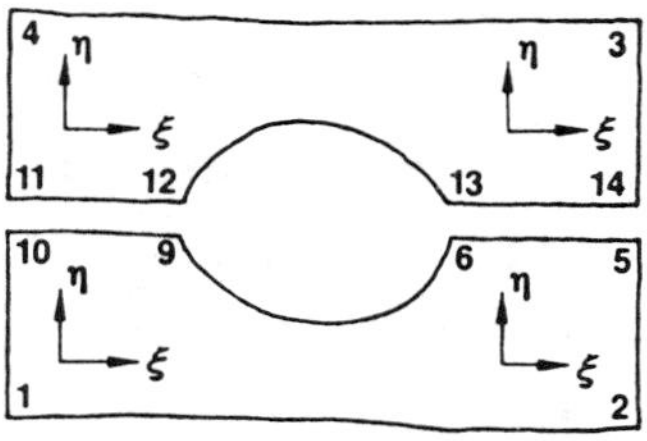

These two regions then deform to rectangles and are fitted to the top and bottom of the rectangle corresponding to the inner system along the contours 12-13 and 9-6 as shown.

Here points at a vertical distance below the segment 11-12 are coincident with points at the same vertical distance below the segment 10-9 on the same vertical line, and vice versa, with similar correspondence for the pair of segments 13-14 and 6-5. Points at a horizontal distance to the left of the segment 8-12, at a vertical distance above point 8, coincide with points at the same horizontal distance to the right of the segment 8-9, at the same vertical distance below point 8. Similar correspondence holds for the pair 7-13 and 7-6. Boundary values are specified on the slit 8-7.

The composite system shown on p. 40 can also be represented as a slit configuration in the transformed region:

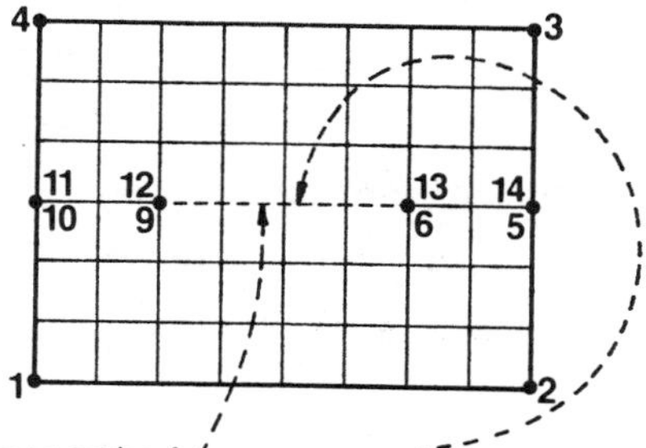

with the inner system represented as

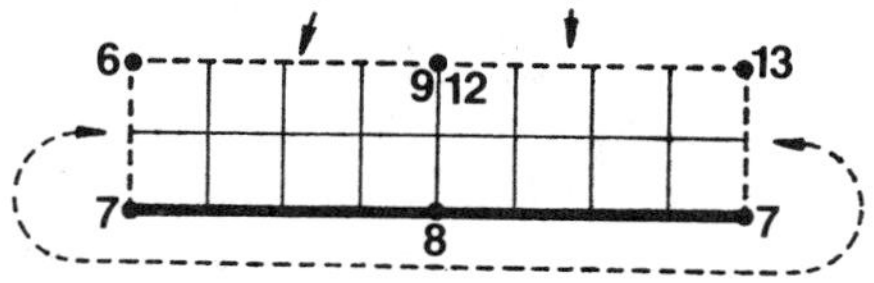

and the lower side of the slit considered re-entrant with the left half of the top boundary of the rectangle corresponding to the inner system, the upper side of the slit being re-entrant with the right half of this top boundary of the inner region. Now the conceptual opening is as follows for the inner region:

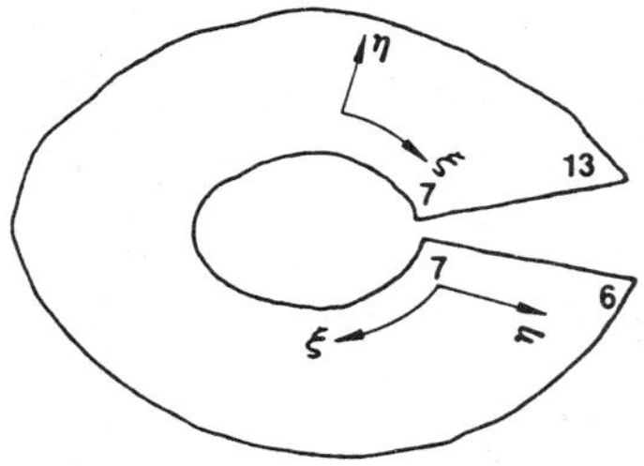

Difference representations made above the slit thus would use points below the right half of the top of the inner region in the transformed region, etc. Similarly, representation made below the left half of the top of the inner region would use points below the slit. The slit is thus a "black hole" into which coordinate lines from the outer system disappear, to reappear as part of the inner system. The slit here, matched with the top of the inner system, is then clearly a branch cut, and passage through the slit onto the inner system is simply passage onto a different sheet.

Note that the embedded system has its own distinctive species and directions for the coordinate lines, entirely separate from the outer system. Thus for the inner region the directions are as follows:

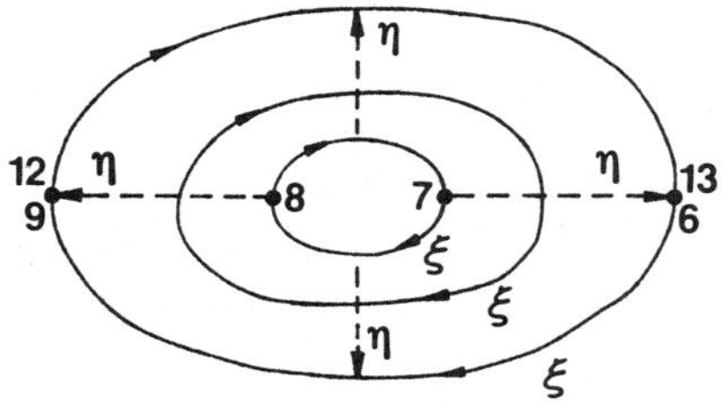

while for the outer region they are as shown below:

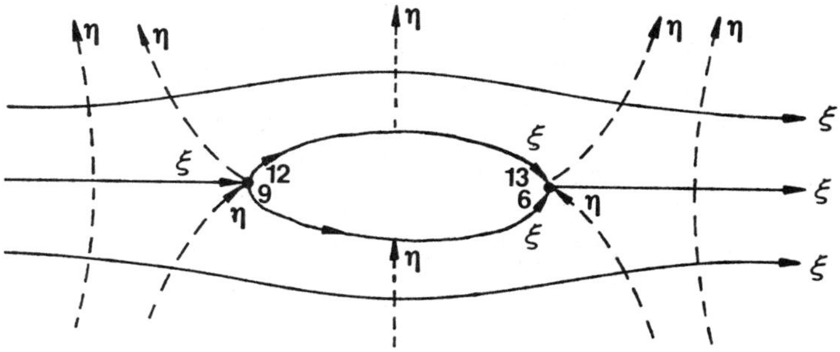

Thus at a point on the upper interface, 12-13, between the systems the lines are as follows:

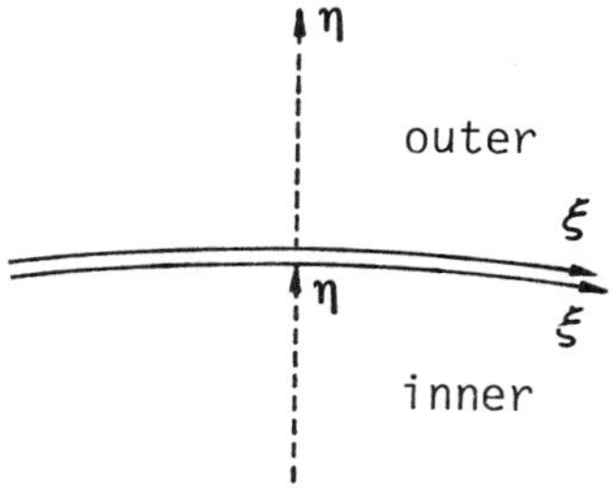

while on the lower interface, 9-6 they are as follows:

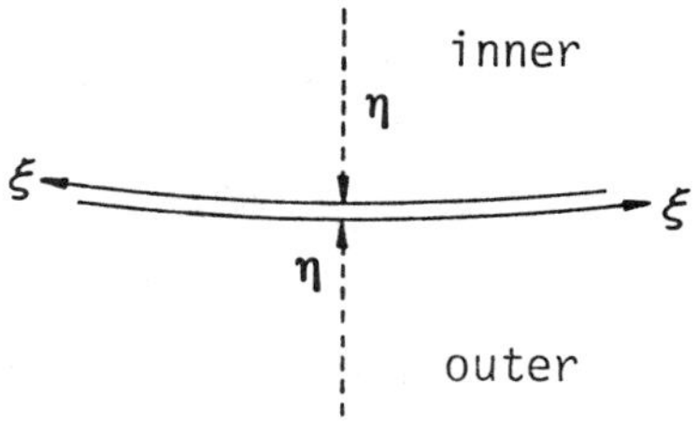

Thus both coordinates reverse direction at the lower interface although the species is continuous, while both the species and directions are continuous across the upper interface. This again corresponds to passage onto a different sheet, for the interface between the inner and outer systems, i.e., the segments 12-13 and 9-6, is actually a branch cut.

The points 9(12) and 6(13) here require special notice. For example, at point 9 the coordinate line configuration is as follows:

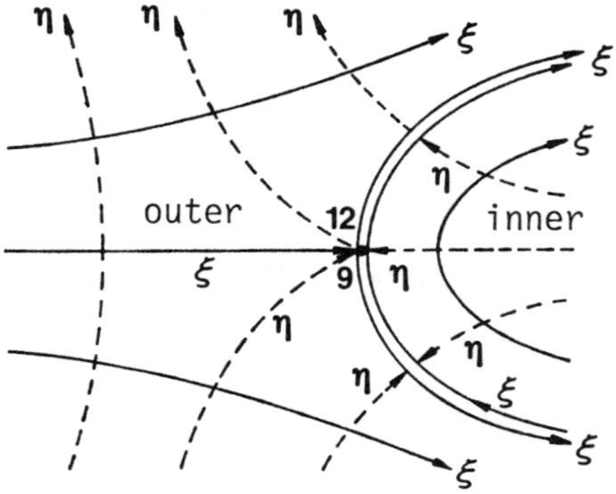

The lines through point 9 are as shown below

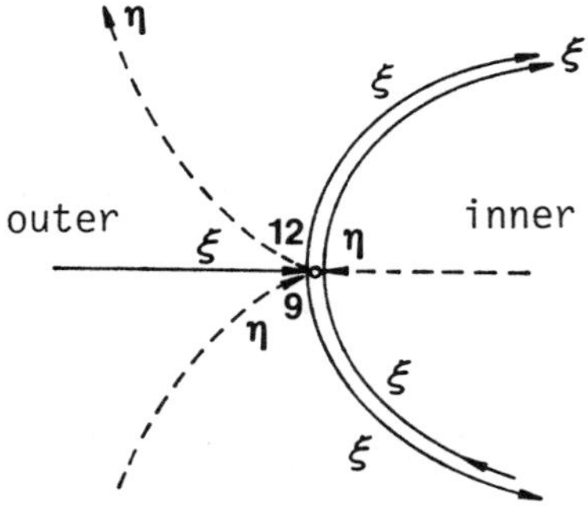

There are thus several changes in species and direction at this point. This type of special point embodies the form which always occurs with the slit configuration, shown on p. 26, and occurs here because the embedded region inside the contour 9-6-13-12 is essentially contained inside a slit defined by the same set of numbers.

The above discussion refers to the slit configuration on p. 41. For the configuration on p. 40, the lines in the outer region are still as diagrammed on p. 43, but the lines in the inner region now are as follows:

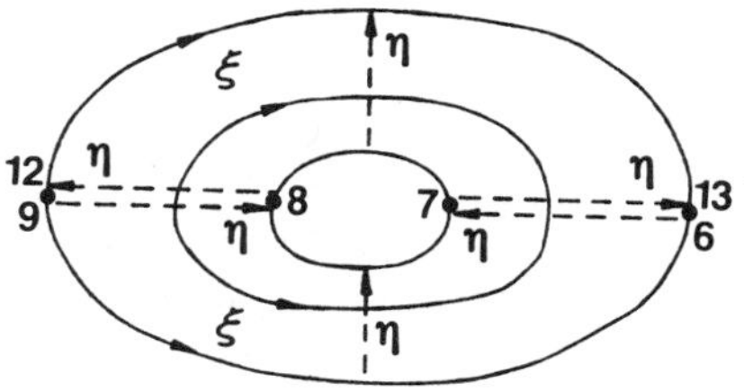

The coordinate line species and direction given on p. 43 for the upper interface, 12-13, thus applies here on the entire interface between the two regions.

An alternative treatment of the two special points is to place them inside cells as shown below:

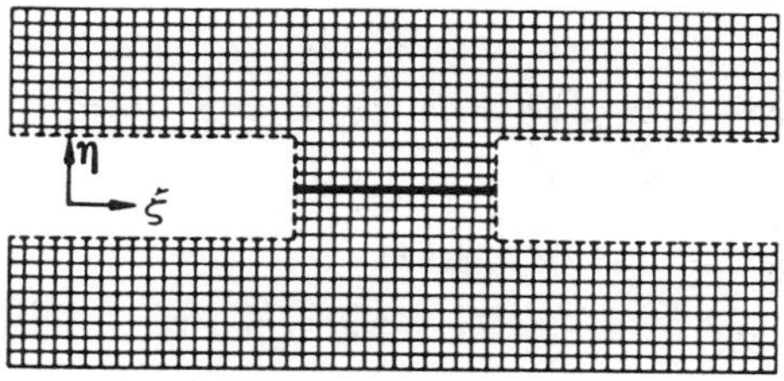

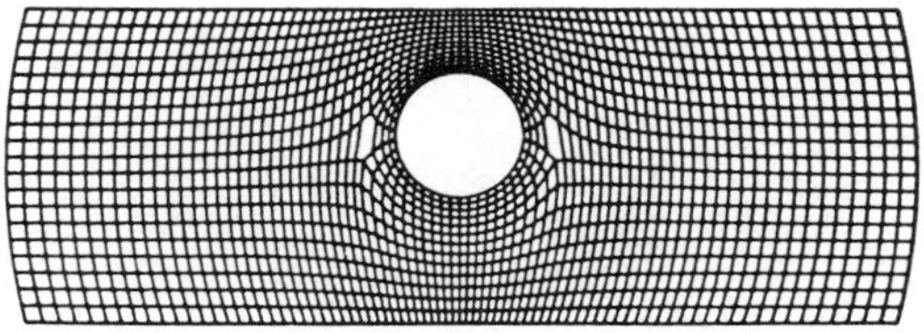

This results in a six-sided cell surrounding each of these two points which requires special treatment as discussed in Chapter IV.

Embedded systems can also be constructed in the block configuration:

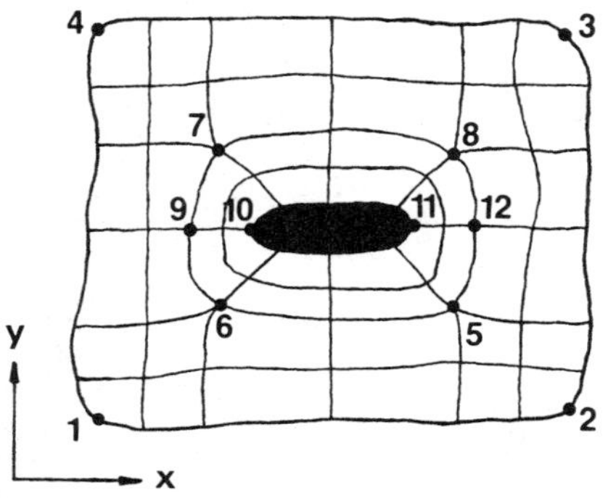

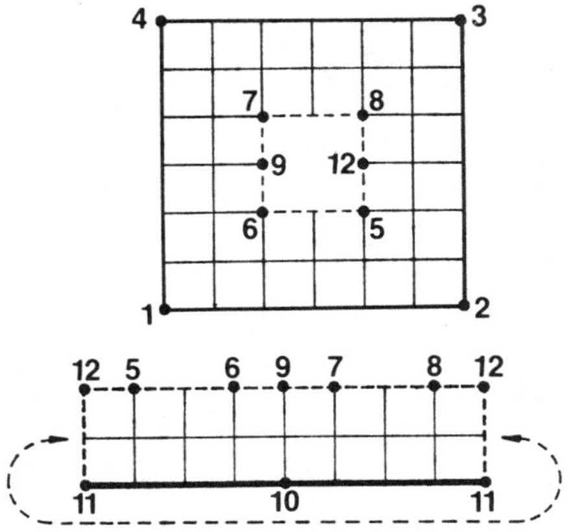

Here the top of the block, 7-8, in the outer system is re-entrant with the corresponding segment, 7-8, on a portion of the top of the inner system. The left side of the block, 6-7, and the bottom of the block, 6-5, are similarly re-entrant with single portions of the top of the inner system. Finally, the right side of the block, 5-12-8, is re-entrant with two portions, 5-12 and 12-8, of the top of the inner system. Points outside one of these segments in one system are thus located at corresponding positions inside the other segment of the re-entrant pair in the other system. The slab sides, matched with the top of the inner system, are thus branch cuts between the inner and outer systems.

Here the coordinate lines proceed as follows for the outer system:

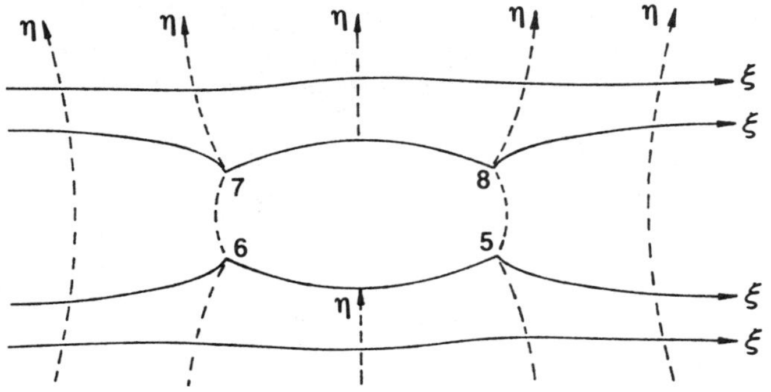

while those for the inner system are the same as before, as shown on p. 42. This means that on the left and right sides of the block, i.e., segments 6-7 and 5-8, the line directions are as follows:

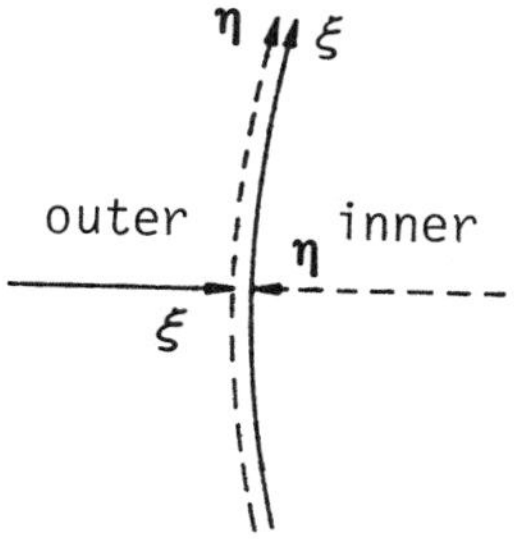

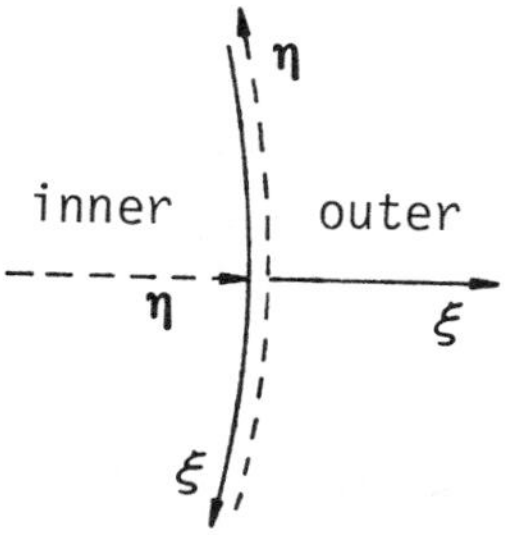

and on the top and bottom, segments 7-8 and 6-5, the directions are as shown below:

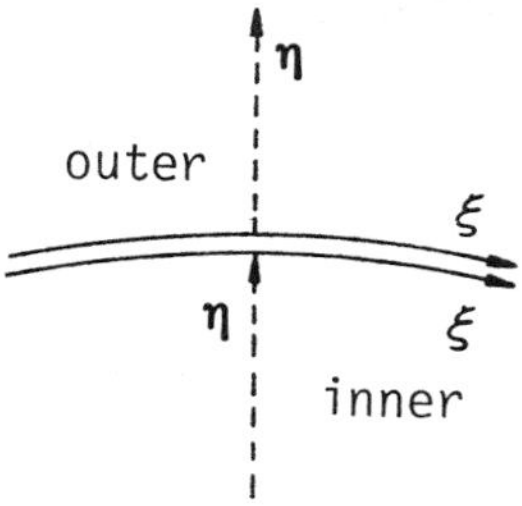

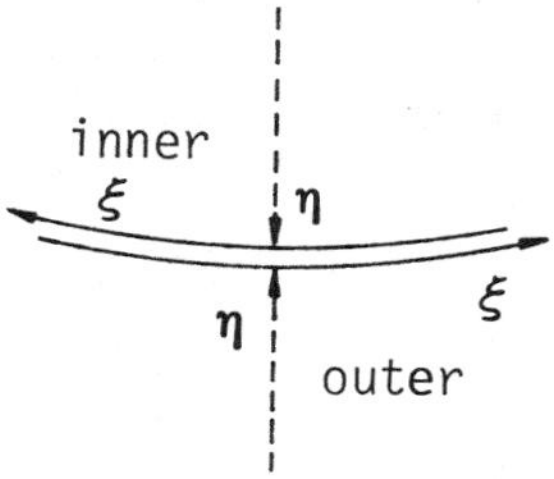

There are thus changes in coordinate species and/or direction that are different on each side of the block.

The point 8 (and points 7,6 and 5) are special points of the following form:

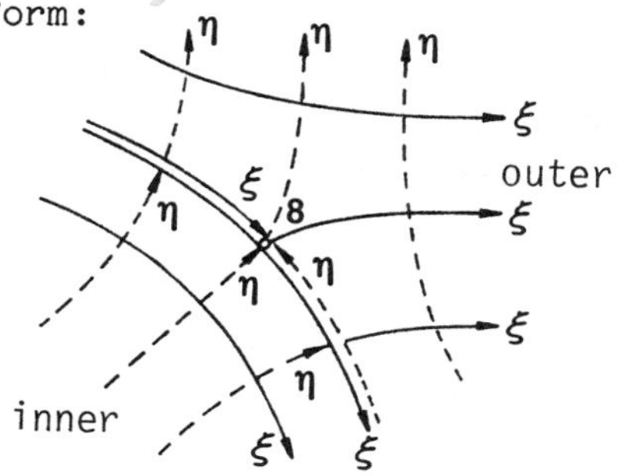

The lines through the point are as shown below:

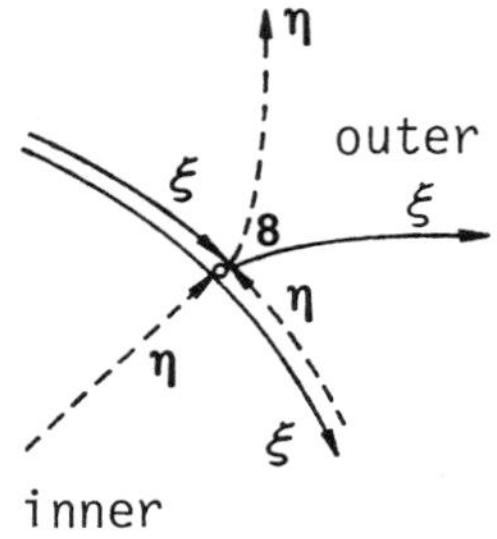

Here the special points occur in the field instead of on the boundary.

An example of a C-type system embedded in another C-type system is given next:

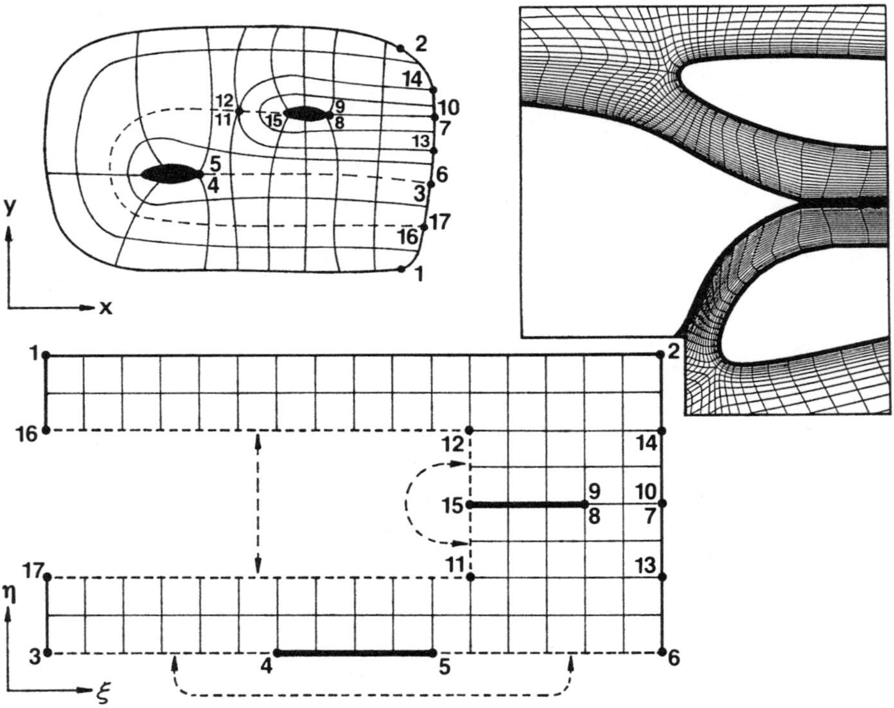

Here the conceptual opening is as follows: First, considering the system about the upper body, we have the following configuration:

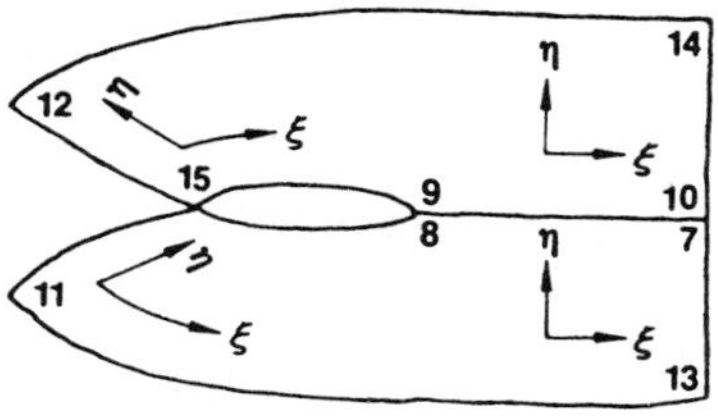

which, with the body collapsed to a slit, opens to the rectangle in the center of the transformed region. Next consider the system about the other body:

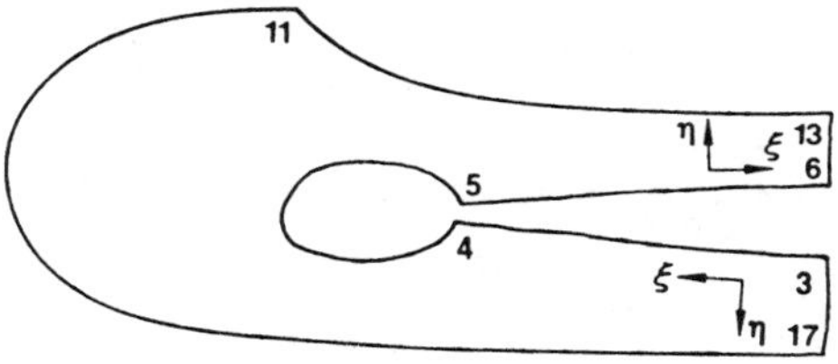

This opens to a rectangle, with the body flattening to a portion of the bottom, which is fitted to the first rectangle along the segment 11-13. Finally, the outermost portion opens as follows:

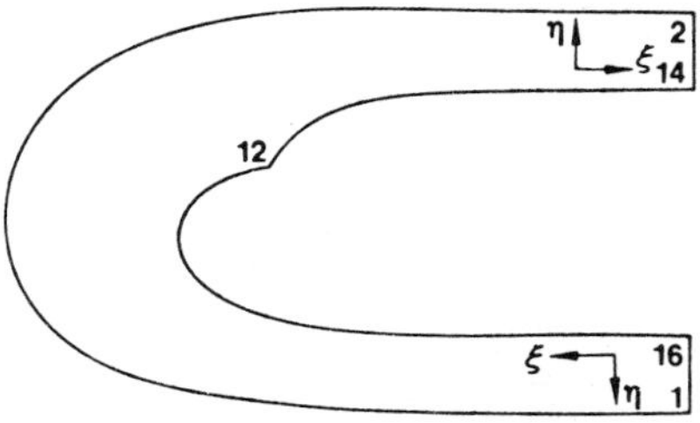

which opens to a rectangle which is fitted to the first one along the segment 12-14.

Again the embedded region inside the contour 14-12-11-13 can be considered to lie inside a slit. This contour, which forms the interface between the inner and outer systems, is actually a branch cut between the two systems, across which there are discontinuities in coordinate species and directon in the same manner as was discussed above for the previous embedded system. Points below segment 16-12 coincide with points below segment 17-11 in this case. Points to the left of segment 15-12, above point 15, are coincident with points to the right of segment 15-11 below point 15. The slit here is formed of the segments 8-15 and 9-15. The coincident points 11 and 12 here must be taken as a point boundary in the physical region, i.e., fixed at a specified value. Several special points of the types discussed above are present here.

An alternative arrangement of the transformed region that corresponds to exactly the same coordinate system in the physical region is as follows:

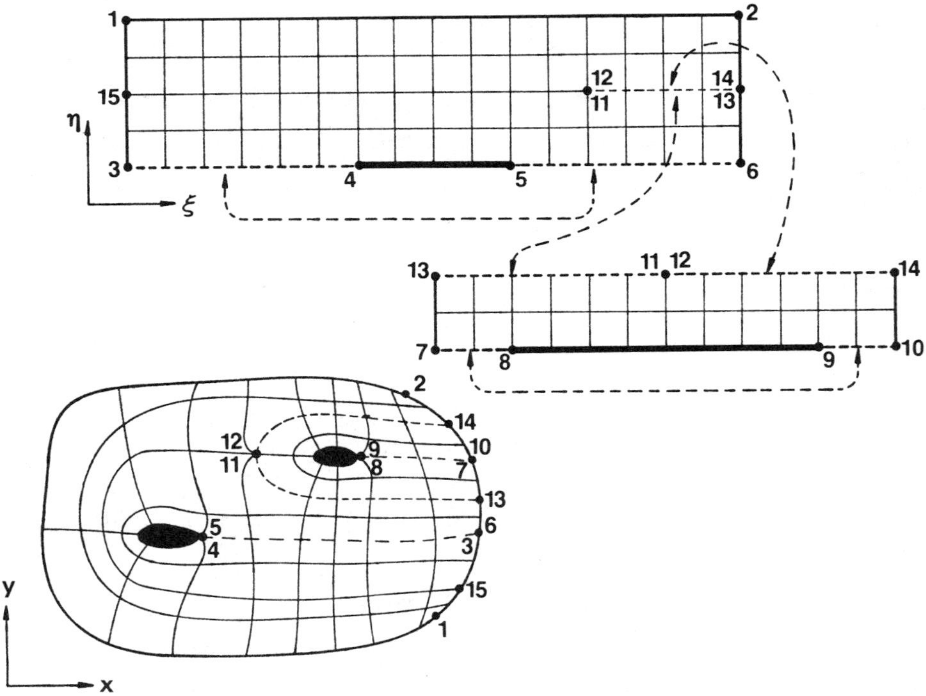

Here points below segment 3-4, to the left of point 4, coincide with points above segment 6-5, to the right of point 5. When calculations are made on or above the segment 12-14 on the larger block, points below this segment coincide with points below the corresponding segment on the smaller block. Similarly, when calculations are made on or below the segment 13-11 on the larger block points above this segment coincide with points below the corresponding segment on the smaller block. Finally, points below the segment 7-8, to the left of point 8, on the smaller block are coincident with points above the segment 10-9, to the right of point 9.

This configuration displays explicitly the correspondence of the embedded region inside the contour 14-12-11-13 to a slit. Conceptually, coordinate lines from the main system disappear into the slit and emerge into the embedded system. These coordinate lines thus are continued from the main system onto another sheet representing the embedded system. This concept of embedded systems, with continuation onto another sheet through a slit adds considerable flexibility to the grid configurations and is of particular importance with multiple boundaries and in three dimensions. The composite structure discussed in Section 4 removes much of the coding complexity associated with systems of this type.

D. Other Configurations

Another arrangement of cuts, where the species of coordinate changes on a continuous line as the cut is crossed, is illustrated below. The transformed region in this case is composed of three blocks connected by the cuts.

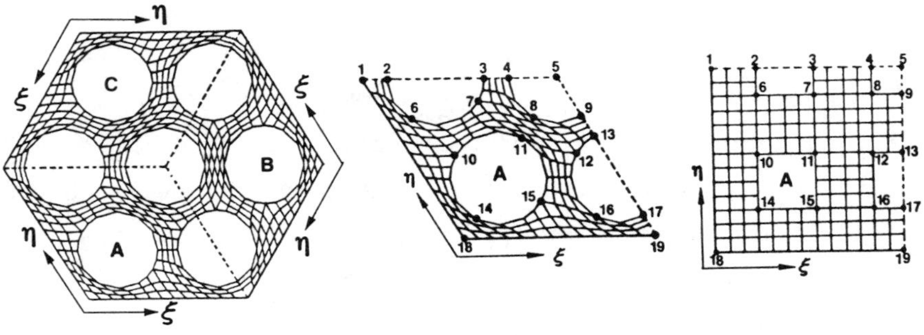

Here points outside one section are coincident with corresponding points inside the adjacent section.

The coordinate line configuration on the interface on the right side of block A here is as follows:

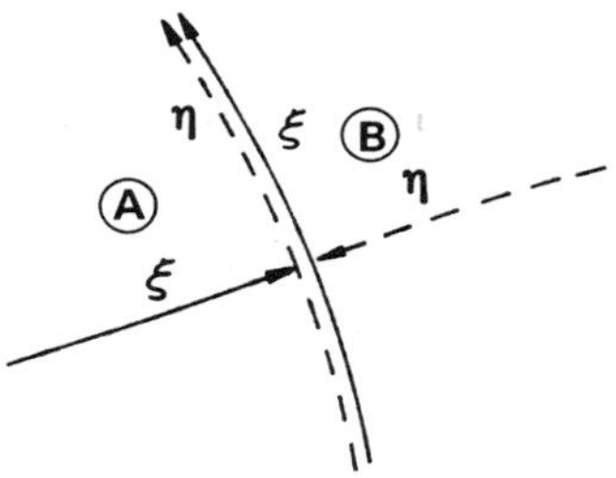

This same type of configuration occurs, in different orientations, on each of the interfaces. These interfaces are branch cuts, so that passage onto the adjacent block amounts to passage onto another sheet in the same manner discussed above.

As a final configuration for consideration in two dimensions, the following example shows a case with fewer lines on one side of a slab than on the other. This does not necessitate the use of different increments of the curvilinear coordinates in the numerical expressions, because, as has been mentioned, these increments always cancel out anyway.

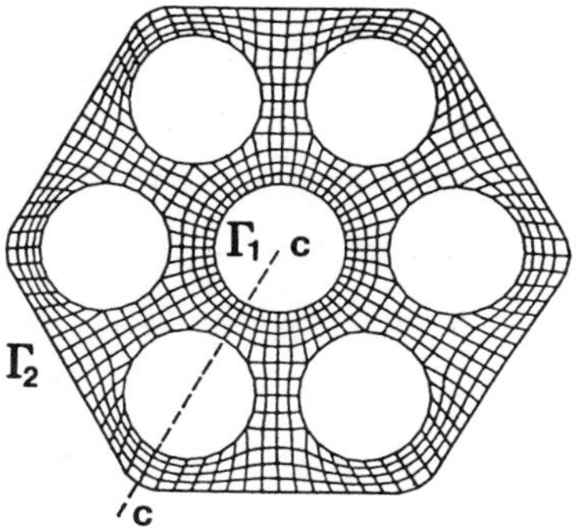

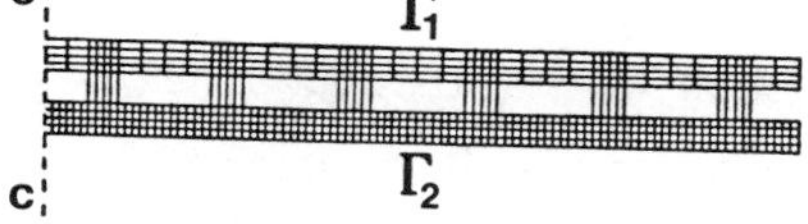

E. Three-dimensional Regions

All the general concepts illustrated in these examples extend directly to three dimensions. Interior boundaries in the transformed region can become rectangular solids and plates, corresponding to the slabs and slits, respectively, illustrated above for two dimensions. Examples of three-dimensional configurations can be found in the surveys given by Ref. [8] and [9].

It is also possible to use branch cuts, as illustrated above for two dimensions, to bring the interior boundaries in the physical region entirely to the exterior boundary of the transformed region:

Physical space

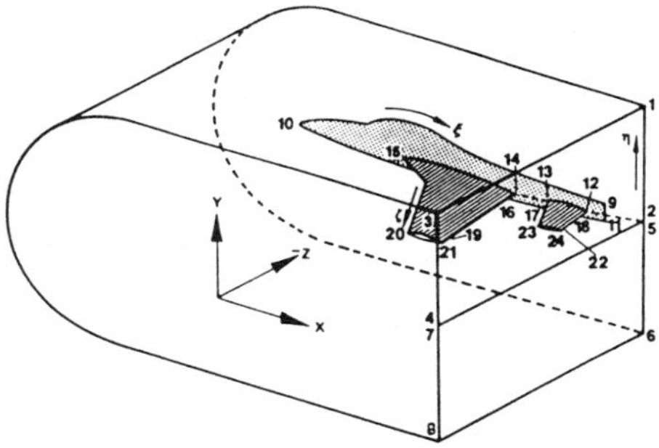

Computational space

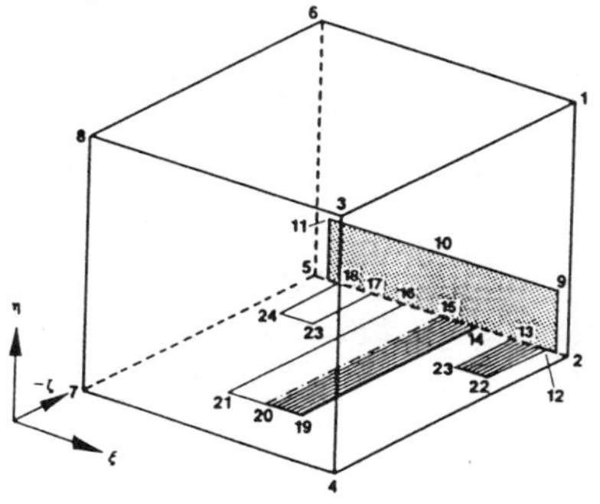

The correspondence between the physical and transformed fields can, however, become much more complicated in three dimensions, and considerable ingenuity may be required to visualize this correspondence. For instance, the simple case of polar coordinates corresponds to a rectangular solid with two opposing sides having the radial coordinate constant thereon, and two re-entrant sides on which the longitudinal coordinate is constant at 0 and 2π, respectively (corresponding to the cut). The remaining two sides correspond to the north and south polar axes, so that an axis opens to cover an entire side. There is thus a line, i.e., the axis, in the physical region that corresponds to an entire side in the transformed region.

Three-dimensional grids may be constructed in some cases by simply connecting corresponding points on two-dimensional grids generated on stacks of planes or curved surfaces:

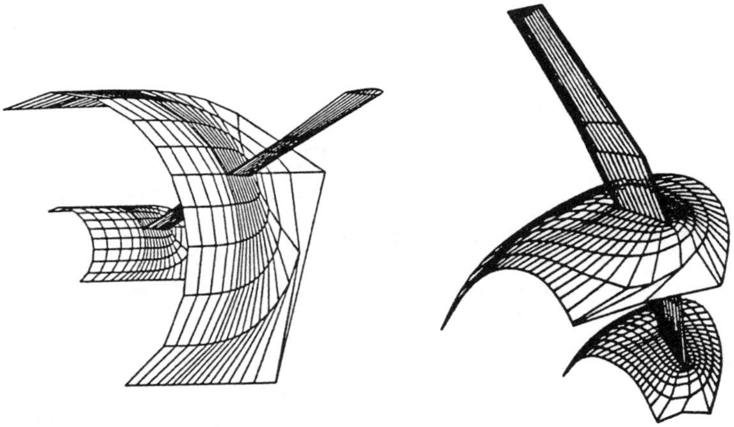

It should be noted, however, that this procedure provides no inherent smoothness in the third direction, except in cases where the stack is formed by an analytical transformation, such as rotation, translation or scaling, of the two-dimensional systems. An example of such an analytical transformation of two-dimensional systems is the construciton of a three-dimensional grid for a curved pipe by rotating and translating (and scaling if the cross-sectional area of the pipe varies) two-dimensional grids generated for the pipe cross-section so as to place these transformed two-dimensional grids normal to the pipe axis at successive locations along the axis:

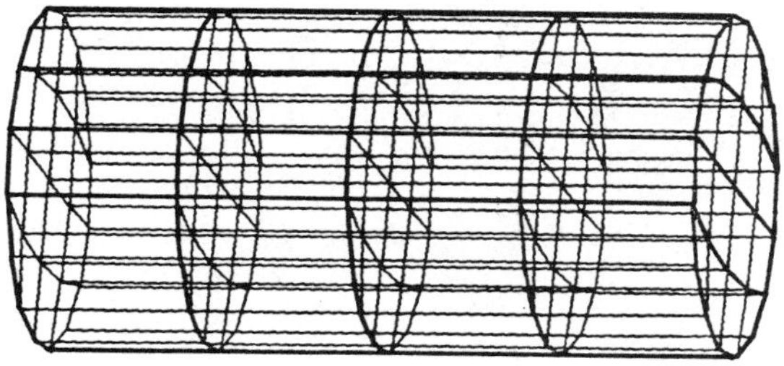

Another example is the rotation of a two-dimensional grid about an axis to produce an axi-symmetric grid:

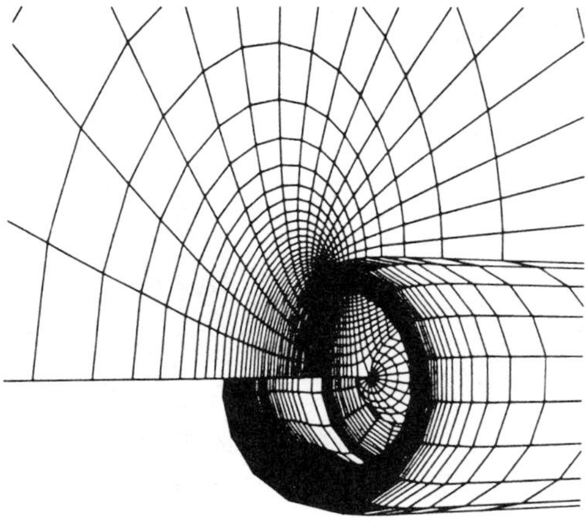

4. Composite Grids

All of the above concepts can be incorporated in a single framework, and the coding complexity can be greatly reduced, by considering the physical field to be segmented into sub-regions, bounded by four (six in 3D) generally curved sides, within each of which an individual coordinate system is generated. The overall coordinate system, covering the entire physical field, is then formed by joining the sub-systems at the sub-region boundaries. The degree of continuity with which this juncture is made is a design consideration in regard to the mode of application intended for the resulting grid.

This segmentation concept is illustrated in the figure below.

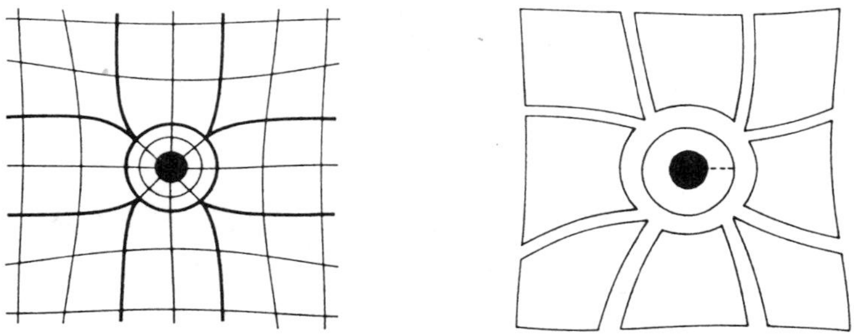

The locations of the interfaces between the sub-regions in the physical region are, of course, arbitrary since these interfaces are not actual boundaries. These interfaces might be fixed, i.e., the location completely specified just as in the case of actual boundaries, or might be left to be located by the grid generation procedure. Also the coordinate lines in adjacent sub-regions might be made to meet at the interface between with complete continuity:

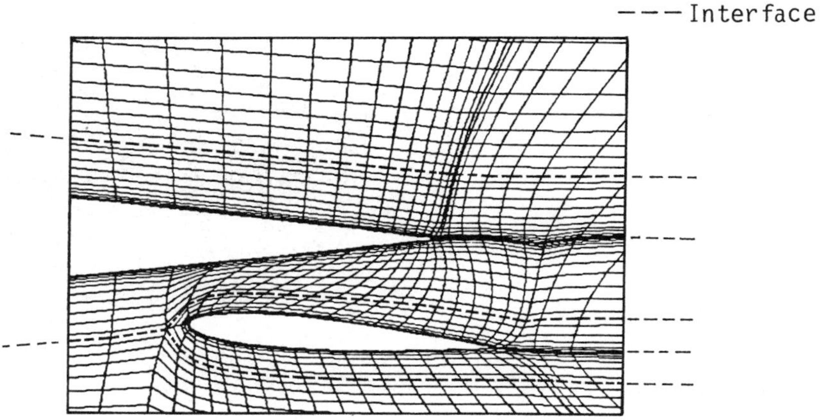

with some lesser degree of continuity, e.g., continuous line slope only:

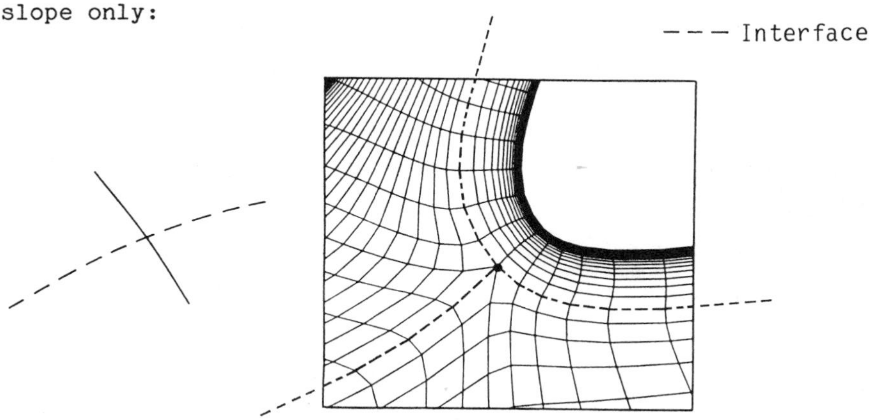

or with a discontinuity in slope:

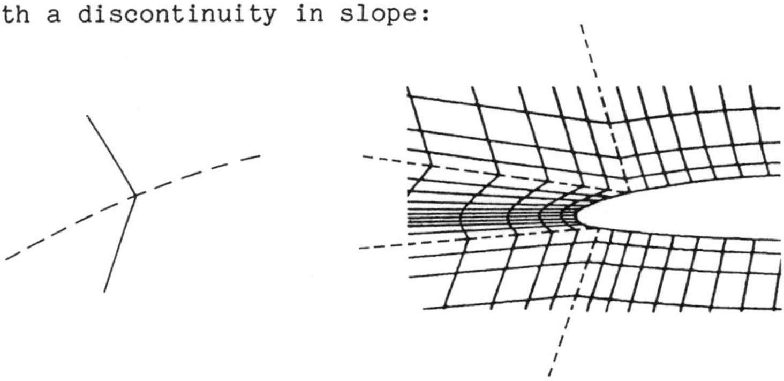

or perhaps not to meet at all:

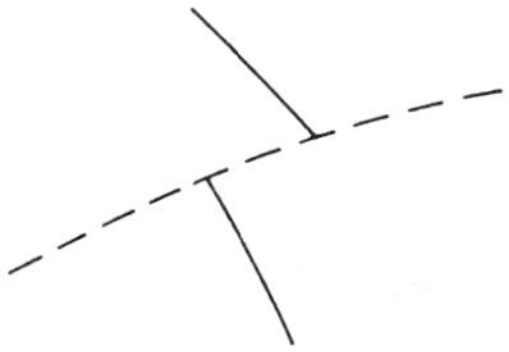

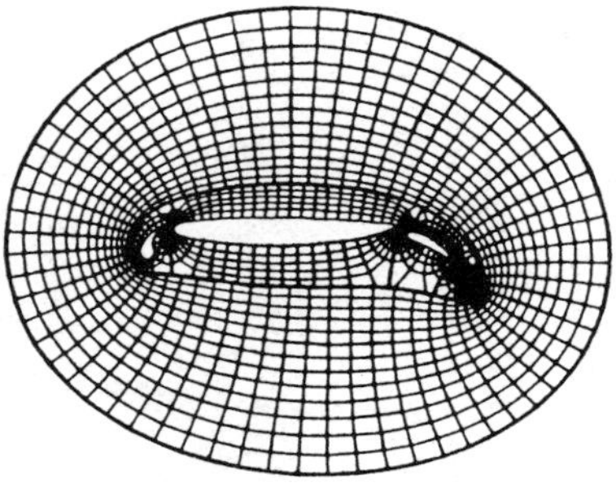

Naturally, progressively more special treatment at the interface will be required in numerical applications as more degrees of line continuity at the interface are lost. Procedures for generating segmented grids with various degrees of interface continuity are discussed later, and conservative interface conditions are given in Ref. [52], [53].

Now, with regard to placing these concepts in the framework of segmentation, the sides of an individual subregion (called a "block" hereafter) can be treated as boundaries on which the coordinate points, and/or the coordinate line intersection angles, are specified, just as is done for actual boundaries, or a side may be treated as one member of a pair of re-entrant boundaries, i.e., one side of a branch cut in the physical region across which complete continuity is established. The other member of the pair may be another side (or portion thereof) of the same block or may be all (or part of) a side of an adjacent block in the physical field. Recall that it is not necessary for a coordinate to remain of the same species across a re-entrant boundary, since the passage is onto a different sheet. This can introduce some coding complexity, but the treatment is straightforward, and in fact the coding can be greatly simplified by using an extra layer of points surrounding each block as is discussed in Section 6.

Some of the general concepts have been embodied in the two-dimensional code discussed in Ref. [19] and in three recent three-dimensional codes, Ref. [13] Ref. [14], and [51]

A. Simply-connected Regions

The first L-shaped simply-connected configuration on p. 21 can be interpreted as being composed of three blocks, with the sides of adjacent blocks forming pairs of re-entrant boundaries:

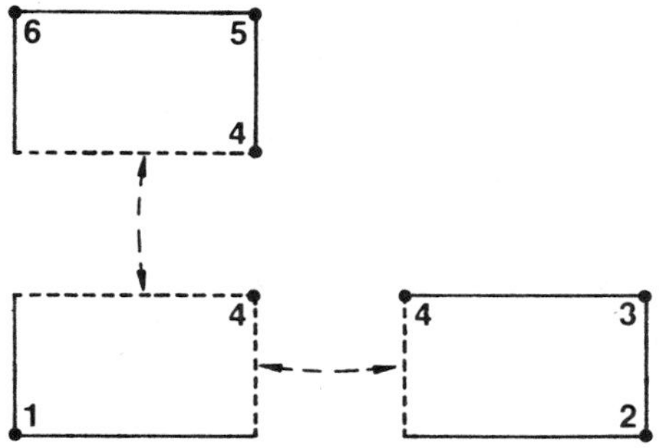

or two blocks with a portion of a side of one block re-entrant with an entire side of another block:

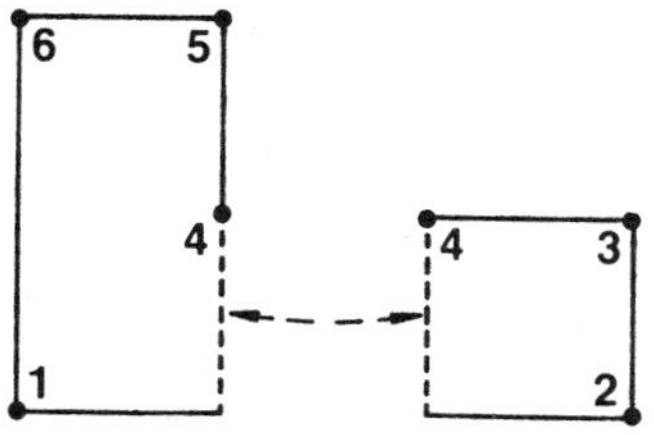

Here, and in the examples to follow, solid lines correspond to physical boundaries, while the dashed lines correspond to the interfaces between the blocks. The dashed arrows indicate the linkage between the interfaces. (Obviously, any single block can be broken into any number of blocks connected by re-entrant boundaries across adjacent sides.) In contrast, the L-shaped configuration on p. 22 corresponds to the use of a single block. Similarly, the configuration on p. 24 can be formed with three blocks:

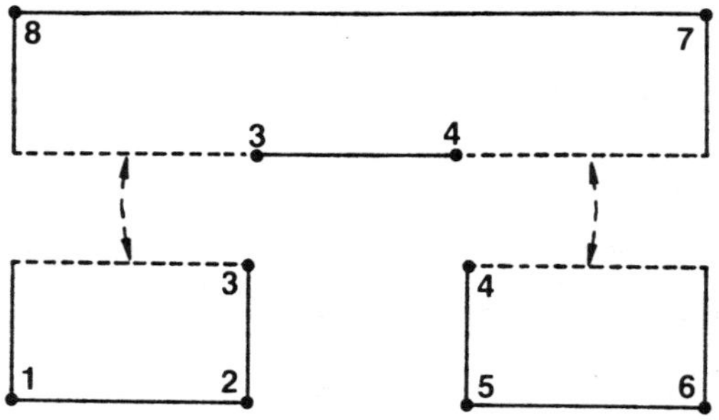

while the first configuration for the same boundary on p. 25 is formed with a single block.

The slit configuration on p. 25 can be formed of three blocks:

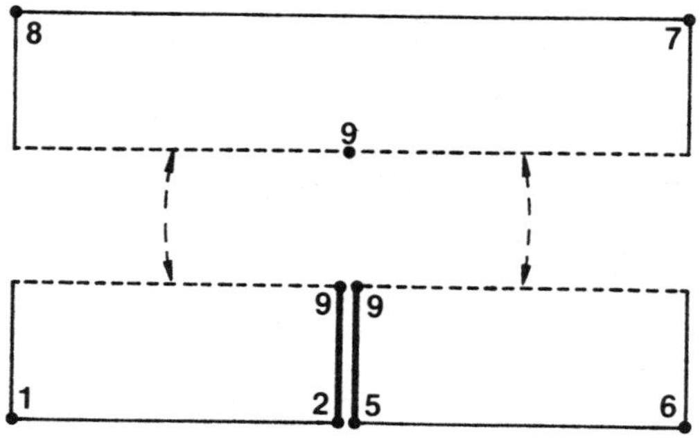

or two blocks with only a portion of the adjacent sides of two blocks forming a re-entrant boundary:

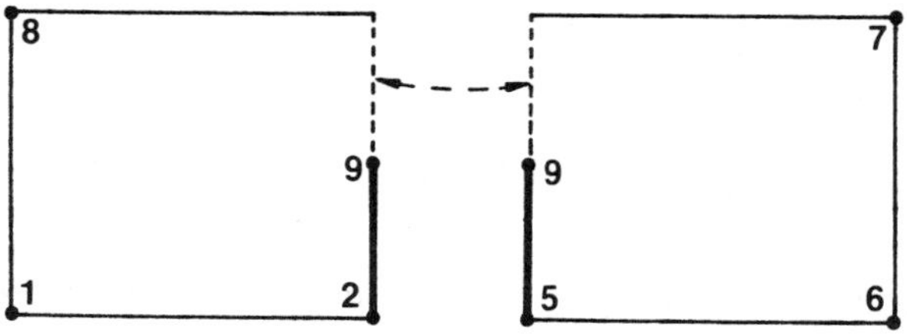

B. Multiply-connected Regions

The configuration with a single cut shown on p. 29 corresponds to the use of a single block with the left and right sides here being the members of a pair of re-entrant boundaries:

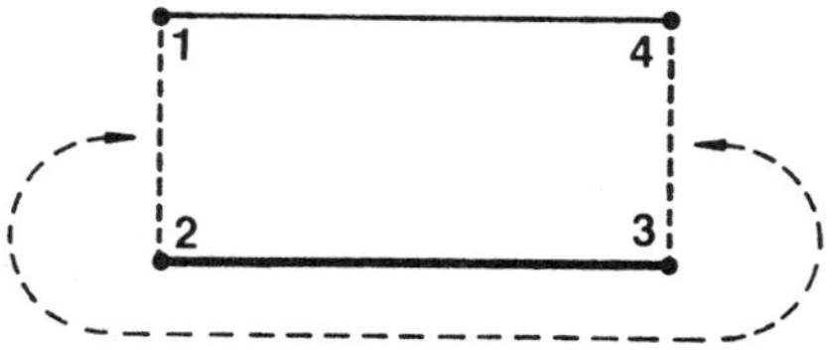

The multiply-connected slab configuration on p. 27 can be broken into four blocks:

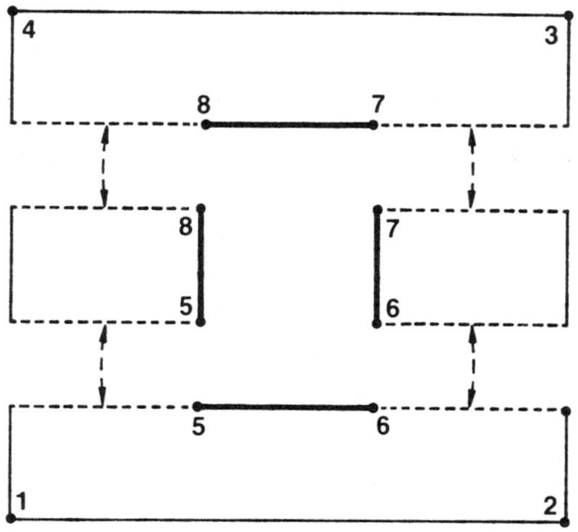

Other decompositions should also be immediately conceivable. The slit configuration on p. 28 can be formed with two blocks, again with only portions of adjacent sides serving as re-entrant boundaries:

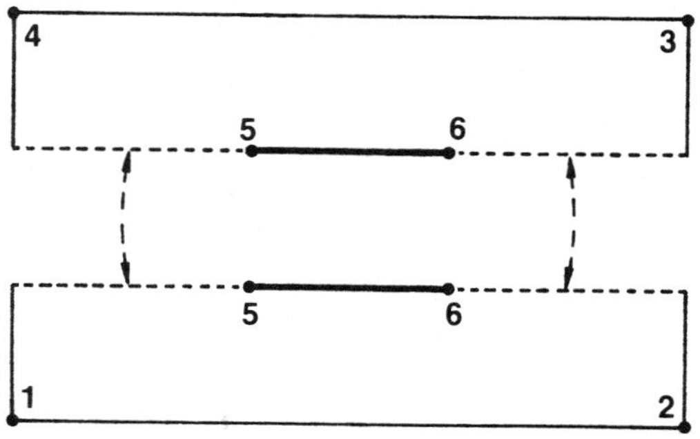

or into four blocks, with entire sides as re-entrant boundaries in all cases:

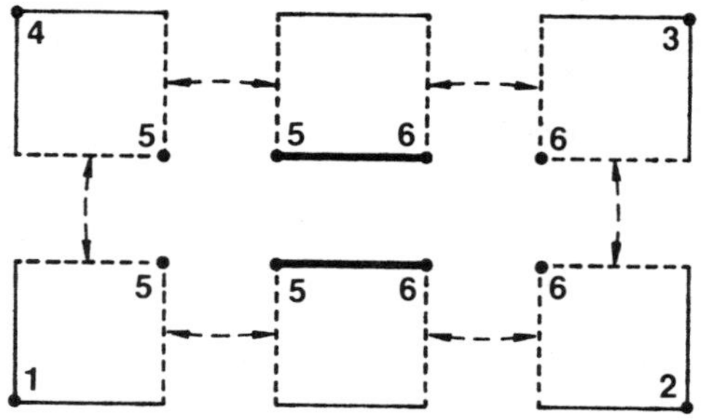

The double-body region on p. 34 opens to a single block as shown there, with portions of sides as re-entrant boundaries. A five block configuration would use only entire sides as re-entrant boundaries, however:

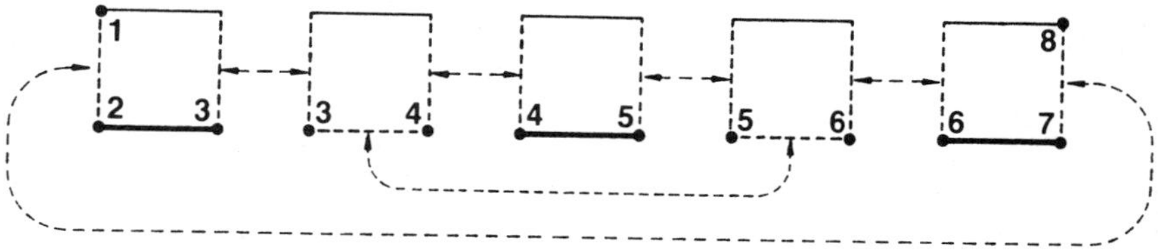

There is no real advantage, however, to the five-block system here.

C. Embedded Regions

The segmentation concept is most useful in the construction of embedded coordinate systems. For instance, the system on p. 40 can be considered to be formed of three blocks as follows:

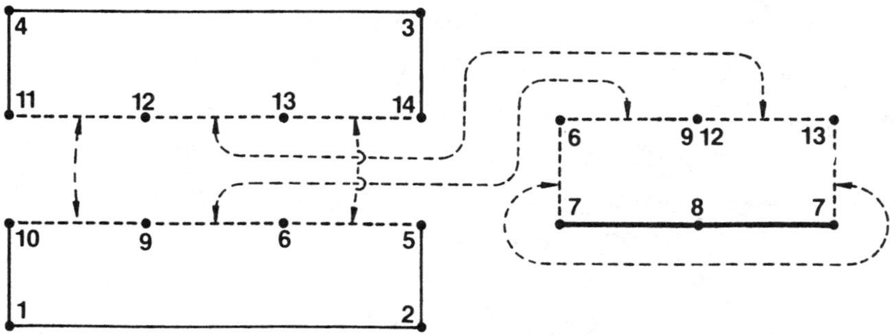

Here portions of adjacent sides of the two larger blocks are re-entrant with each other, while each of the remaining portions of these sides is re-entrant with half of one side of the smaller block. The left and right sides of the smaller block are re-entrant with each other. This configuration could also have been constructed with eight blocks:

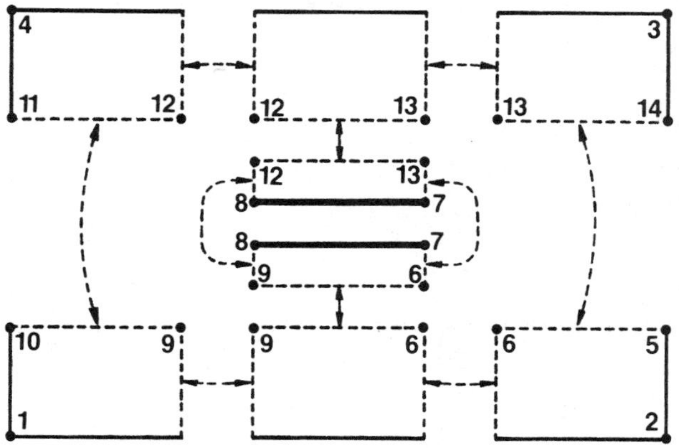

with only entire sides being involved in re-entrant pairs as shown.

With embedded systems the coordinate species often changes as the re-entrant boundary is crossed. These systems also show that the blocks need be physically adjacent only in the physical field, and it is in this sense that "adjacent" is always to be interpreted. The transformed (computational) field should always be viewed as only a bookkeeping structure. Various constructions are possible for the configurations on p. 48 and 50, and a two block structure was actually used on p. 50. A further example follows:

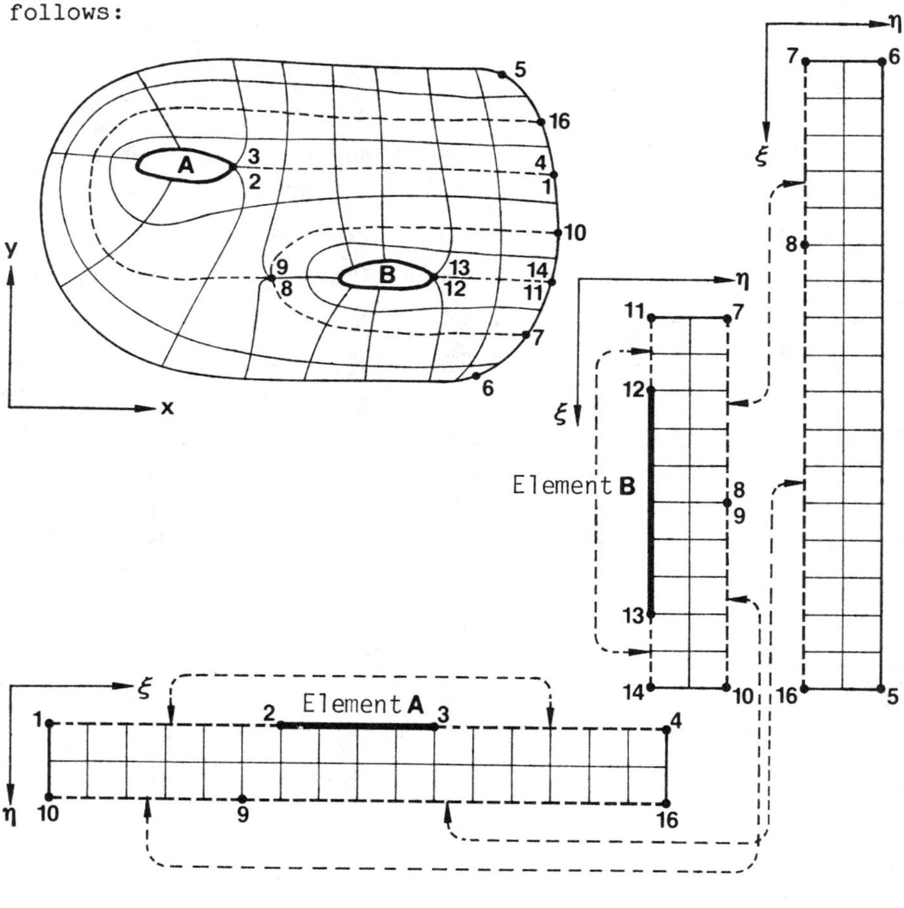

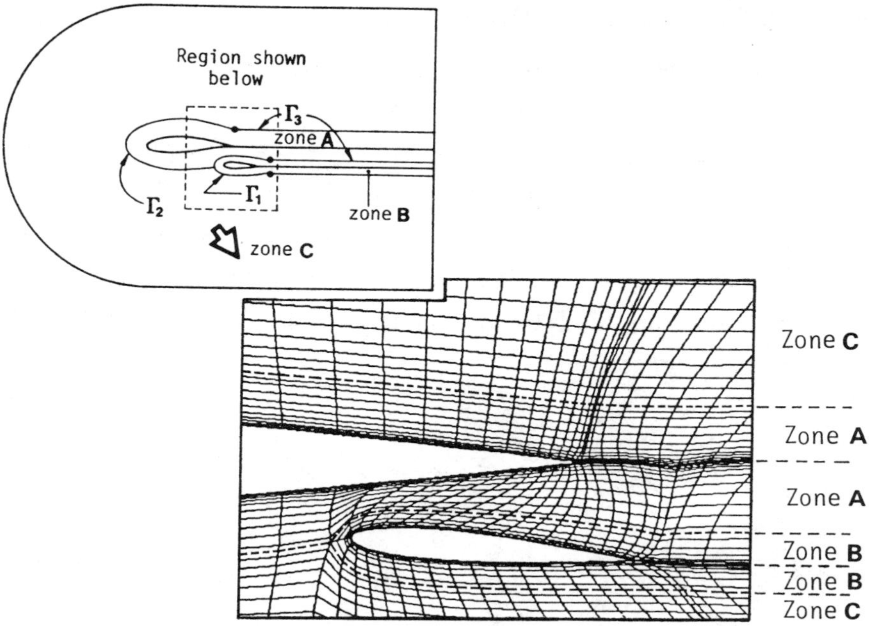

Another example follows:

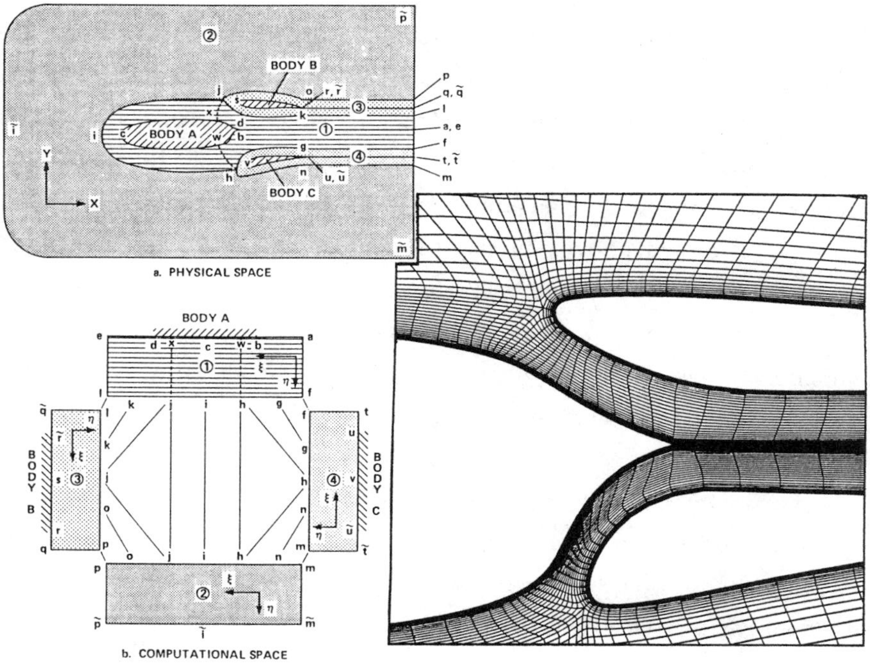

D. Three-dimensional Regions

For general three-dimensional configurations, it is usually very difficult to obtain a reasonable grid with the entire physical region transformed to a single rectangular block. A better approach in most cases is to segment the physical region into contiguous sub-regions, each bounded by six curved surfaces, with each sub-region being transformed into a rectangular block. An individual grid is generated in each sub-region:

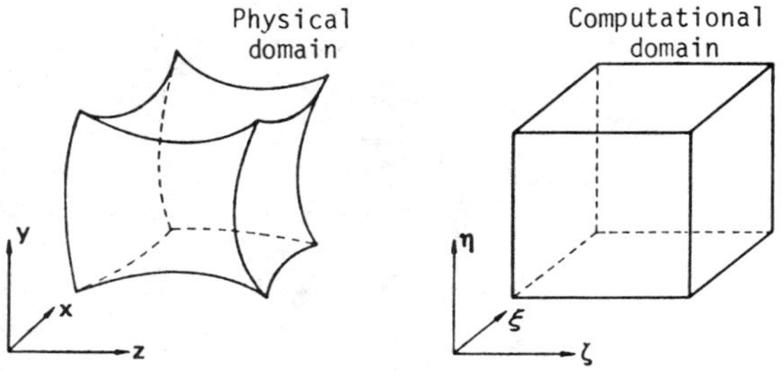

These sub-region grids are patched together to form the overall grids, as in the two-dimensional cases discussed above. Examples of the use of this segmentation in three dimensions are found, in particular, in Ref. [11] and [12]. Others are noted in the survey given by Ref. [9].

As noted above, complete continuity can be achieved at the sub-region interfaces by noting the correspondence of points exterior to one sub-region with points interior to another. The necessary bookkeeping can be accomplished, and the coding complexity can be greatly reduced, by using an auxiliary layer of points just outside each of the six sides of the computational region, analogous to the procedure mentioned above for two dimensions. A correspondence is then established in the code between the auxiliary points and the appropriate points just inside other sub-regions. This approach has recently been incorporated in an internal region code, Ref. [13], and in two codes for general regions, Ref. [14] and [51]. This is discussed in more detail in Section 6.

General three-dimensional regions can be built up using sub-regions as follows: First, point distributions are specified on the edges of a curved surface forming one boundary of a sub-region:

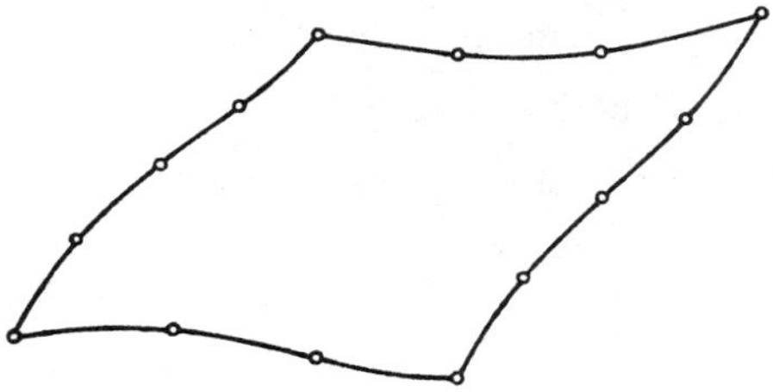

and a two-dimensional coordinate system is generated on the surface:

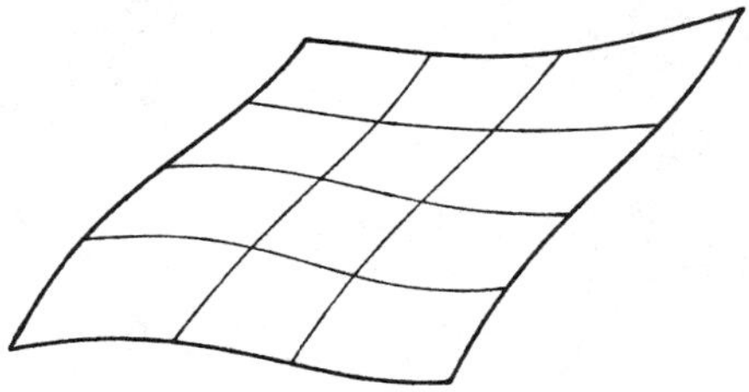

When this has been done for all surfaces bounding the sub-region, the three-dimensional system within the sub-region is generated using the points on the surface grids as boundary conditions:

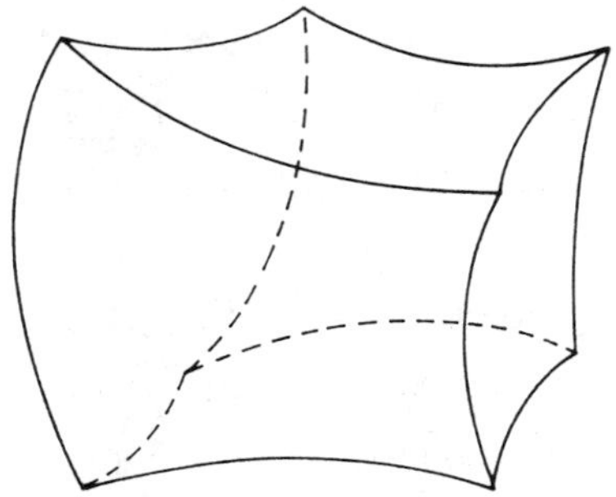

In three dimensions it is possible for a line, e.g., a polar axis, in the physical region to map to an entire side of the computational region as in the illistration below, where the axis corresponds to the entire left side of the block:

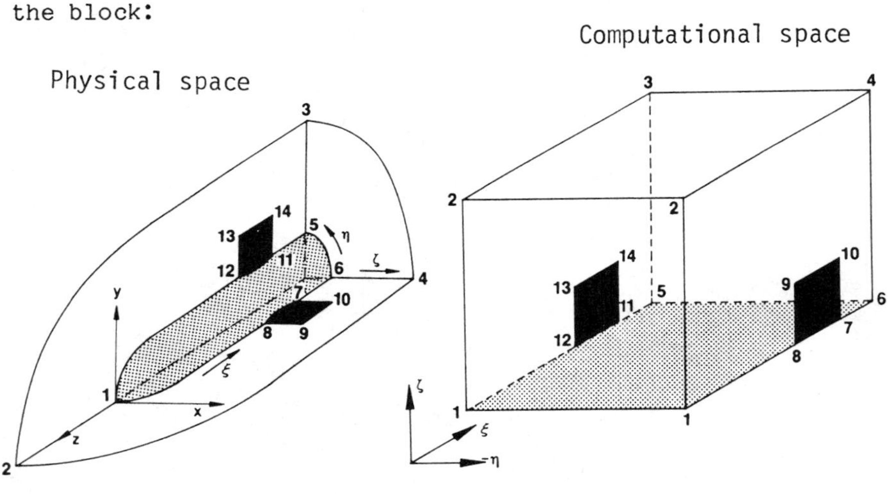

The system illustrated here could be one of several identical blocks joining together to form a complete system around the axis.

It is illustrated by an exercise that the occurence of a polar axis can be avoided, and this facilitates the construction of a block structure. Thus a surface grid, having eight "corners", analogous to the four "corners" on the circle in the 2D grid on p. 23, can be constructed on the surface of a sphere. This serves much better than a latitude-longitude type system for joining to adjacent regions. Similarly, the use of the four "corner" system, rather than a cylindrical system, in a circular pipe allows T-sections and bifurcations to be treated easily by a composite structure, c.f. Ref. [13].

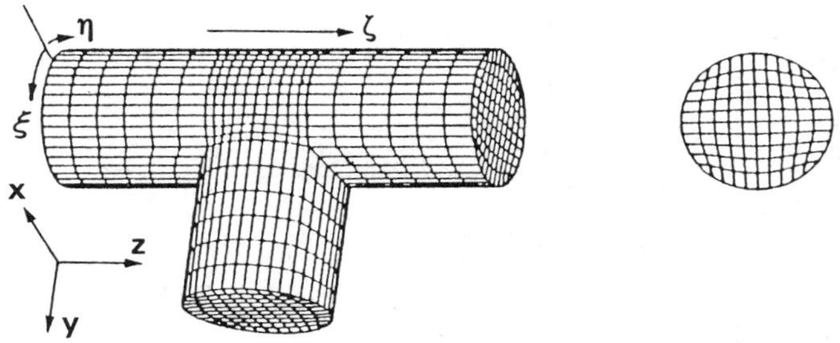

Generally, grid configurations with polar axis should not be used in composite grid structures.

E. Overlaid Grids

Another approach to complicated configurations is to overlay coordinate systems of different types, or those generated for different sub-regions:

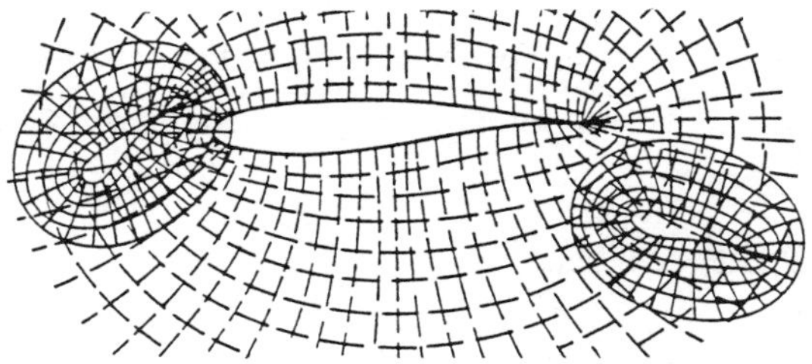

Here an appropriate grid is generated to fit each individual component of the configuration, such that each grid has several lines of overlap with an adjacent grid. Interpolation is then used in the region of overlap when solutions are done on the composite grid, with iteration among the various grids. This approach has the advantage of simplicity in the grid generation, in that the various sub-region grids are only required to overlap, not to fit. However, there would appear to be problems if regions of strong gradients fall on the overlap regions. Also the interpolation may have to be constructed differently for different configurations, so that a general code may be hard to produce. Some applications of such overlaid grids are noted in Ref. [5].

5. Branch Cuts

As has been noted in the above discussion of transformed field configurations, it is possible for discontinuities in coordinate species and/or direction to occur at branch cuts, in the sense of passage onto another sheet. Continuity can be maintained, however, by conceptually remaining on the same overlapping sheet as the cut is crossed. All derivatives thus do exist at the cut, but careful attention to difference formulations is necessary to represent derivatives correctly across the cut. Although the correct representation can be accomplished directly by surrounding the computational region with an extra layer of points, as is discussed in Section 6, it is instructive to consider what is required of a correct representation further here.

A. Point Correspondence

Points on re-entrant boundaries in the transformed region, i.e., on branch cuts in the physical region, are not special points in the sense used above. Points on re-entrant boundaries, in fact, differ no more from the other field points than do the points on the 0 and 2π lines in a cylindrical coordinate system. Care must be taken, however, to identify the interior points coinciding with the extensions from such points beyond the field in the transformed space. This correspondence was noted above in each of the configurations shown above, being indicated by the dashed connecting lines joining the two members of a pair of re-entrant boundaries. There are essentially four types of pairs of re-entrant boundaries, as illustrated in the following discussion of derivative correspondence. In these illustrations one exterior point, and its corresponding interior point, are shown for each case. The converse of the correspondence should be evident in each configuration.

For the configurations involving a change in the coordinate species at the cut, not only must the coordinate directions be taken into account as the cut is crossed, but also the coordinate species may need to be interpreted differently from that established across the cut in order to remain on the same sheet as the cut is crossed. For example, points on an η-line belonging to section A in the figure on p. 52, but located outside the right side of this region, are coincident with points on a ξ-line of region B at a corresponding distance (in the transformed region) below the top of this region.

B. Derivative Correspondence

Care must be taken at branch cuts to represent derivatives correctly in relation to the particular side of the cut on which the derivative is to be used. The existence of branch cuts indicates that the transformed region is multi-sheeted, and computations must remain on the same sheet as the cut is crossed. Remaining on the same sheet means continuing the coordinate lines across the cut coincident with those of the adjacent region, but keeping the same interpretation of coordinate line species and directions as the cut is crossed, rather than adopting those of the adjacent region. As noted above, points outside a region across a cut in the transformed space are coincident with points inside the region across the other member of the pair of re-entrant boundary segments corresponding to the cut in the transformed space. The positive directions of the curvilinear coordinates to be used at these points inside the region across the other member of the pair in some cases are the same as the defined directions there, but in other cases are the opposite directions. As noted above, the coordinate species may change also.

For cuts located on opposing sides of the transformed region, the proper form is simply a continuation across the cut. Thus in the configuration on p. 29, with a computation site on the right side of the transformed region, i.e., on the upper side of the cut in the physical plane, we have points to the right of the site (below the cut in the physical plane) coinciding with points to the right of the left side of the transformed region (below the cut in the physical plane) as noted above. When ξ-derivatives and η-derivatives for use outside the right side of the transformed region are represented inside the left side, the positive directions of ξ and η to be used there are to the right and upward, respectively, as is illustrated below. (In this and the following figures of the section, the dotted arrows indicate the proper directions to be used at the interior points coincident with the required exterior points, i.e., on the same sheet across the cut, while solid arrows indicate the locally established directions for the coordinate lines, i.e., on a different sheet.)

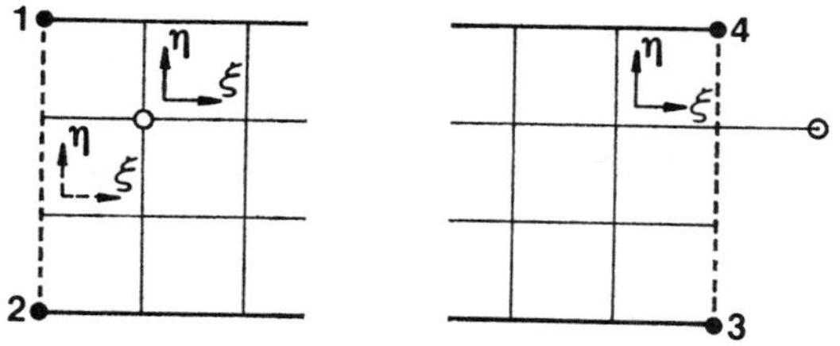

With the two sides of the cut both located on the same coordinate line, i.e., on the same side of the transformed region as in the configuration on p. 30, however, the situation is not as simple as the above. In this case, when the computation site is on the left branch of the cut in the transformed region (on the lower branch in the physical region), the points below this boundary in the transformed region coincide with points located above the right branch of the cut (above the cut in the physical region) at mirror-image positions, as has been noted earlier. The η-derivatives for use at such points below the left branch thus must be represented at these corresponding points above the right branch. The positive direction of η for purposes of this calculation of derivatives above the right branch, for use below the left branch, must be taken as downward, not upward. There is a similar reversal in the interpretation of the positive direction of ξ. This is in accordance with the discussion on p. 31. These interpretations are illustrated below:

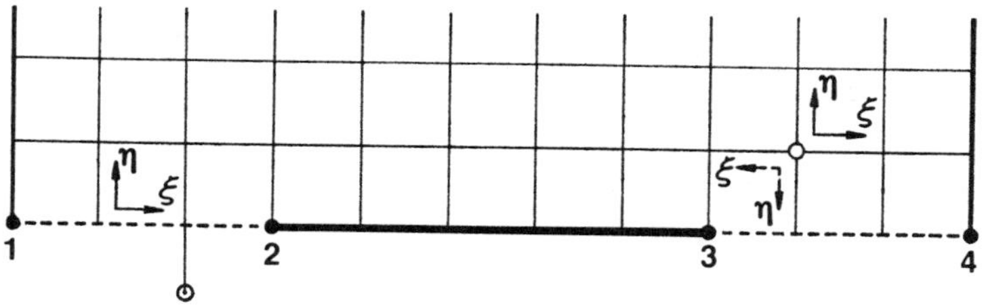

In the configuration on p. 40, where two sides of a cut face each other across a void, there is really no problem of interpretation, since the directions in the configuration are treated simply as if the void did not exist. This correspondence is as shown below:

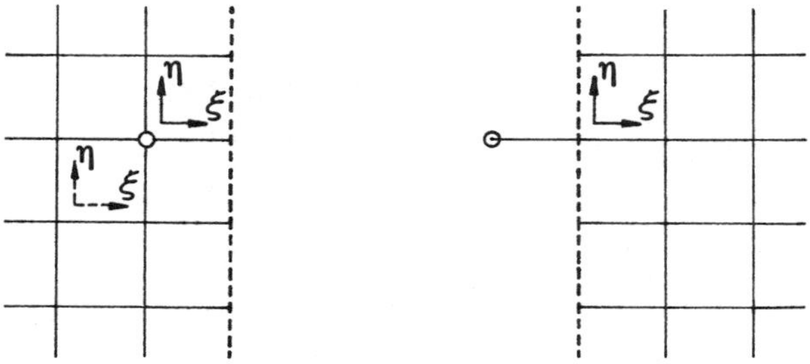

In all cases the interpretation of the positive directions of the curvilinear coordinates must be such as to preserve the direction in the physical region, i.e., on the same sheet, as the cut is crossed. In the cases where the coordinate species change at the cut, the situation is even more complicated. Thus on the left side, segment 6-7, of the slab interface between the inner and outer systems in the embedded configuration on p. 46, where the species changes across the cut, the correspondence is as follows:

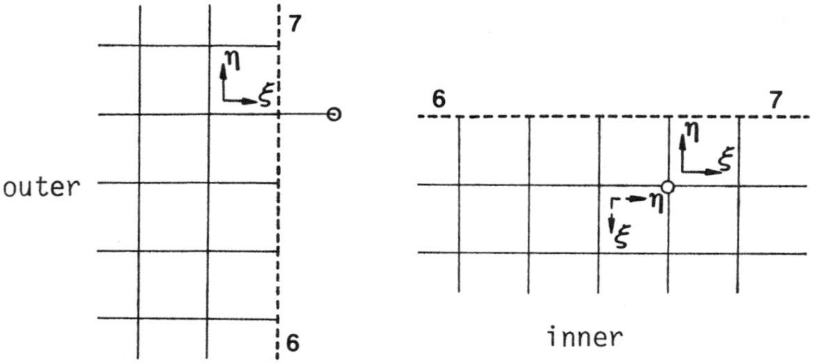

Thus, when a ξ-derivative is needed outside the outer sytem, for use inside the left slab interface, the positive ξ-direction at the corresponding points inside the inner system must be taken to coincide with the negative η-direction of the inner system. Similarly, an η-derivative would be represented taking the positive η-direction to coincide with the positive ξ-direction of the inner system. In an analogous fashion, a ξ-derivative needed outside the inner system, for use inside the segment 6-7, would be represented at the corresponding point inside the outer system, i.e., to the left of the left slab side, but with the positive ξ-direction taken to be the positive η-direction of the outer system. An η-derivative would be represented similarly, taking the positive η-direction to be the negative ξ-direction of the outer system.

A ξ-derivative to the left of the right side of the slab in the outer system would be represented below segment 12-5 or 8-12, as the case may be, but with the positive ξ-direction taken to be the positive η-direction of the inner system. Similarly, an η-derivative would be represented taking the positive η-direction to be the negative ξ-direction of the inner system. For a ξ-derivative above the bottom of the slab in the outer system, the correspondence is to below the segment 5-6 inside the inner system, with the positive ξ-direction taken to be the negative ξ-direction of the inner system. The η-derivative is represented taking the positive η-direction to be the negative η-direction of the inner sytem. Finally, for derivatives below the top of the slab in the outer system, the correspondence is to below the segment 7-8 inside the inner system, with both the species and direction of the coordinates unchanged.

The proper interpretation of coordinate species and

direction across branch cuts for all the other configurations discussed above can be inferred directly from these examples. A conceptual joining of the two members of a pair of re-entrant boundaries in accordance with the dashed line notation used on the configurations given in this chapter will always show exactly how to interpret both the coordinate species and directions in order to remain on the same sheet and thus to maintain continuity in derivative representation across the cut. Examples of the proper difference representation are given in the following section. The complexities of this correspondence can be completely avoided, however, by using surrounding layers around each block in a segmented structure as discussed in the next section.

6. Implementation

As discussed above, the transformed region is always comprised of contiguous rectangular blocks by construction. This occurs because of the essential fact that one of the curvilinear coordinates is defined as constant on each segment of the physical boundary. Consequently, each segment of the physical boundary corresponds to a plane segment of the boundary of the transformed region that is parallel to a coordinate plane there. The complete boundary of the transformed region then is composed of plane segments, all intersecting at right angles. Although the transformed region may not be a simple six-sided rectangular solid, it can be broken up into a contiguous collection of such solids, here called blocks.

Now it is noted in Chapter III that the increments $\Delta\xi^i$ cancel from all difference expressions, and that the actual values of the curvilinear coordinates ξ^i are immaterial. The coordinates in the transformed region can thus be considered simple counters identifying the points on the grid. This being the case, and the transformed region being comprised of a collection of rectangular blocks, it is convenient to identify the grid points with integer values of the curvilinear coordinates in each block, and thus to place the cartesian coordinates of a grid point in $\underline{r}_{ijk}$, where the subscripts (i,j,k) here indicate position (ξ^1,ξ^2,ξ^3) in the transformed region. (In coding, a fourth index may be added to identify the block.) In each block, the curvilinear coordinates are then taken to vary as $\xi^i = 1,2,\ldots,I^i$ over the grid points, where I^i is the number of points in the ξ^i-direction. Grid points on a boundary segment of the transformed region will be placed in $\underline{r}_{ijk}$ with one index fixed.

Now each block has six exterior boundaries, and may also have any number of interior boundaries (cf. the slab and slit configurations of Section 3), all of which will always be plane segments intersecting at right angles, although the occurence of interior boundaries can be avoided if desired by breaking the block up into a collection of smaller blocks as discussed in Section 4. The boundary segments in the transformed plane may correspond to actual segments of the physical boundary, or may correspond to cuts in the physical region. As discussed in Section 5, these cuts are not physical boundaries, but rather are interfaces across which the field is re-entrant on itself. A boundary segment in the transformed region corresponding to such a cut then is an interface across which one block is connected with complete continuity to another block, or to another side of itself, several examples having been given above in this chapter.

Depending on the type of grid generation system used (cf., the later chapters), the cartesian coordinates of the grid points on a physical boundary segment may either be specified or may be free to move over the boundary in order to satisfy a condition, e.g., orthogonality, or the angle at which coordinate lines intersect the boundary.

To set up the configuration of the transformed region, a correspondence is established between each (exterior or interior) segment of the boundary of the transformed region and either a segment of the physical boundary or a segment of a cut in the physical region. This is best illustrated by a series of examples using the configurations of this chapter. The first step in general is to position points on the physical boundary, or on a cut, which are to correspond to corners of the transformed region (exterior or interior). As noted in Section 3, these points do not have to be located at actual corners (slope discontinuities) on the physical boundary.

For example, considering the two-dimensional simply-connected region on p. 23, four points on the physical boundary are selected to correspond to the four corners of the empty rectangle that forms the transformed region here:

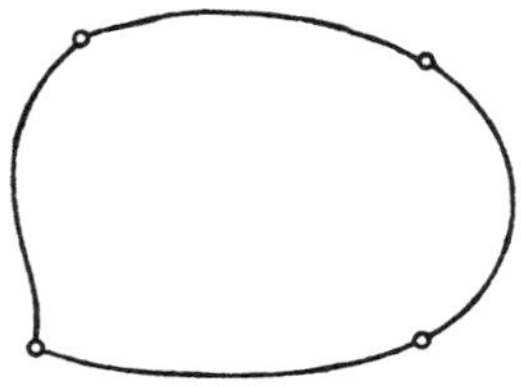

Now, considering any one of these four points, one species of curvilinear coordinate will run from that point to one of the two neighboring corner points, while the other species will run to the other neighbor:

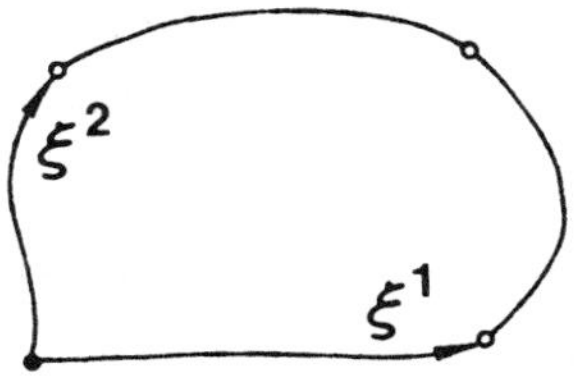

The corresponding species of coordinates will run to connect opposite pairs of corner points:

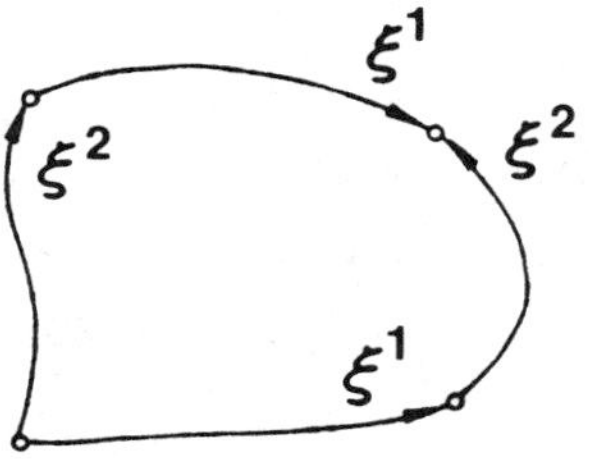

Since the curvilinear coordinates are to be assigned integer values at the grid points, ξ^i is to vary from 1 at one corner to a maximum value, I^i, at the next corner, where I^i is the number of grid points on the boundary segment between these two corners. Thus, proceeding clockwise from the lower left corner, the cartesian coordinates of the four corner points are placed in $\underline{r}_{1,1}$, $\underline{r}_{1,J}$, $\underline{r}_{I,J}$, and $\underline{r}_{I,1}$, where I^1 = I and I^2 = J.

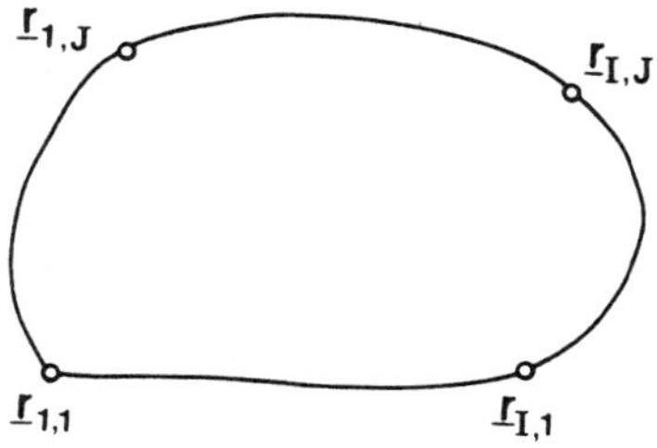

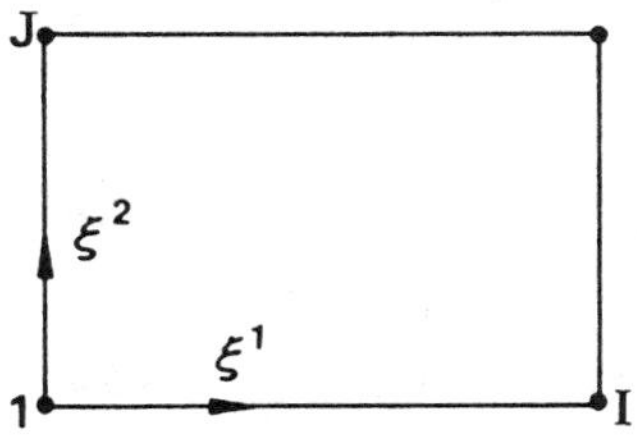

The boundary specification is then completed by positioning

I-2 points on the lower and upper boundary segments of the physical region as desired, and J-2 points on the left and right segments. The cartesian coordinates of these points on the lower and upper segments are placed in $\underset{\sim}{r}_{i,1}$ and $\underset{\sim}{r}_{i,J}$, respectively, for i from 2 to I-1, and those on the left and right segments are placed, respectively, in $\underset{\sim}{r}_{1,j}$ and $\underset{\sim}{r}_{I,j}$ for j from 2 to J-1.

This process of boundary specification can be most easily understood by viewing the rectangular boundary of the transformed region, with I equally-spaced points along two opposite sides and J equally-spaced points along the other two sides, conceptually, as being deformed to fit on the physical boundary. The corners can be located anywhere on the physical boundary, of course. Here the point distribution on the sides can be conceptually stretched and compressed to position points as desired along the physical boundary. The cartesian coordinates of all the selected point locations on the physical boundary are then placed in $\underset{\sim}{r}_{ij}$, as described above.

This conceptual deformation of the rectangular boundary of the transformed region to fit on the physical boundary serves to quickly illustrate the boundary specification for the doubly-connected physical field shown on p. 29, which involes a cut. Thus I points are positioned as desired clockwise around the inner boundary of the physical region from 2 to 3, and I points are positioned as desired, also in clockwise progression, around the outer boundary from 1 to 4. The cartesian coordinates of these points on the inner boundary are placed in $\underset{\sim}{r}_{i,1}$, and those on the outer boundary in $\underset{\sim}{r}_{i,J}$, with i from 1 to I. Note that here the first and last points must coincide on each boundary, i.e., $\underset{\sim}{r}_{I,1} = \underset{\sim}{r}_{1,1}$ and $\underset{\sim}{r}_{I,J} = \underset{\sim}{r}_{1,J}$. The left and right sides of the transformed region (i = 1 and i = I) are re-entrant boundaries, corresponding to the cut, and hence values on these boundaries are not set but will be determined by the generation system. The system must provide that the same value appears on both of these sides, i.e., $\mathbf{r}_{I,j} = \mathbf{r}_{1,j}$ for all j from 2 to J-1.

The conceptual deformation of the rectangle for a C-type configuration is illustrated on p. 31. Here, with I1 the number of points on the segments 1-2 and 3-4 (which must have the same number of points), I2 points are positioned as desired around the inner boundary in the physical region in a clockwise sense from 2 to 3, and the cartesian coordinates of these points are placed in $\mathbf{r}_{i,1}$ for i from I1 to I1+I2-1.

The first and last of these points must be coincident, i.e., $\mathbf{r}_{I1,1} = \mathbf{r}_{I1+I2-1,1}$. Now the top, and the left and right sides, of the rectangle are deformed here to fit on the outer boundary of the physical region. (In the illustration given, the two top corners are placed on the two corners that occur in the physical boundary, a selection that is logical but not mandatory.) The cartesian coordinates of the J points(positioned as desired on the segment 4-5 of the physical boundary) are placed in $\mathbf{r}_{I,j}$, proceeding upward on the physical boundary from 4 to 5 for j = 1 to J, and those on the segment 1-6 are placed in $\mathbf{r}_{1,j}$, but proceeding downward on the physical boundary from 1 to 6 for j = 1 to J. Finally, the cartesian coordinates of the I selected points on the physical boundary segment 6-5 are placed in $\mathbf{r}_{i,J}$, proceeding clockwise from 6 to 5 for i = 1 to I. Since the same number of points must occur on the top and bottom of the rectangle, we must have I = 2(I1-1)+I2. Here the portions of the lower side of the rectangle, i.e., i from 2 to I1-1, and from I1+I2 to I-1 with j=1, are re-entrant boundaries corresponding to the cut, and hence no values are to be specified on these segments. The generation system must make the correspondence $\mathbf{r}_{i,1} = \mathbf{r}_{I-i+1,1}$ for i=2 to I1-1 on these segments.

The conceptual deformation of the boundary of the transformed regions also serves for the slab configuration on p. 27, where the interior rectangle deforms to fit the interior physical boundary, while the outer rectangle deforms to fit the outer physical boundary. On the inner boundary, the cartesian coordinates of J2-J1+1 selected points on the segment 5-8 of the physical boundary are placed in $\mathbf{r}_{I1,j}$ for j from J1 to J2, proceeding upward on the physical boundary from 5 to 8, where J1 and J2 are the j-indices of the lower and upper sides, respectively, of the interior rectangle and I1 is the i-index of the left side of this rectangle. Similarly, J2-J1+1 points are positioned as desired on the segment 6-7 of the physical boundary and are placed in $\mathbf{r}_{I2,j}$, where I2 is the i-index of the right side of the inner rectangle. Also I2-I1+1 points on the segments 5-6 and 8-7 of the physical boundary are placed in $\mathbf{r}_{i,J1}$ and $\mathbf{r}_{i,J2}$, respectively, for i from I1 to I2, proceeding to the right on each segment. The outer boundary is treated as has been described for an empty rectangle. Here there will be J1-1 coordinate lines running from left to right below the inner boundary, and J-J2 lines running above the inner boundary. Similarly, there will be I1-1 lines running upward to the left of the interior boundary and I-I2 lines to the right. Thus the specifications of the desired number of coordinate lines running on each side of the inner boundary serves to determine the indicies I1, I2, J1, and J2. Note that the points inside the slab, i.e., $I1 < i < I2$ and $J1 < j < J2$ are simply excluded from the calculation.

The slit configuration, illustrated on p. 28, can also be treated via the conceptual deformation, but now with a portion of a line inside the rectangle opening to fit the interior boundary of the physical region. This requires that provision be made in coding for two values of the cartesian coordinates to be stored on the slit. If the i-indices of the slit ends, 5 and 6, are I1 and I2, respectively, then the cartesian coordinates of I2-I1+1 points positioned as desired on the lower portion of the physical interior boundary, again proceeding from 5 to 6, are placed in a one-dimensional array, while the coordinates of the same number of points selected on the upper portion of the physical interior boundary, again proceeding from 5 to 6, are placed in another one-dimensional array. The first and last points in one of these arrays must, of course, coincide with those in the other. Then the generation system must read values into $\underset{\sim}{r}_{i,J1}$ for i from I1 to I2 (J1 being the j-index of the slit) from the former array for use below the slit, or values from the latter array for use above. (As has been noted, the use of a composite structure eliminates the need for these two auxiliary arrays.) Note that the index values I1 and I2 are determined by the number of lines desired to run upward to the left and right of the interior boundary, respectively, i.e., I1-1 lines on the left and I-I2 on the right. Similarly, there will be J1-1 lines below the interior boundary, and J-J1 above.

Configurations, such as those illustrated on pp. 24-25, which involve slabs or slits that intersect the outer boundary are treated similarly, with points inside the slab again being simply excluded from the calculations. Also multiple slab or slit arrangements are treated by obvious extensions of the above procedures. Here the indices corresponding to each slab or slit will be determined by the number of points on the interior boundary segments and the number of coordinate lines specified to run between the various boundaries. For example, in the slit configuration shown on p. 33, the ends of the horizontal slit would be at i-indices I1 and I2, where I1-1 lines run vertically to the left of the slit and there are I2-I1+1 points on the slit. The vertical slit would be at i=I3 where there are I3-I2-1 vertical lines between this slit and the horizontal slit (and I-I2 lines to the right). Similarly, if the j-indices of the ends of the vertical slit or J1 and J2, there will be J1-1 horizontal lines below this slit and J-J2 lines above. With the j-index of the horizontal slit as J3, there will be J3-1 horizontal lines below this slit and J-J3 above. Provision will now have to be made in coding for two one-dimensional arrays for each slit to hold the cartesian coordinates of the points on the segments of the physical interior boundaries corresponding to the two sides of each slit. Again this coding complexity is avoided in the composite structure.

The use of the conceptual deformation of the rectangle to set up the boundary configuration for the case with multiple interior boundaries on p. 34 should follow with little further explanation. Here there must be the same number of points on the pair of segments 2-3 and 6-7, which correspond to the two segments forming the interior boundary on the right. There must also be the same number of points on the pair, 3-4 and 5-6, corresponding to the cut connecting the two interior boundaries. Finally the number of points on the outer boundary must, of course, be the same as that on the bottom boundary. Note also that the values of the cartesian coordinates placed at 2 must be the same as are placed at 7; those at 3 must be the same as those at 6, and those at 4 the same as at 5. Values are not set on the cuts, of course, but the generation system must provide that values at points on the segment from 3 to 4 are the same as those on the segment 5-6, but proceeding from 6 to 5. Also values on the segments 2-1 and 7-8 must be the same, proceeding upward in each case.

Following the conceptual deformation of the rectangular boundaries of the transformed region and the indexing system illustrated above, it now should be possible to set up the more complicated configurations such as the embedded regions shown in Section 3C. As noted there, however, the most straightforward and general approach to such more complicated configurations is to divide the field into contiguous rectangular blocks, each of which has its own intrinsic set of curvilinear coordinates and hence its own (i,j,k) indexing system. The necessary correspondence between the individual coordinate systems across the block interfaces was discussed in some detail in Section 3C. This block structure greatly simplifies the setup of the configuration. For example, consider the 3-block structure shown on p. 49 for the physical field shown on p. 48, for which the blocks are as follows:

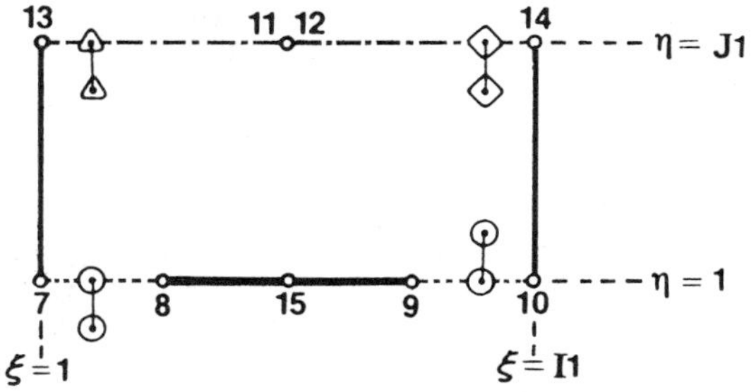

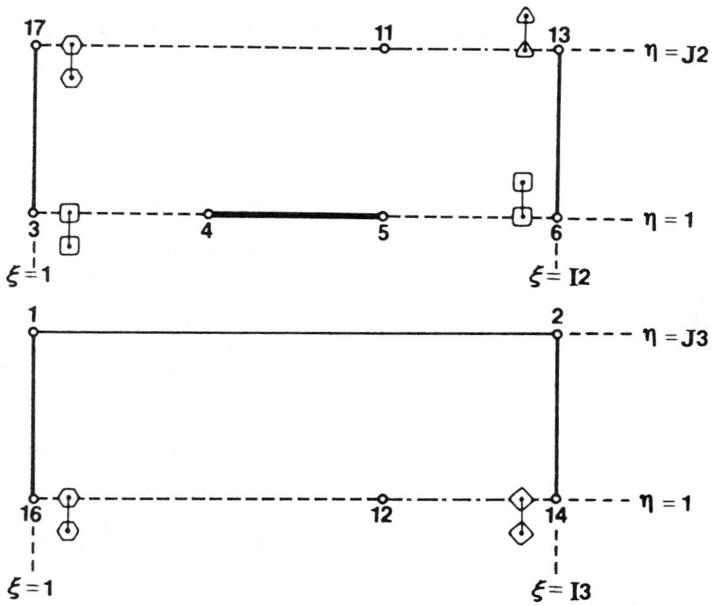

Here the selected points on the right interior boundary (segment 8-15-9) are placed in $\mathbf{r}_{i,1}$ of the first block, for i from the i-index at 8 to that at 9, proceeding clockwise from 8 to 9 on the physical boundary. (The difference between these two i-indices here is equal to the number of points on this interior boundary, less one.) Similary, the selected points on the left interior boundary (segment 4-5) are placed in $\mathbf{r}_{i,1}$ of the second block for i from the i-index at 4 to that at 5, proceeding clockwise from 4 to 5 on the physical boundary. The selected points on the outer boundary of the physical region are placed in $\mathbf{r}_{1,j}$ of the third block for j from 1 to J3, in $\mathbf{r}_{i,J3}$ for i from 1 to I3, and in $\mathbf{r}_{I3,j}$ for j from J3 to 1, proceeding from 16 to 1 to 2 to 14 on the physical boundary. Points on the remainder of the physical outer boundary are placed in $\mathbf{r}_{1,j}$ of the second block for j from 1 to J2 and in $\mathbf{r}_{I2,j}$ for j from J2 to 1, proceeding from 3 to 17 for the former and from 13 to 6 for the latter, and in $\mathbf{r}_{1,j}$ of the first block for j from

1 to J1 and in $r_{I1,j}$ for j from J1 to 1, proceeding from 7 to 13 for the former and from 14 to 10 for the latter.

Since the three blocks must fit together we have I3=I2, (I1+1)/2 equal to the difference in i-indices between 11 and 13 in the second block and to that between 12 and 14 of the third block. The quantities J1, J2, and J3 determine how many C-type lines occur in each block, and can be chosen independently. Here the segment 11-13 on the top of the first block interfaces with the corresponding segment on the top of the second block. The segment 12-14, which forms the remainder of the top of the first block, interfaces with the corresponding segment on the bottom of the third block. Finally, the segment 12-16, which forms the remainder of the bottom of the third block, interfaces with segment 11-17, which forms the remainder of the top of the second block. The segments 3-4 and 6-5 on the bottom of the second block interface with each other in the order indicated, as do also the segments 7-8 and 10-9 on the bottom of the first block.

In coding, this block structure can be handled by using a fourth index to identify the block, placing an extra layer around each block, (i=0 and I+1, j=0 and J+1) and providing an image-point array by which any point of any block can be paired with any point of any other, or the same, block. Such pairs of points are coincident in the physical region, being on or across block interfaces, and consequently are to be given the same values of the cartesian coordinates by the generation system. This imaging extends to the extra layer surrounding each block, so that appropriate points inside other blocks can be identified for use in difference representations on the block interfaces that require points outside the block, (cf. Section 5).

Interface correspondence then can be established by

input by setting the image-point correspondence on the appropriate block sides, i.e., placing the (i,j,k) indices and block number of one member of a coincident pair of points in the image-point array at the indices and block number of the other member of the pair. This correspondence is indicated on the block diagram on pp. 85-86 by the points enclosed in certain geometric symbols.

Thus, for the 3-block configuration considered above, the indices (I1-i+1, 1) and block number 1, corresponding to a point on the segment 9-10 of the first block, would be placed in the image-point array at the point (i,1) on the segment 7-8 of this block, and vice versa. A similar pairing occurs for points on the segments 3-4 and 5-6 of the second block. The indices (I2-i+1, J2) and block number 2, corresponding to a point on the segment 11-13 of the second block would be placed in the image-point array at the point (i,J1) of the first block on the segment 13-11 of that block, and vice versa. The indices (I3-I1+i, 1) and block number 3 (a point on the segment 12-14 of the third block) would be placed in the array at the point (i, J1) of the first block for a point on the segment 12-14 of that block. Finally, the indices (i,1) and block number 3 (point on segment 16-12 of the third block) would be placed in the array at the point (i,J2) of the second block for a point on the segment 17-11 of that block. The remaining segments all correspond at portions of the physical boundary and hence do not have image points.

In the same manner the following image correspondence can be set between interior points and points on the surrounding layers in order to establish difference representations across the block interfaces: (This correspondence is indicated symbolically on the block diagram on pp. 85-86 by geometric symbols.):

Object Point			Image Point		
Segment	Block	Indices	Segment	Block	Indices
○ 7-8	1	i,0	9-10	1	I1-i+1,2
7-8	1	i,2	9-10	1	I1-i+1,0
□ 3-4	2	i,0	5-6	2	I2-i+1,2
3-4	2	i,2	5-6	2	I2-i+1,0
△ 13-11	1	i,J1-1	11-13	2	I2-i+1,J2+1
13-11	1	i,J1+1	11-13	2	I2-i+1,J2-1
◇ 12-14	1	i,J1-1	12-14	3	I3-I1+i,0
12-14	1	i,J1+1	12-14	3	I3-I1+i,2
⬠ 17-11	2	i,J2-1	16-12	3	i,0
17-11	2	i,J2+1	16-12	3	i,2

As noted, all of this information would be input into the image-point array. Then with values of the cartesian coordinates at the image points on the surrounding layer set equal to those at the corresponding object point inside one of the blocks, it is possible to use the same difference representations on the interfaces that are used in the interior.

The discussion given in this chapter should now allow the image-point input to be constructed for any configuration of interest. As noted, it is not necessary that the coordinate species remain the same as the interface is crossed. Thus, for instance, a point on the right side of one block could be paired with one on the bottom of another block. In such a case the image point of the point (I+1, j) outside the first block would be the point (j,2) inside the second block. Similarly the image of the point (i,0) below the second block would be the point (I-1, i) inside the first block. The correct difference representation across interfaces is thus automatically established, eliminating the need for the concern with passage onto different sheets discussed in detail earlier in this chapter.

This greatly simplifies the coding, since with the surrounding layers and the use of the image points, all of the derivative correspondences are automatic and do not have to be specified for each configuration. It is only necessary to specify the point correspondence by input. This construction also allows codes for the numerical solution of partial differential equations on the grid to be written to operate on rectangular blocks. Then any configuration can be treated by sweeping over all the blocks. The surrounding layers of points and the image correspondence provide the proper linkage across the block interfaces. In an implicit solution the values on the interfaces would have to be updated iteratively in the course of the solution. The solution for the generation of the grid would similarly keep the interface and surrounding layer values updated during the course of the iterative solution.

This, of course, maintains complete continuity across the block interfaces. If complete continuity is not required, then the surrounding layer is not required and the interfaces would be treated in the same manner as are physical boundaries. However, the surrounding layer and the point correspondence thereon discussed above might still be needed for the numerical solution to be done on the grid.

The extension of all of the above concepts and structures to three dimensions is direct, the illustrations having been given in two dimensions only for economy of presentation.

Exercises

1. Sketch the grid when the physical region shown below is transformed to an empty rectangle as indicated.

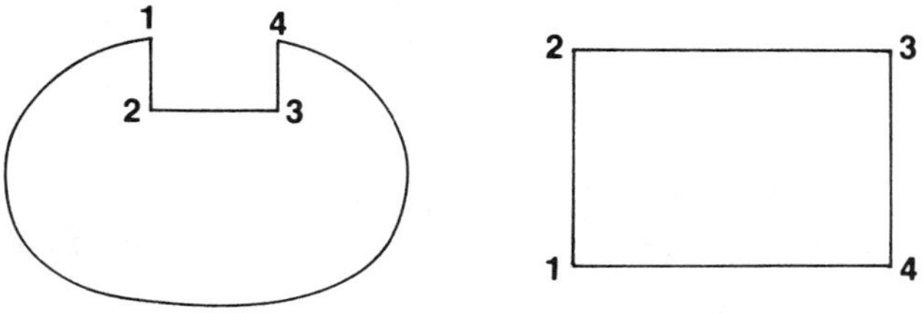

2. Locate the cuts on the grid in the physical region for the grids shown on pp. 35-37.

3. Sketch the grid for a O-grid, a C-grid, and a slit configuration for a cascade arrangement (a periodic stack of bodies, e.g., turbine blades.)

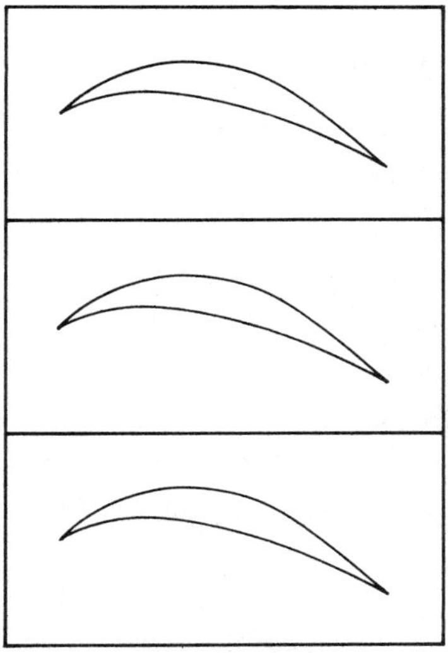

4. For the configuration shown below, let body II trans- from form to a slit in a C-type system about I. Sketch the grid lines and show the transformed region configuration.

5. Sketch the surface grid on a sphere with eight "corners" on the sphere. This is analogous to the 2D grid for the circle with four "corners" on p. 23. This configuration would be more appropriate for embedding a spherical region in a composite structure than would the usual polar system.

6. Diagram the transformed region configuration for a polar coordinate system between two concentric spheres. Here the polar axis will map to an entire side of the transformed region.

7. Sketch the surface grid on a circular cylinder with a hemispherical cap for two cases: (1) with a cylindrical-coordinate type grid on the hemisphere and (2) with four "corners" on the intersection of the hemisphere with the cylinder. The latter case would be more appropriate for a composite system.

8. For the two-slit configuration on p. 33, diagram a composite system composed of empty blocks. Show all cuts and the correspondence between pairs thereof.

9. Sketch the grid and diagram the blocks for a composite two-block system for a circular pipe T-section. It is

necessary to use a cross-sectional grid of the type shown on p. 23, having four "corners" on the circle, since cylindrical grids would not join with line continuity.

10. Diagram the block structure and grid for a six-block composite system for the region between two concentric spheres, based on the surface grid of Exercise 10. Note that no polar axis occurs with this configuration.

11. Consider a region between two boundaries, both of which are formed of cylinders with hemi-spherical caps, these being coaxial with one inside the other. Sketch the grid and diagram the blocks for a three-block system, with one block corresponding to the annular region between the caps, for the following two configurations: (1) with the polar axis connecting the caps and (2) with no polar axis. In the latter case each of four sides of one block will correspond to one of four portions of one side of the other block.

12. Diagram the point correspondence across all the cuts in the two-body O-grid on p. 34. Also give the relation between the indices of corresponding points on the cuts. Finally give the relation between the indices of points on a surrounding layer of points and points inside the field inside the cuts.

13. For one block of the system on p. 52, give the correspondence between indices of points on the surrounding layers and points inside adjacent blocks for the cuts.

14. Sketch the grid for a 2-D composite system having two circular regions embedded in a grid which is generally rectangular. Let one of the circular regions have a

cylindrical-coordinate type of grid and the other have a grid of the type with four "corners" on the circle as on p. 23.

15. Show that is is not possible to handle the point correspondence across the cuts in the embedded slab type system shown on p. 46 (a 2-block system) by using an extra layer of points just inside the slab in the outer system. Also show that it is possible to represent the correspondence across the cuts using surrounding layers if a 4-block composite system is used.

III. TRANSFORMATION RELATIONS

The transformation relations from cartesian coordinates to a general curvilinear system are developed here using certain concepts from differential geometry and tensor analysis, which are introduced only as needed. Warsi [15] has given an extensive collection of concepts from tensor analysis and differential geometry applicable to the generation of curvilinear coordinate systems. Another discussion is given in Eiseman [16], where these concepts are developed as part of a general survey on the generation and use of curvilinear coordinate systems. Eiseman includes a discussion on differential forms, which is a fundamental part of modern differential geometry, but primarily restricts his development to Euclidean space. In contrast, Warsi has given a classical development that includes curved space, but not differential forms.

Partial derivatives with respect to cartesian coordinates are related to partial derivatives with respect to curvilinear coordinates by the chain rule which may be written in either of two ways. If A is a scalar-valued function, then

$$A_{x_i} = \sum_{j=1}^{3} A_{\xi^j} (\xi^j)_{x_i} \qquad (i = 1,2,3) \tag{1}$$

or, equivalently,

$$A_{\xi^i} = \sum_{j=1}^{3} A_{x_j} (x_j)_{\xi^i} \qquad (i = 1,2,3) \tag{2}$$

Either formulation may be used to relate the cartesian and curvilinear derivatives of the function A. However, there is a difference in the transformation derivatives which must be inserted in these relations. In the first case one must be able to evaluate (or approximate) the vectors

$$\nabla\xi^{i} \qquad (i = 1,2,3)$$

whereas, the second case requires

$$\underset{\sim}{r}_{\xi^i} \qquad (i = 1,2,3)$$

Thus all the transformation relations may be based on either of these two sets of vectors. Various properties of, and relationships between, these vectors are developed and applied in this chapter to provide the necessary transformation relations.

1. Base Vectors

The curvilinear coordinate lines of a three-dimensional system are space curves formed by the intersection of surfaces on which one coordinate is constant. One coordinate varies along a coordinate line, of course, while the other two are constant thereon. The tangents to the coordinate line and the normals to the coordinate surface are the base vectors of the coordinate system.

A. Covariant

Consider first a coordinate line along which only the coordinate ξ varies:

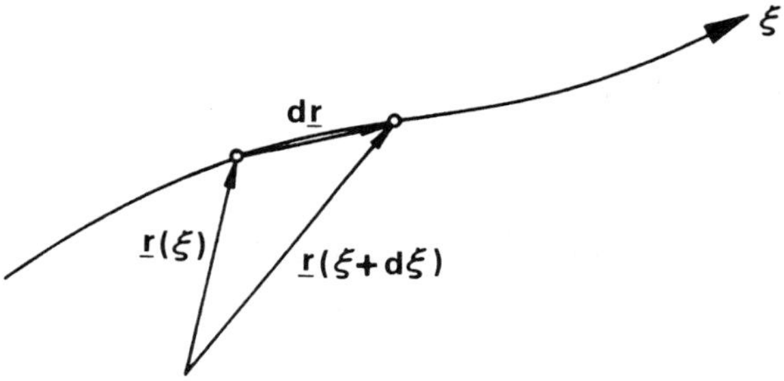

Clearly a tangent vector to the coordinate line is given by

$$\lim_{d\xi\to 0} \frac{\underset{\sim}{r}(\xi + d\xi) - \underset{\sim}{r}(\xi)}{d\xi} = \underset{\sim}{r}_\xi$$

(Coordinates appearing as subscripts will always indicate partial differentiation.) These tangent vectors to the three coordinate lines are the three covariant base vectors of the curvilinear coordinate system, designated

$$\underset{\sim}{a}_i \equiv \underset{\sim}{r}_{\xi^i} \quad (i = 1,2,3) \tag{3}$$

where the three curvilinear coordinates are represented by ξ^i $(i = 1,2,3)$, and the subscript $\underline{i}$ indicates the base vector corresponding to the ξ^i coordinate, i.e., the tangent to the coordinate line along which only ξ^i varies.

B. Contravariant

A normal vector to a coordinate surface on which the coordinate ξ is constant is given by $\underset{\sim}{\nabla}\xi$:

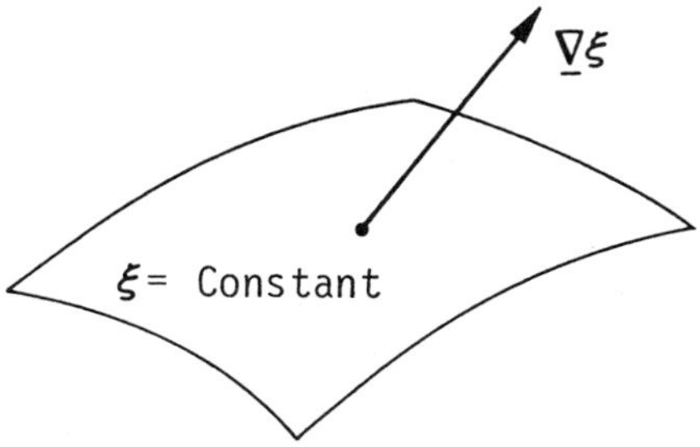

These normal vectors to the three coordinate surfaces are the three <u>contravariant</u> base vectors of the curvilinear coordinate system, designated

$$\underset{\sim}{a}^i \equiv \underset{\sim}{\nabla}\xi^i \qquad (i = 1,2,3) \tag{4}$$

Here the coordinate index $\underline{i}$ appears as a superscript on the base vector to differentiate these contravariant base vectors from the covariant base vectors. The two types of base vectors are illustrated in the following figure, showing an element of volume with six sides, each of which lies on some coordinate surface.

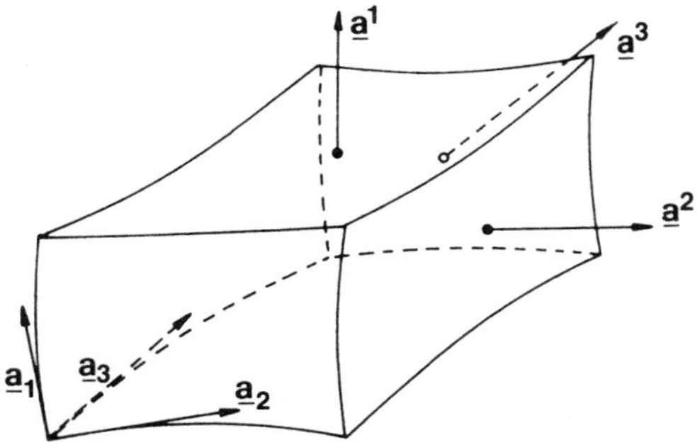

C. Orthogonality

Only in an orthogonal coordinate system are the two types of base vectors parallel, since for a non-orthogonal system, the normal to a coordinate surface does not necessarily coincide with the tangent to a coordinate line crossing that surface:

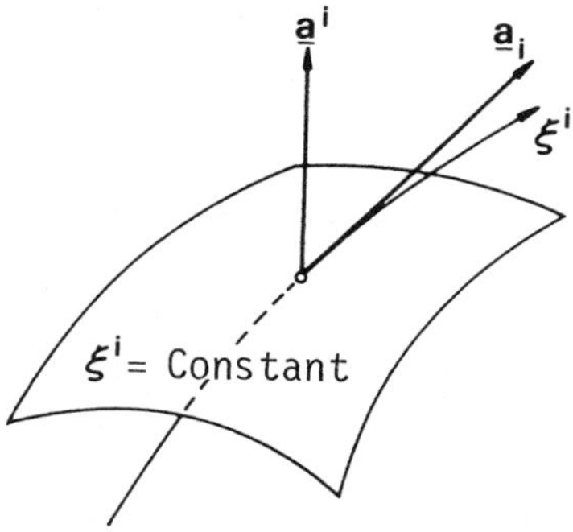

Also for an orthogonal system the three base vectors of each type are obviously mutually perpendicular.

2. Differential Elements

The differential increments of arc length, surface, and volume, which are needed for the formulation of the respective integrals, can be generated directly from the covariant base vectors. The general arc length increment leads also to the definition of a fundamental metric tensor.

A. Covariant metric tensor

The general differential increment (not necessarily along a coordinate line) of a position vector is given by

$$d\underset{\sim}{r} = \sum_{i=1}^{3} \underset{\sim}{r}_{\xi^i}\, d\xi^i = \sum_{i=1}^{3} \underset{\sim}{a}_i\, d\xi^i$$

An increment of arc length along a general space curve then is given by

$$(ds)^2 = |d\underset{\sim}{r}|^2 = \sum_{i=1}^{3} \sum_{j=1}^{3} \underset{\sim}{a}_i \cdot \underset{\sim}{a}_j\, d\xi^i d\xi^j$$

The general arc length increment thus depends on the nine dot products, $\underset{\sim}{a}_i \cdot \underset{\sim}{a}_j$, $(i = 1,2,3)$ and $(j = 1,2,3)$, which form a symmetric tensor. These quantities are the covariant metric tensor components:

$$g_{ij} = \underset{\sim}{a}_i \cdot \underset{\sim}{a}_j = g_{ji} \qquad (i = 1,2,3), \ (j = 1,2,3) \tag{5}$$

Thus the general arc length increment can be written as

$$(ds)^2 = \sum_{i=1}^{3} \sum_{j=1}^{3} g_{ij} \, d\xi^i d\xi^j \tag{6}$$

B. Arc length element

An increment of arc length on a coordinate line along which ξ^i varies is given by

$$ds^i = |\underset{\sim}{r}_{\xi^i}| d\xi^i = |\underset{\sim}{a}_i| d\xi^i = \sqrt{g_{ii}} \, d\xi^i \tag{7}$$

C. Surface area element

Also an increment of area on a coordinate surface of constant ξ^i is given by

$$dS^i = |\underset{\sim}{r}_{\xi^j} \times \underset{\sim}{r}_{\xi^k}| d\xi^j d\xi^k = |\underset{\sim}{a}_j \times \underset{\sim}{a}_k| d\xi^j d\xi^k \tag{8}$$

$$(i = 1,2,3) \qquad (i,j,k) \text{ cyclic}$$

Using the vector identity

$$(\underset{\sim}{A} \times \underset{\sim}{B}) \cdot (\underset{\sim}{C} \times \underset{\sim}{D}) = (\underset{\sim}{A} \cdot \underset{\sim}{C})(\underset{\sim}{B} \cdot \underset{\sim}{D}) - (\underset{\sim}{A} \cdot \underset{\sim}{D})(\underset{\sim}{B} \cdot \underset{\sim}{C}) \tag{9}$$

we have

$$|\underset{\sim}{a}_j \times \underset{\sim}{a}_k|^2 = (\underset{\sim}{a}_j \cdot \underset{\sim}{a}_j)(\underset{\sim}{a}_k \cdot \underset{\sim}{a}_k) - (\underset{\sim}{a}_j \cdot \underset{\sim}{a}_k)^2$$

$$= g_{jj} g_{kk} - g_{jk}^2 \tag{10}$$

so that the increment of surface area can be written as

$$dS^i = \sqrt{g_{jj}g_{kk} - g_{jk}^2}\, d\xi^j d\xi^k \qquad (i = 1,2,3)$$

$$(i,j,k) \text{ cyclic} \tag{11}$$

D. Volume element

An increment of volume is given by

$$dV = \underset{\sim}{r}_{\xi^i} \cdot (\underset{\sim}{r}_{\xi^j} \times \underset{\sim}{r}_{\xi^k})d\xi^i d\xi^j d\xi^k \quad (i,j,k)\text{cyclic}$$

$$= \underset{\sim}{a}_1 \cdot (\underset{\sim}{a}_2 \times \underset{\sim}{a}_3)d\xi^1 d\xi^2 d\xi^3 \tag{12}$$

But, by the identity (9),

$$[\underset{\sim}{a}_1 \cdot (\underset{\sim}{a}_2 \times \underset{\sim}{a}_3)]^2 = (\underset{\sim}{a}_1 \cdot \underset{\sim}{a}_1)[(\underset{\sim}{a}_2 \times \underset{\sim}{a}_3) \cdot (\underset{\sim}{a}_2 \times \underset{\sim}{a}_3)]$$

$$- |\underset{\sim}{a}_1 \times (\underset{\sim}{a}_2 \times \underset{\sim}{a}_3)|^2$$

Also from (9),

$$(\underset{\sim}{a}_2 \times \underset{\sim}{a}_3) \cdot (\underset{\sim}{a}_2 \times \underset{\sim}{a}_3) = (\underset{\sim}{a}_2 \cdot \underset{\sim}{a}_2)(\underset{\sim}{a}_3 \cdot \underset{\sim}{a}_3) - (\underset{\sim}{a}_2 \cdot \underset{\sim}{a}_3)^2$$

and by the vector identity

$$\underset{\sim}{A} \times (\underset{\sim}{B} \times \underset{\sim}{C}) = (\underset{\sim}{A} \cdot \underset{\sim}{C})\underset{\sim}{B} - (\underset{\sim}{A} \cdot \underset{\sim}{B})\underset{\sim}{C} \tag{13}$$

we have

$$\underset{\sim}{a}_1 \times (\underset{\sim}{a}_2 \times \underset{\sim}{a}_3) = (\underset{\sim}{a}_1 \cdot \underset{\sim}{a}_3)\underset{\sim}{a}_2 - (\underset{\sim}{a}_1 \cdot \underset{\sim}{a}_2)\underset{\sim}{a}_3$$

so that with the dot products replaced according to the definition (5),

$$[\underset{\sim}{a}_1 \cdot (\underset{\sim}{a}_2 \times \underset{\sim}{a}_3)]^2 = g_{11}(g_{22}g_{33} - g_{23}^2) - g_{13}^2 g_{22}$$

$$- g_{12}^2 g_{33} + 2g_{13}g_{12}g_{23}$$

$$= g_{11}(g_{22}g_{33} - g_{23}^2)$$

$$- g_{12}(g_{12}g_{33} - g_{13}g_{23})$$

$$+ g_{13}(g_{12}g_{23} - g_{13}g_{22})$$

This last expression is simply the determinant of the (symmetric) covariant metric tensor expanded by cofactors. Therefore

$$[\underset{\sim}{a}_1 \cdot (\underset{\sim}{a}_2 \times \underset{\sim}{a}_3)]^2 = \det|g_{ij}| = g \tag{14}$$

so that the volume increment can be written

$$dV = \sqrt{g}\, d\xi^1 d\xi^2 d\xi^3 \tag{15}$$

where $\sqrt{g}$ (called the Jacobian of the transformation) can be evaluated by either of the following expressions:

$$\sqrt{g} = \sqrt{\det|g_{ij}|} = \underset{\sim}{a}_1 \cdot (\underset{\sim}{a}_2 \times \underset{\sim}{a}_3) \tag{16}$$

3. Derivative Operators

Expressions for the derivative operators, such as gradient, divergence, curl, Laplacian, etc., are obtained by applying the Divergence Theorem to a differential volume increment bounded by coordinate surfaces. The gradient operator then leads to the expression of contravariant base vectors in terms of the covariant base vectors, and to the contravariant metric tensor as the inverse of the covariant metric tensor.

By the Divergence Theorem,

$$\iiint_V \underset{\sim}{\nabla} \cdot \underset{\sim}{A} \, dV = \oiint_S \underset{\sim}{A} \cdot \underset{\sim}{n} \, dS \tag{17}$$

for any tensor $\underset{\sim}{A}$, where $\underset{\sim}{n}$ is the outward-directed unit normal to the closed surface S enclosing the volume V. For a differential surface element lying on a coordinate surface we have, by Eq. (8),

$$\underset{\sim}{n} \, dS^i = \pm \underset{\sim}{a}_j \times \underset{\sim}{a}_k \, d\xi^j d\xi^k \tag{18}$$

with the choice of sign being dependent on the location of the volume relative to the surface. Then considering a differential element of volume, δV, bounded by six faces lying on coordinate surfaces, as shown in the figure on p. 98, we have, using Eq. (15) and (18),

$$\iiint_{\delta V} (\underset{\sim}{\nabla} \cdot \underset{\sim}{A}) \sqrt{g} \, d\xi^1 d\xi^2 d\xi^3$$

$$= \sum_{i=1}^{3} [\iint_{\delta S^i_+} \underset{\sim}{A} \cdot (\underset{\sim}{a}_j \times \underset{\sim}{a}_k) d\xi^j d\xi^k$$

$$- \iint_{\delta S^i_-} \underset{\sim}{A} \cdot (\underset{\sim}{a}_j \times \underset{\sim}{a}_k) d\xi^j d\xi^k] \tag{19}$$

where the notation δS^i_+ and δS^i_- indicates the element on two sides of the which ξ^i is constant and which are located at larger and smaller values, respectively, of ξ^i. Here, as usual, the indices (i,j,k) are cyclic.

A. Divergence

Proceeding to the limit as the element of volume shrinks to zero we then have an expression for the divergence:

$$\underset{\sim}{\nabla} \cdot \underset{\sim}{A} = \frac{1}{\sqrt{g}} \sum_{i=1}^{3} [(\underset{\sim}{a}_j \times \underset{\sim}{a}_k) \cdot \underset{\sim}{A}]_{\xi^i} \tag{20}$$

where, as noted, the subscript ξ^i on the bracket indicates partial differentiation.

A basic metric identity is involved here, since

$$\sum_{i=1}^{3} (\underset{\sim}{a}_j \times \underset{\sim}{a}_k)_{\xi^i} = \sum_{i=1}^{3} (\underset{\sim}{r}_{\xi^j} \times \underset{\sim}{r}_{\xi^k})_{\xi^i}$$

$$= \sum_{i=1}^{3} \underset{\sim}{r}_{\xi^j \xi^i} \times \underset{\sim}{r}_{\xi^k} + \sum_{i=1}^{3} \underset{\sim}{r}_{\xi^j} \times \underset{\sim}{r}_{\xi^k \xi^i}$$

The indices (i,j,k) are cyclic, and therefore the last summation may be written equivalently as

$$\sum_{i=1}^{3} \underset{\sim}{r}_{\xi^j} \times \underset{\sim}{r}_{\xi^k \xi^i} = \sum_{i=1}^{3} \underset{\sim}{r}_{\xi^k} \times \underset{\sim}{r}_{\xi^i \xi^j}$$

Since this is then the negative of the first summation we have the identity,

$$\sum_{i=1}^{3} (\underset{\sim}{a}_j \times \underset{\sim}{a}_k)_{\xi^i} = 0 \tag{21}$$

This is a fundamental metric identity which will be used several times in the developments that follow. This identity also follows directly from Eq. (20) for uniform $\underset{\sim}{A}$. It then follows that the divergence can also be written as

$$\underset{\sim}{\nabla} \cdot \underset{\sim}{A} = \frac{1}{\sqrt{g}} \sum_{i=1}^{3} (\underset{\sim}{a}_j \times \underset{\sim}{a}_k) \cdot \underset{\sim}{A}_{\xi^i} \tag{22}$$

Although the equations (20) and (22) are equivalent expressions for the divergence, because of the identity (21), the numerical representations of these two forms may not be equivalent. The form given by Eq. (20) is called the conservative form, and that of Eq. (22), where the product derivative has been expanded and Eq. (21) has been used, is called the non-conservative form. Recalling that the quantity ($a_j \times a_k$) represents an increment of surface area (cf. Eq. (8)), so that ($a_j \times a_k$) $\cdot$ A is a flux through this area, it is clear that the difference between the two forms is that the area used in numerical representation of the flux in the conservative form, Eq. (20), is the area of the individual sides of the volume element, but in the non-conservative form, a common area evaluated at the center of the volume element is used. The conservative form thus gives the telescopic collapse of the flux terms when the difference equations are summed over the field, so that this summation then involves only the boundary fluxes. This would seem to favor the conservative form as the better numerical representation of the net flux through the volume element.

It is important to note that since the conservative form of the divergence, and of the gradient, curl, and Laplacian to follow, is obtained directly from the closed surface integral in the Divergence Theorem, the use of the conservative difference forms for these derivative operators is equivalent to using difference forms for that closed surface integral. Therefore the finite volume difference formulation can be implemented by using these conservative forms directly in the differential equations of motion without the necessity of returning to the integral form of the equations of motion.

B. Curl

Since Eq. (17) is also valid with the dot products replaced by cross products, the conservative and non-conservative expressions for the curl follow immediately from Eq. (20) and (22):

$$\underset{\sim}{\nabla} \times \underset{\sim}{A} = \frac{1}{\sqrt{g}} \sum_{i=1}^{3} [(\underset{\sim}{a}_j \times \underset{\sim}{a}_k) \times \underset{\sim}{A}]_{\xi^i} \tag{23}$$

and

$$\underset{\sim}{\nabla} \times \underset{\sim}{A} = \frac{1}{\sqrt{g}} \sum_{i=1}^{3} (\underset{\sim}{a}_j \times \underset{\sim}{a}_k) \times \underset{\sim}{A}_{\xi^i} \tag{24}$$

These expressions can also be written, using Eq. (13), as

$$\underset{\sim}{\nabla} \times \underset{\sim}{A} = \frac{1}{\sqrt{g}} \sum_{i=1}^{3} [(\underset{\sim}{a}_j \cdot \underset{\sim}{A})\underset{\sim}{a}_k - (\underset{\sim}{a}_k \cdot \underset{\sim}{A})\underset{\sim}{a}_j]_{\xi^i} \tag{25}$$

and

$$\underset{\sim}{\nabla} \times \underset{\sim}{A} = \frac{1}{\sqrt{g}} \sum_{i=1}^{3} [(\underset{\sim}{a}_j \cdot \underset{\sim}{A}_{\xi^i})\underset{\sim}{a}_k - (\underset{\sim}{a}_k \cdot \underset{\sim}{A}_{\xi^i})\,\underset{\sim}{a}_j] \tag{26}$$

C. Gradient

Eq. (17) is also valid with $\underset{\sim}{A}$ replaced by a scalar, and the dot product replaced by simple operation on the left and multiplication on the right. Therefore the conservative and non-conservative expressions for the gradient also follow directly from Eq. (20) and (22) as

$$\underset{\sim}{\nabla} A = \frac{1}{\sqrt{g}} \sum_{i=1}^{3} [(\underset{\sim}{a}_j \times \underset{\sim}{a}_k)A]_{\xi^i} \tag{27}$$

and

$$\underset{\sim}{\nabla} A = \frac{1}{\sqrt{g}} \sum_{i=1}^{3} (\underset{\sim}{a}_j \times \underset{\sim}{a}_k) A_{\xi^i} \tag{28}$$

D. Laplacian

The expressions for the Laplacian then follow from Eq. (20) or (22), with $\underset{\sim}{A}$ replaced by $\underset{\sim}{\nabla} A$ from Eq. (27) or (28). Thus the conservative form is

$$\nabla^2 A = \underset{\sim}{\nabla} \cdot (\underset{\sim}{\nabla} A)$$

$$= \frac{1}{\sqrt{g}} \sum_{i=1}^{3} \sum_{l=1}^{3} \{\frac{1}{\sqrt{g}}[(\underset{\sim}{a}_j \times \underset{\sim}{a}_k) \cdot [(\underset{\sim}{a}_m \times \underset{\sim}{a}_n)A]_{\xi^l}\}_{\xi^i} \tag{29}$$

(i,j,k) cyclic (l,m,n) cyclic

and the non-conservative is

$$\nabla^2 A = \frac{1}{\sqrt{g}} \sum_{i=1}^{3} \sum_{l=1}^{3} (\underset{\sim}{a}_j \times \underset{\sim}{a}_k) \cdot [\frac{1}{\sqrt{g}}(\underset{\sim}{a}_m \times \underset{\sim}{a}_n) A_{\xi^l}]_{\xi^i} \tag{30}$$

(i,j,k) cyclic (l,m,n) cyclic

With the product derivative expanded, the non-conservative form, Eq. (30), can also be written as

$$\nabla^2 A = \frac{1}{g} \sum_{i=1}^{3} \sum_{l=1}^{3} (\underset{\sim}{a}_j \times \underset{\sim}{a}_k) \cdot (\underset{\sim}{a}_m \times \underset{\sim}{a}_n) A_{\xi^l \xi^i}$$

$$+ \frac{1}{\sqrt{g}} \sum_{i=1}^{3} \sum_{l=1}^{3} (\underset{\sim}{a}_j \times \underset{\sim}{a}_k) \cdot [\frac{1}{\sqrt{g}}(\underset{\sim}{a}_m \times \underset{\sim}{a}_n)]_{\xi^i} A_{\xi^l} \tag{31}$$

4. Relations Between Covariant and Contravariant Metrics

A. Base vectors

The expression (28) for the gradient allows the contravariant base vectors to be expressed in terms of the covariant base vectors as follows. With $A=\xi^m$ in (28), we have

$$\underset{\sim}{\nabla}\xi^m = \frac{1}{\sqrt{g}} \sum_{i=1}^{3} (\underset{\sim}{a}_j \times \underset{\sim}{a}_k)\delta_i^m$$

since the three curvilinear coordinates are independent of each other. Then

$$\underset{\sim}{\nabla}\xi^i = \frac{1}{\sqrt{g}} \underset{\sim}{a}_j \times \underset{\sim}{a}_k \qquad (i = 1,2,3) \qquad (i,j,k) \text{ cyclic} \tag{32}$$

This gives a relation between the derivatives of the curvilinear coordinates $(\xi^i)_{x_1}$ and the derivatives $(x_r)_\xi$s of the cartesian coordinates. By Eq. (4) the contravariant base vectors may be written in terms of the covariant base vectors as

$$\underset{\sim}{a}^i = \underset{\sim}{\nabla}\xi^i = \frac{1}{\sqrt{g}}\underset{\sim}{a}_j \times \underset{\sim}{a}_k \tag{33}$$

$$(i = 1,2,3) \qquad (i,j,k) \text{ cyclic}$$

By Eq. (33),

$$\underset{\sim}{a}_i \cdot \underset{\sim}{a}^j = \frac{1}{\sqrt{g}} \underset{\sim}{a}_i \cdot (\underset{\sim}{a}_k \times \underset{\sim}{a}_l)$$

where here (j,k,l) are cyclic. If $j \neq i$, either k or l must be i, and in that case the right-hand side vanishes since the three vectors in the triple product may be in any cyclic order and the cross product of any vector with itself vanishes. When j = i, the right-hand side is simply unity. Therefore, in general

$$\underset{\sim}{a}_i \cdot \underset{\sim}{a}^j = \delta_i^j \tag{34}$$

Because of this relation, any vector $\underset{\sim}{A}$ can be expressed in terms of either set of base vectors as

$$\underset{\sim}{A} = \sum_{i=1}^{3} (\underset{\sim}{a}^i \cdot \underset{\sim}{A})\, \underset{\sim}{a}_i \tag{35}$$

and

$$\underset{\sim}{A} = \sum_{i=1}^{3} (\underset{\sim}{a}_i \cdot \underset{\sim}{A})\, \underset{\sim}{a}^i \tag{36}$$

Here the quantities $A^i = \underset{\sim}{a}^i \cdot \underset{\sim}{A}$ and $A_i = \underset{\sim}{a}_i \cdot \underset{\sim}{A}$, are the contravariant and covariant components, respectively, of the vector $\underset{\sim}{A}$.

B. Metric Tensors

The components of the contravariant metric tensor are the dot products of the contravariant base vectors:

$$g^{ij} = \underset{\sim}{a}^i \cdot \underset{\sim}{a}^j = g^{ji} \qquad (i = 1,2,3) \qquad (j = 1,2,3) \tag{37}$$

The relation between the covariant and contravariant metric tensor components is obtained by use of Eq. (33) in (37). Thus, with (i,j,k) cyclic and (l,m,n) cylic,

$$g^{il} = \underset{\sim}{a}^i \cdot \underset{\sim}{a}^l = \frac{1}{g} (\underset{\sim}{a}_j \times \underset{\sim}{a}_k) \cdot (\underset{\sim}{a}_m \times \underset{\sim}{a}_n)$$

$$= \frac{1}{g} [(\underset{\sim}{a}_j \cdot \underset{\sim}{a}_m)(\underset{\sim}{a}_k \cdot \underset{\sim}{a}_n) - (\underset{\sim}{a}_j \cdot \underset{\sim}{a}_n)(\underset{\sim}{a}_k \cdot \underset{\sim}{a}_m)]$$

by the identity (9). Then from the definition (5),

$$g^{il} = \frac{1}{g}(g_{jm}g_{kn} - g_{jn}g_{km}) \qquad (38)$$

$$(i = 1,2,3) \qquad (l = 1,2,3)$$

$$(i,j,k) \text{ cyclic} \qquad (l,m,n) \text{ cyclic}$$

Since the quantity in parentheses in the above equation is the signed cofactor of the il component of the covariant metric tensor, the right-hand side above is the li component of the inverse of this tensor. Then, since the metric tensor is symmetric we have immediately that the contravariant metric tensor is simply the inverse of the covariant metric tensor. It then follows that

$$\det|g^{ij}| = \frac{1}{\det|g_{ij}|} = \frac{1}{g}$$

so that, in terms of the contravariant base vectors, the Jacobian is

$$\sqrt{g} = (\det|g^{ij}|)^{-1/2} = \frac{1}{\underset{\sim}{a}^1 \cdot (\underset{\sim}{a}^2 \times \underset{\sim}{a}^3)} \qquad (39)$$

The identity (21) can be given, using Eq. (33), as

$$\sum_{i=1}^{3} (\sqrt{g}\, \underset{\sim}{a}^i)_{\xi^i} = 0 \qquad (40)$$

5. Restatement of Derivative Operators

In view of Eq. (33), the cross products of the covariant base vectors in the expressions given above for the gradient, divergence, curl, and Laplacian can be replaced directly by the contravariant base vectors (multiplied by the Jacobian). The components of these contravariant base vectors $\underset{\sim}{a}^i$ in the expressions are the derivatives of the curvilinear coordinates with respect to the cartesian coordinates, and this notation, rather than the cross-products, often appears in the literature. Thus, by Eq. (4), the x_j-component of $\underset{\sim}{a}^i$ can be written as

$$(\underset{\sim}{a}^i)_j = (\xi^i)_{x_j} \tag{41}$$

The expressions for the gradient, divergence, curl, Laplacian, etc., given above in terms of the cross products of the covariant base vectors, $\underset{\sim}{a}_i$, involve the derivatives of the cartesian coordinates with respect to the curvilinear coordinates, e.g. $(x_i)_{\xi^j}$. The expressions given below in terms of the contravariant base vectors, $\underset{\sim}{a}^i$, involve the derivatives $(\xi^i)_{x_j}$ when $\sqrt{g}$ is evaluated from (39). From a coding standpoint, however, the contravariant base vectors $\underset{\sim}{a}^i$ in these expressions would be evaluated from the covariant base vectors using Eq. (33).

A. Conservative

The conservative forms are as follows:

$$\underset{\sim}{\nabla} A = \frac{1}{\sqrt{g}} \sum_{i=1}^{3} (\sqrt{g}\, \underset{\sim}{a}^i A)_{\xi^i} \tag{42}$$

$$\underset{\sim}{\nabla} \cdot \underset{\sim}{A} = \frac{1}{\sqrt{g}} \sum_{i=1}^{3} (\sqrt{g}\, \underset{\sim}{a}^i \cdot \underset{\sim}{A})_{\xi^i} \tag{43}$$

$$\underset{\sim}{\nabla} \times \underset{\sim}{A} = \frac{1}{\sqrt{g}} \sum_{i=1}^{3} (\sqrt{g}\, \underset{\sim}{a}^i \times \underset{\sim}{A})_{\xi^i} \tag{44}$$

$$\nabla^2 A = \frac{1}{\sqrt{g}} \sum_{i=1}^{3} \sum_{j=1}^{3} [\underset{\sim}{a}^i \cdot (\sqrt{g}\, \underset{\sim}{a}^j A)_{\xi^j}]_{\xi^i} \tag{45}$$

By expanding the inner derivative, the Laplacian can be expressed as

$$\nabla^2 A = \frac{1}{\sqrt{g}} \sum_{i=1}^{3} \sum_{j=1}^{3} (\sqrt{g}\, g^{ij} A_{\xi^j})_{\xi^i} \tag{46}$$

For $\nabla \cdot (\alpha \underset{\sim}{\nabla} A)$ we have

$$\underset{\sim}{\nabla} \cdot (\alpha \underset{\sim}{\nabla} A) = \frac{1}{\sqrt{g}} \sum_{i=1}^{3} \sum_{j=1}^{3} [\alpha \underset{\sim}{a}^i \cdot (\sqrt{g}\, \underset{\sim}{a}^j A)_{\xi^j}]_{\xi^i} \tag{47}$$

or, with the inner derivative expanded,

$$\underset{\sim}{\nabla} \cdot (\alpha \underset{\sim}{\nabla} A) = \frac{1}{\sqrt{g}} \sum_{i=1}^{3} \sum_{j=1}^{3} (\alpha \sqrt{g}\, g^{ij} A_{\xi^j})_{\xi^i} \tag{48}$$

In the expressions for the divergence, $\underset{\sim}{A}$ may be a tensor, in which case we have

$$(\underset{\sim}{\nabla} \cdot \underset{\sim}{A})_k = \sum_{l=1}^{3} (A_{kl})_{x_l}$$

$$= \frac{1}{\sqrt{g}} \sum_{i=1}^{3} \sum_{l=1}^{3} [\sqrt{g}\, (\underset{\sim}{a}^i)_l A_{kl}]_{\xi^i} \qquad (k = 1,2,3) \tag{49}$$

From Eq. (42) we have the conservative expressions for the first derivative:

$$A_{x_j} = (\underset{\sim}{\nabla} A)_j = \frac{1}{\sqrt{g}} \sum_{i=1}^{3} [\sqrt{g}\, (\underset{\sim}{a}^i)_j A]_{\xi^i} \tag{50}$$

where $(\cdot)_j$ is the component in the x_j-direction. Also, for the second derivative,

$$A_{x_j x_k} = [\nabla(A_{x_j})]_k = \frac{1}{\sqrt{g}} \sum_{i=1}^{3} \sum_{l=1}^{3} \{(\underset{\sim}{a}^i)_k [\sqrt{g}\,(\underset{\sim}{a}^l)_j A]_{\xi^l}\}_{\xi^i} \tag{51}$$

or, with the inner derivative expanded,

$$A_{x_j x_k} = \frac{1}{\sqrt{g}} \sum_{i=1}^{3} \sum_{l=1}^{3} [\sqrt{g}\,(\underset{\sim}{a}^i)_k (\underset{\sim}{a}^l)_j A_{\xi^l}]_{\xi^i} \tag{52}$$

It then follows that all of the above conservative expressions can be written in the form

$$\frac{1}{\sqrt{g}} \sum_{i=1}^{3} (A^i)_{\xi^i} \tag{53}$$

where the quantity A^i takes the following form for the various operations, with i = 1,2,3,

$$\underset{\sim}{\nabla} A : \quad A^i = \sqrt{g}\, \underset{\sim}{a}^i A \tag{54}$$

$$\underset{\sim}{\nabla} \cdot \underset{\sim}{A} \text{ (vector } \underset{\sim}{A}) : \quad A^i = \sqrt{g}\, \underset{\sim}{a}^i \cdot \underset{\sim}{A} \tag{55}$$

$$\underset{\sim}{\nabla} \cdot \underset{\sim}{A} \text{ (tensor } \underset{\sim}{A}) : \quad \underset{\sim}{A}^i = \sqrt{g}\, \underset{\sim}{A}\, \underset{\sim}{a}^i \tag{56}$$

(matrix product of square matrix $\underset{\sim}{A}$ and column vector $\underset{\sim}{a}^i$. Here $\underset{\sim}{A}^i$ is a vector)

$$\underset{\sim}{\nabla} \times \underset{\sim}{A} : \quad \underset{\sim}{A}^i = \sqrt{g}\, \underset{\sim}{a}^i \times \underset{\sim}{A} \tag{57}$$

$$\nabla^2 A : \quad A^i = \underset{\sim}{a}^i \cdot \sum_{j=1}^{3} (\sqrt{g}\, \underset{\sim}{a}^j A)_{\xi^j} \quad \text{(for Eq. (45))} \tag{58}$$

$$A^i = \sqrt{g} \sum_{j=1}^{3} g^{ij} A_{\xi^j} \quad \text{(for Eq. (46))} \tag{59}$$

$$\underset{\sim}{\nabla} \cdot (\alpha \underset{\sim}{\nabla} A) : \quad A^i = \alpha \sum_{j=1}^{3} [\underset{\sim}{a}^i \cdot (\sqrt{g}\, \underset{\sim}{a}^j A)_{\xi^j}] \text{ (for Eq. (47))} \quad (60)$$

$$A^i = \alpha\sqrt{g} \sum_{j=1}^{3} g^{ij} A_{\xi^j} \quad \text{(for Eq. (48))} \quad (61)$$

$$A_{x_j} : \quad A^i = \sqrt{g}\, (\underset{\sim}{a}^i)_j A \quad (62)$$

$$A_{x_j x_k} : \quad A^i = (\underset{\sim}{a}^i)_k \sum_{l=1}^{3} [\sqrt{g}\, (\underset{\sim}{a}^l)_j A]_{\xi^l} \quad \text{(for Eq. (51))} \quad (63)$$

$$A^i = \sqrt{g}\, (\underset{\sim}{a}^i)_k \sum_{l=1}^{3} (\underset{\sim}{a}^l)_j A_{\xi^l} \quad \text{(for Eq. (52))} \quad (64)$$

It is computationally more efficient to evaluate the product $\sqrt{g}\, \underset{\sim}{a}^i$ as an entity from Eq. (33) when the conservative forms are used, in order to avoid the extra multiplication by $\sqrt{g}$. Another alternative is to include $\sqrt{g}$ with $\underset{\sim}{A}$.

B. Non-conservative

The non-conservative relations are as follows:

$$\underset{\sim}{\nabla} A = \sum_{i=1}^{3} \underset{\sim}{a}^i A_{\xi^i} \quad (65)$$

From Eq. (65) the $\underset{\sim}{\nabla}$ operator can be represented by

$$\underset{\sim}{\nabla} = \sum_{i=1}^{3} \underset{\sim}{a}^i \frac{\partial}{\partial \xi^i} \quad (66)$$

and

$$\nabla \cdot A = \sum_{i=1}^{3} a^i \cdot A_{\xi^i} \tag{67}$$

$$\nabla \times A = \sum_{i=1}^{3} a^i \times A_{\xi^i} \tag{68}$$

$$\nabla^2 A = \sum_{i=1}^{3} \sum_{j=1}^{3} a^i \cdot a^j \; A_{\xi^i \xi^j}$$

$$+ \sum_{i=1}^{3} \sum_{j=1}^{3} a^i \cdot (a^j)_{\xi^i} \, A_{\xi^j} \tag{69}$$

Since

$$\nabla^2 \xi^l = \nabla \cdot (\nabla \xi^l) = \nabla \cdot a^l = \sum_{i=1}^{3} a^i \cdot (a^l)_{\xi^i} \tag{70}$$

by Eq. (37) and (69), the Laplacian can also be written as

$$\nabla^2 A = \sum_{i=1}^{3} \sum_{j=1}^{3} g^{ij} \, A_{\xi^i \xi^j} + \sum_{j=1}^{3} (\nabla^2 \xi^j) A_{\xi^j} \tag{71}$$

Using Eq. (67) and (65) we also have

$$\nabla \cdot (\alpha \nabla A) = \sum_{i=1}^{3} a^i \cdot (\alpha \nabla A)_{\xi^i}$$

$$= \sum_{i=1}^{3} \sum_{j=1}^{3} a^i \cdot (\alpha a^j A_{\xi^j})_{\xi^i}$$

$$= \sum_{i=1}^{3} \sum_{j=1}^{3} a^i \cdot [a^j (\alpha A_{\xi^j})_{\xi^i} + \alpha (a^j)_{\xi^i} \, A_{\xi^j}]$$

Thus, by Eq. (70), the non-conservative expression is

$$\nabla \cdot (\alpha \nabla A) = \sum_{i=1}^{3} \sum_{j=1}^{3} g^{ij} (\alpha A_{\xi^j})_{\xi^i} + \alpha \sum_{j=1}^{3} (\nabla^2 \xi^j) A_{\xi^j} \tag{72}$$

A more practical equation than Eq. (70) for the evaluation of $\nabla^2 \xi^j$ in these expressions can be obtained as follows.

Since $\nabla^2 \mathbf{r} = 0$ it follows from Eq. (71) that

$$\sum_{i=1}^{3} \sum_{j=1}^{3} g^{ij} \mathbf{r}_{\xi^i \xi^j} + \sum_{j=1}^{3} (\nabla^2 \xi^j) \mathbf{r}_{\xi^j} = 0 \tag{73}$$

But $\mathbf{r}_{\xi^j} = \mathbf{a}_j$. Then dotting $\mathbf{a}^l$ into this equation and using Eq. (34), we have

$$\sum_{i=1}^{3} \sum_{j=1}^{3} g^{ij} \mathbf{a}^l \cdot \mathbf{r}_{\xi^i \xi^j} + \sum_{j=1}^{3} (\nabla^2 \xi^j) \delta_j^l = 0$$

so that $\nabla^2 \xi^l$ is given by

$$\nabla^2 \xi^l = - \sum_{i=1}^{3} \sum_{j=1}^{3} g^{ij} \mathbf{a}^l \cdot \mathbf{r}_{\xi^i \xi^j} \qquad (l = 1,2,3) \tag{74}$$

The non-conservative form of the divergence of a tensor is, by expansion in Eq. (49),

$$(\nabla \cdot \mathbf{A})_k = \sum_{l=1}^{3} (A_{kl})_{x_l} = \sum_{i=1}^{3} \sum_{l=1}^{3} (\mathbf{a}^i)_l (A_{kl})_{\xi^i} \tag{75}$$

$$(k = 1,2,3)$$

From Eq.(65) the non-conservative expressions for the first and second derivatives are

$$A_{x_j} = (\nabla A)_j = \sum_{i=1}^{3} (\mathbf{a}^i)_j A_{\xi^i} \tag{76}$$

and

$$A_{x_j x_k} = [\nabla(A_{x_j})]_k = \sum_{i=1}^{3} \sum_{l=1}^{3} (\underset{\sim}{a}^i)_k \, [(\underset{\sim}{a}^l)_j A_{\xi^l}]_{\xi^i}$$

$$= \sum_{i=1}^{3} \sum_{l=1}^{3} (\underset{\sim}{a}^i)_k \, [(\underset{\sim}{a}^l)_j A_{\xi^l \xi^i} + (\underset{\sim}{a}^l_{\xi^i})_j A_{\xi^l}] \qquad (77)$$

This non-conservative form in terms of the contravariant base vectors is referred to by some as the "chain-rule conservation" form (Eq. (76) is equivalent to Eq. (1)). In any case only the conservative form gives the telescopic collapse over the field that characterizes conservative numerical representations, and it is necessary to substitute for the contravariant base vectors from Eq. (33) in implementation, since it is the covariant base vectors that are directly calculated from the grid point locations.

6. Normal and Tangential Derivatives

Expressions for derivatives normal and tangential to coordinate surfaces are needed in boundary conditions and are obtained from the base vectors as follows.

A. Tangent to coordinate lines

Since the covariant base vectors are tangent to the coordinate lines, the tangential derivative on a coordinate line along which ξ^i varies is given by

$$(A)^i_\tau = \frac{\underset{\sim}{a}_i}{|\underset{\sim}{a}_i|} \cdot \nabla A = \frac{1}{|\underset{\sim}{a}_i|} \sum_{j=1}^{3} (\underset{\sim}{a}_i \cdot \underset{\sim}{a}^j) A_{\xi^j}$$

using Eq. (65). In view of Eq. (34), this reduces to

$$(A)^i_\tau = \frac{A_{\xi^i}}{\sqrt{g_{ii}}} \qquad (i = 1,2,3) \qquad (78)$$

B. Normal to coordinate surfaces

Also, since the contravariant base vectors are normal to the coordinate surfaces, the normal derivative to a coordinate surface on which ξ^i is constant is given by

$$(A)_n^i = \frac{\underset{\sim}{a}^i}{|\underset{\sim}{a}^i|} \cdot \nabla A = \frac{1}{|\underset{\sim}{a}^i|} \sum_{j=1}^{3} (\underset{\sim}{a}^i \cdot \underset{\sim}{a}^j) A_{\xi^j}$$

$$= \frac{1}{\sqrt{g^{ii}}} \sum_{j=1}^{3} g^{ij} A_{\xi^j} \quad (i = 1,2,3) \tag{79}$$

C. Normal to coordinate lines and tangent to coordinate surfaces

The vector $\underset{\sim}{a}^i \times \underset{\sim}{a}_i$ is normal to the coordinate line on which ξ^i varies and is also tangent to the coordinate surface on which ξ^i is constant:

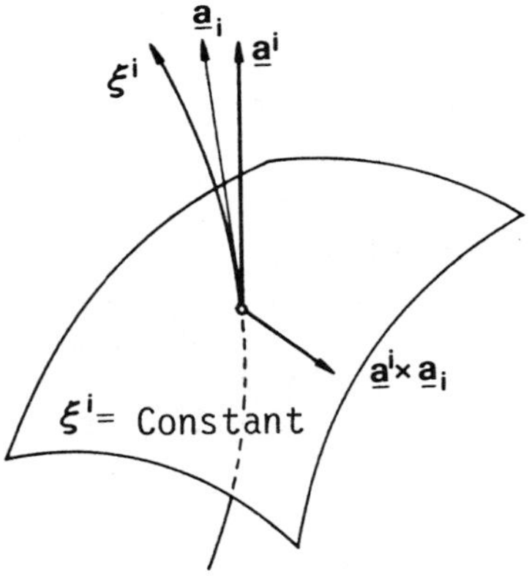

Using Eq. (33) and the identity (13), this vector is given by

$$\underset{\sim}{a}^i \times \underset{\sim}{a}_i = \frac{1}{\sqrt{g}} (\underset{\sim}{a}_j \times \underset{\sim}{a}_k) \times \underset{\sim}{a}_i$$

$$= - \frac{1}{\sqrt{g}}[(\underset{\sim}{a}_i \cdot \underset{\sim}{a}_k)\underset{\sim}{a}_j - (\underset{\sim}{a}_i \cdot \underset{\sim}{a}_j) \underset{\sim}{a}_k]$$

$$= - \frac{1}{\sqrt{g}}(g_{ik}\underset{\sim}{a}_j - g_{ij}\underset{\sim}{a}_k) \qquad (80)$$

and the magnitude is given by

$$|\underset{\sim}{a}^i \times \underset{\sim}{a}_i|^2 = \frac{1}{g} (g_{ik}^2 g_{jj} + g_{ij}^2 g_{kk} - 2g_{ik}g_{ij}g_{jk})$$

$$= \frac{1}{g}[g_{ik}(g_{ik}g_{jj} - g_{ij}g_{jk}) + g_{ij}(g_{ij}g_{kk} - g_{ik}g_{jk})]$$

The bracket is the negative of the second and third terms of the determinant, $g = \det|g_{ij}|$, expanded by cofactors. Therefore, we have

$$|\underset{\sim}{a}^i \times \underset{\sim}{a}_i|^2 = \frac{1}{g} [g_{ii}(g_{jj} g_{kk} - g_{jk}^2) - g]$$

$$= \frac{g_{ii}}{g} (g_{jj}g_{kk} - g_{jk}^2) - 1 \qquad (81)$$

The derivative normal to the coordinate line on which ξ^i varies and in the coordinate surface on which ξ^i is constant then, using Eq. (65) and (80), is given by

$$(A)_T^i = \frac{\underset{\sim}{a}^i \times \underset{\sim}{a}_i}{|\underset{\sim}{a}^i \times \underset{\sim}{a}_i|} \cdot \underset{\sim}{\nabla} A$$

$$= -\frac{1}{\sqrt{g}|\underset{\sim}{a}^i \times \underset{\sim}{a}_i|} \sum_{l=1}^{3} (g_{ik}\underset{\sim}{a}_j \cdot \underset{\sim}{a}^l A_{\xi^l} - g_{ij}\underset{\sim}{a}_k \cdot \underset{\sim}{a}^l A_{\xi^l})$$

$$= -\frac{1}{\sqrt{g}|\underset{\sim}{a}^i \times \underset{\sim}{a}_i|} (g_{ik}A_{\xi^j} - g_{ij}A_{\xi^k})$$

by Eq. (34). Thus, using Eq. (81),

$$(A)_T^i = \frac{g_{ij}A_{\xi^k} - g_{ik}A_{\xi^j}}{\sqrt{g_{ii}(g_{jj}g_{kk} - g_{jk}^2) - g}} \qquad (i,j,k) \text{ cyclic} \qquad (82)$$

7. Integrals

Expressions for surface, volume, and line integrals are easily developed from the base vectors as follow.

A. Surface integral

Returning to the Divergence Theorem, Eq. (17), and its counterparts with the dot product replaced by a cross product or simple operation (the latter with A replaced by a scalar), we have approximate expressions for the surface integrals over the surface of the volume element given, using Eq. (15), by

$$\oiint_{\delta S} \underset{\sim}{A} \circ \underset{\sim}{n}\, dS \cong \sqrt{g}\, \underset{\sim}{\nabla} \circ \underset{\sim}{A}\, d\xi^1 d\xi^2 d\xi^3 \qquad (83)$$

with the open circle indicating the product operation, and using the appropriate expression for $\underset{\sim}{\nabla} \circ \underset{\sim}{A}$ from those given in the developments above. This emphasizes again that difference representations based on integral formulations, e.g. finite volume, can be obtained by using conservative expressions for the derivative operators directly in the partial differential equations.

B. Volume integral

The approximate expression for the volume integral over the volume element, again using Eq. (15), is simply

$$\iiint_{\delta V} \underset{\sim}{A}\, dV \cong \sqrt{g}\, \underset{\sim}{A}\, d\xi^1 d\xi^2 d\xi^3 \tag{84}$$

C. Line integrals

Using Eq. (3), the line integral on a coordinate line element on which ξ^i varies is simply

$$\int_{\delta s} \underset{\sim}{A} \circ d\underset{\sim}{r} = \underset{\sim}{A} \circ \underset{\sim}{a}_i\, d\xi^i \tag{85}$$

where again the open circle indicates any type of operation, and $\underset{\sim}{A}$ is any tensor. Also since

$$\oint_{c} \underset{\sim}{A} \circ d\underset{\sim}{r} = \iint_{S} (\underset{\sim}{n} \times \underset{\sim}{\nabla}) \circ \underset{\sim}{A}\, dS$$

we have for a closed circuit lying on a coordinate surface on which ξ^i is constant,

$$\oint_{c^i} \underset{\sim}{A} \circ d\underset{\sim}{r} = \sqrt{g} \sum_{l=1}^{3} [(\underset{\sim}{a}^i \times \underset{\sim}{a}^l)\frac{\partial}{\partial \xi^l}] \circ \underset{\sim}{A}\, d\xi^m d\xi^n$$

$$= \sqrt{g} \sum_{l=1}^{3} (\underset{\sim}{a}^i \times \underset{\sim}{a}^l) \circ \underset{\sim}{A}_{\xi^l}\, d\xi^m d\xi^n$$

using Eq. (83), where (l,m,n) are cyclic. But

$$\mathbf{a}^i \times \mathbf{a}^l = \mathbf{a}^i \times (\mathbf{a}_m \times \mathbf{a}_n)/\sqrt{g}$$

Then using the identity (13), we have

$$\mathbf{a}^i \times \mathbf{a}^l = (\mathbf{a}^i \cdot \mathbf{a}_n)\mathbf{a}_m - (\mathbf{a}^i \cdot \mathbf{a}_m)\mathbf{a}_n$$

$$= \delta^i_n\, \mathbf{a}_m - \delta^i_m\, \mathbf{a}_n \qquad (86)$$

by using Eq. (34). With (i,j,k) cyclic we have for the circuit integral,

$$\oint_{c^i} \mathbf{A} \circ d\mathbf{r} = \mathbf{a}_k \circ \mathbf{A}_{\xi^j}\, d\xi^j - \mathbf{a}_j \circ \mathbf{A}_{\xi^k}\, d\xi^k \qquad (87)$$

8. Two-Dimensional Forms

In two dimensions, let the x_3 direction be the direction of invariance, and let the ξ^3 curvilinear coordinate be identical with x_3. Also, for convenience of notation, let the other coordinates be identified as

$$x_1 = x, \qquad x_2 = y, \qquad \xi^1 = \xi, \qquad \text{and } \xi^2 = \eta$$

A. Metric elements

Then

$$\mathbf{a}_3 = \mathbf{a}^3 = \mathbf{k},$$

and the other base vectors are

$$\mathbf{a}_1 = \mathbf{r}_\xi = \mathbf{i}x_\xi + \mathbf{j}y_\xi$$

$$\qquad (88)$$

$$\mathbf{a}_2 = \mathbf{r}_\eta = \mathbf{i}x_\eta + \mathbf{j}y_\eta$$

The covariant metric components then are

$$g_{33} = \underset{\sim}{k} \cdot \underset{\sim}{k} = 1$$

$$g_{13} = g_{31} = \underset{\sim}{a}_1 \cdot \underset{\sim}{k} = 0$$

$$g_{23} = g_{32} = \underset{\sim}{a}_2 \cdot \underset{\sim}{k} = 0$$

$$g_{11} = x_\xi^2 + y_\xi^2$$

$$g_{22} = x_\eta^2 + y_\eta^2$$

$$g_{12} = g_{21} = x_\xi x_\eta + y_\xi y_\eta \tag{89}$$

From Eq. (16), the Jacobian is given by

$$\sqrt{g} = \sqrt{\det|g_{ij}|} = \sqrt{g_{11}g_{22} - g_{12}^2}$$

$$= \underset{\sim}{k} \cdot (\underset{\sim}{a}_1 \times \underset{\sim}{a}_2) = |\underset{\sim}{a}_1 \times \underset{\sim}{a}_2|$$

$$= x_\xi y_\eta - x_\eta y_\xi \tag{90}$$

The other contravariant base vectors are, from Eq. (33),

$$\underset{\sim}{a}^1 = \frac{1}{\sqrt{g}}(\underset{\sim}{a}_2 \times \underset{\sim}{k}) = \frac{1}{\sqrt{g}}(\underset{\sim}{i}y_\eta - \underset{\sim}{j}x_\eta)$$

$$\underset{\sim}{a}^2 = \frac{1}{\sqrt{g}}(\underset{\sim}{k} \times \underset{\sim}{a}_1) = \frac{1}{\sqrt{g}}(-\underset{\sim}{i}y_\xi + \underset{\sim}{j}x_\xi) \tag{91}$$

and the contravariant metric components are, from Eq. (37) or (38),

$$g^{33} = \underset{\sim}{k} \cdot \underset{\sim}{k} = 1$$

$$g^{13} = g^{31} = \underset{\sim}{a}^1 \cdot \underset{\sim}{k} = 0$$

$$g^{23} = g^{32} = \underset{\sim}{a}^2 \cdot \underset{\sim}{k} = 0$$

$$g^{11} = \frac{g_{22}}{g}, \qquad g^{22} = \frac{g_{11}}{g}$$

$$g^{12} = g^{21} = -\frac{g_{12}}{g} \tag{92}$$

From Eq. (4) we have $\underset{\sim}{\nabla}\xi = \underset{\sim}{a}^1$ and $\underset{\sim}{\nabla}\eta = \underset{\sim}{a}^2$, so that by Eq. (91),

$$\xi_x = \frac{y_\eta}{\sqrt{g}}, \quad \xi_y = -\frac{x_\eta}{\sqrt{g}}, \qquad \eta_x = -\frac{y_\xi}{\sqrt{g}}, \quad \eta_y = \frac{x_\xi}{\sqrt{g}} \tag{93}$$

B. Transformation relations

Divergence (conservative), Eq. (43):

$$\underset{\sim}{\nabla} \cdot \underset{\sim}{A} = \frac{1}{\sqrt{g}}[(y_\eta A_1 - x_\eta A_2)_\xi + (-y_\xi A_1 + x_\xi A_2)_\eta] \tag{94}$$

(non-conservative), Eq. (67):

$$\underset{\sim}{\nabla} \cdot \underset{\sim}{A} = \frac{1}{\sqrt{g}}[y_\eta (A_1)_\xi - x_\eta (A_2)_\xi - y_\xi (A_1)_\eta + x_\xi (A_2)_\eta] \tag{95}$$

Gradient (conservative), Eq., (42):

$$f_x = \frac{1}{\sqrt{g}}[(y_\eta f)_\xi - (y_\xi f)_\eta] \tag{96}$$

$$f_y = \frac{1}{\sqrt{g}}[-(x_\eta f)_\xi + (x_\xi f)_\eta] \tag{97}$$

(non-conservative), Eq. (65):

$$f_x = \frac{1}{\sqrt{g}}(y_\eta f_\xi - y_\xi f_\eta) \tag{98}$$

$$f_y = \frac{1}{\sqrt{g}}(-x_\eta f_\xi + x_\xi f_\eta) \tag{99}$$

By Eq. (93), or directly from Eq. (76), these non-conservative forms may be given as

$$f_x = f_\xi \xi_x + f_\eta \eta_x$$

$$f_y = f_\xi \xi_y + f_\eta \eta_y$$

which are the so-called "chain-rule conservative" forms. This form, however, is not conservative and the relations given by Eq. (93) must be substituted in the implementation in any case, since it is x_ξ, etc., rather than ξ_x, that is directly calculated from the grid point locations.

Curl (conservative), Eq. (44):

$$\underset{\sim}{\nabla} \times \underset{\sim}{A} = \frac{\underset{\sim}{k}}{\sqrt{g}}[(y_\eta A_2 + x_\eta A_1)_\xi - (y_\xi A_2 + x_\xi A_1)_\eta] \tag{100}$$

(non-conservative), Eq. (68):

$$\underset{\sim}{\nabla} \times \underset{\sim}{A} = \frac{\underset{\sim}{k}}{\sqrt{g}}[y_\eta (A_2)_\xi + x_\eta (A_1)_\xi - y_\xi (A_2)_\eta - x_\xi (A_1)_\eta] \tag{101}$$

Laplacian (conservative), Eq. (45):

$$\sqrt{g}\,\nabla^2 f = \{[\frac{1}{\sqrt{g}} y_\eta[(y_\eta f)_\xi - (y_\xi f)_\eta]$$

$$- \frac{1}{\sqrt{g}} x_\eta[-(x_\eta f)_\xi + (x_\xi f)_\eta]\}_\xi$$

$$+ \{- \frac{1}{\sqrt{g}} y_\xi[(y_\eta f)_\xi - (y_\xi f)_\eta]$$

$$+ \frac{1}{\sqrt{g}} x_\xi[-(x_\eta f)_\xi + (x_\xi f)_\eta]\}_\eta \qquad (102)$$

(non-conservative), Eq. (69):

$$\nabla^2 f = \frac{1}{g}[(x_\eta^2 + y_\eta^2)f_{\xi\xi} - 2(x_\xi x_\eta + y_\xi y_\eta)f_{\xi\eta}$$

$$+ (x_\xi^2 + y_\xi^2)f_{\eta\eta}] + (\nabla^2\xi)f_\xi + (\nabla^2\eta)f_\eta \qquad (103)$$

<u>Second derivatives</u> (non-conservative):

$$f_{xx} = (y_\eta^2 f_{\xi\xi} - 2y_\xi y_\eta f_{\xi\eta} + y_\xi^2 f_{\eta\eta})/g$$

$$+ [(y_\eta^2 y_{\xi\xi} - 2y_\xi y_\eta y_{\xi\eta} + y_\xi^2 y_{\eta\eta})(x_\eta f_\xi - x_\xi f_\eta)$$

$$+ (y_\eta^2 x_{\xi\xi} - 2y_\xi y_\eta x_{\xi\eta} + y_\xi^2 x_{\eta\eta})(y_\xi f_\eta - y_\eta f_\xi)]/g^{3/2} \qquad (104)$$

$$f_{yy} = (x_\eta^2 f_{\xi\xi} - 2x_\xi x_\eta f_{\xi\eta} + x_\xi^2 f_{\eta\eta})/g$$

$$+ [(x_\eta^2 y_{\xi\xi} - 2x_\xi x_\eta y_{\xi\eta} + x_\xi^2 y_{\eta\eta})(x_\eta f_\xi - x_\xi f_\eta)$$

$$+ (x_\eta^2 x_{\xi\xi} - 2x_\xi x_\eta x_{\xi\eta} + x_\xi^2 x_{\eta\eta})(y_\xi f_\eta - y_\eta f_\xi)]/g^{3/2} \tag{105}$$

$$f_{xy} = [(x_\xi y_\eta + x_\eta y_\xi)f_{\xi\eta} - x_\xi y_\xi f_{\eta\eta} - x_\eta y_\eta f_{\xi\xi}]/g$$

$$+ \{(x_\xi y_{\eta\eta} - x_\eta y_{\xi\eta})/g + [x_\eta y_\eta(\sqrt{g})_\xi - x_\xi y_\eta(\sqrt{g})_\eta]/g^{3/2}\}f_\xi$$

$$+ \{(x_\eta y_{\xi\xi} - x_\xi y_{\xi\eta})/g + [x_\xi y_\xi(\sqrt{g})_\eta - x_\eta y_\xi(\sqrt{g})_\xi]/g^{3/2}\}f_\eta \tag{106}$$

Normal derivative (conservative):

$$f_{n(\xi)} = \frac{1}{\sqrt{g}\sqrt{x_\eta^2 + y_\eta^2}}\{y_\eta[(y_\eta f)_\xi - (y_\xi f)_\eta]$$

$$- x_\eta[-(x_\eta f)_\xi + (x_\xi f)_\eta]\} \tag{107}$$

$$f_{n(\eta)} = \frac{1}{\sqrt{g}\sqrt{x_\xi^2 + y_\xi^2}}\{-y_\xi[(y_\eta f)_\xi - (y_\xi f)_\eta]$$

$$+ x_\xi[-(x_\eta f)_\xi + (x_\xi f)_\eta]\} \tag{108}$$

(non-conservative):

$$f_{n(\xi)} = \frac{1}{\sqrt{g}\sqrt{x_\eta^2 + y_\eta^2}}[(x_\eta^2 + y_\eta^2)f_\xi - (x_\xi x_\eta + y_\xi y_\eta)f_\eta] \tag{109}$$

$$f_{n(\eta)} = \frac{1}{\sqrt{g}\sqrt{x_\xi^2 + y_\xi^2}}[-(x_\xi x_\eta + y_\xi y_\eta)f_\xi + (x_\xi^2 + y_\xi^2)f_\eta] \tag{110}$$

Tangential derivative (conservative):

$$f_{\tau(\xi)} = \frac{1}{\sqrt{g}\sqrt{x_\eta^2 + y_\eta^2}} \{x_\eta[(y_\eta f)_\xi - (y_\xi f)_\eta]$$

$$- y_\eta[(x_\eta f)_\xi - (x_\xi f)_\eta]\} \qquad (111)$$

$$f_{\tau(\eta)} = \frac{1}{\sqrt{g}\sqrt{x_\xi^2 + y_\xi^2}} \{x_\xi[(y_\eta f)_\xi - (y_\xi f)_\eta]$$

$$- y_\xi[(x_\eta f)_\xi - (x_\xi f)_\eta]\} \qquad (112)$$

(non-conservative):

$$f_{\tau(\xi)} = \frac{1}{\sqrt{x_\eta^2 + y_\eta^2}} f_\eta \qquad (113)$$

$$f_{\tau(\eta)} = \frac{1}{\sqrt{x_\xi^2 + y_\xi^2}} f_\xi \qquad (114)$$

Surface integral:

$$\iint_S f dS = \sqrt{g}\, f d\xi d\eta \qquad (115)$$

9. Time derivatives

A. First Derivative

With moving grids the time derivatives must be transformed also. For the first derivative we have

$$\left(\frac{\partial A}{\partial t}\right)_{\xi} = \left(\frac{\partial A}{\partial t}\right)_{x} + \nabla A \cdot \left(\frac{\partial x}{\partial t}\right)_{\xi} \tag{116}$$

where here, and in Eq. (117) below, the subscripts indicate the variable being held constant in the partial differentiation. Here the time derivative on the left side is at a fixed position in the transformed space, i.e., at a given grid point. The time derivative on the right is at a fixed position in the physical space, i.e., the time derivative that appears in the physical equations of motion. The quantity $\partial x/\partial t_{\xi}$ is the grid point speed, to be written $\dot{x}$ hereafter. Thus we have, for substitution into the physical equations of motion, the relation

$$\left(\frac{\partial A}{\partial t}\right)_{x} = \left(\frac{\partial A}{\partial t}\right)_{\xi} - \dot{x} \cdot \nabla A \tag{117}$$

with ∇A to come from the transformation relations given previously. With the time derivatives transformed, only time derivatives at fixed points in the transformed space will appear in the equations and, therefore, all computation can be done on the fixed uniform grid in the transformed field without interpolation, even though the grid points are in motion in the physical space. The last term in Eq. (117) resembles a convective term and accounts for the motion of the grid.

B. Convective terms

Consider the generic convective terms

$$C = A_t + \nabla \cdot (uA) \tag{118}$$

where $\underset{\sim}{u}$ is a velocity, which occur in many conservation equations. Using Eq. (117) we have

$$C = A_t - \dot{\underset{\sim}{x}} \cdot \underset{\sim}{\nabla} A + \underset{\sim}{\nabla} \cdot (\underset{\sim}{u} A)$$

where now the time derivative is understood to be at a fixed point in the transformed space. Then using Eq. (42) and (43) for the gradient and divergence, this becomes

$$C = A_t - \frac{1}{\sqrt{g}} \dot{\underset{\sim}{x}} \cdot \sum_{i=1}^{3} (\sqrt{g}\, \underset{\sim}{a}^i A)_{\xi^i} + \frac{1}{\sqrt{g}} \sum_{i=1}^{3} (\sqrt{g}\, \underset{\sim}{a}^i \cdot \underset{\sim}{u} A)_{\xi^i} \qquad (119)$$

By Eq. (16),

$$(\sqrt{g})_t = [\underset{\sim}{a}_1 \cdot (\underset{\sim}{a}_2 \times \underset{\sim}{a}_3)]_t$$

$$= (\underset{\sim}{a}_1)_t \cdot (\underset{\sim}{a}_2 \times \underset{\sim}{a}_3) + (\underset{\sim}{a}_2)_t \cdot (\underset{\sim}{a}_3 \times \underset{\sim}{a}_1)$$

$$+ (\underset{\sim}{a}_3)_t \cdot (\underset{\sim}{a}_1 \times \underset{\sim}{a}_2)$$

$$= \sqrt{g} \sum_{i=1}^{3} (\underset{\sim}{a}_i)_t \cdot \underset{\sim}{a}^i$$

by Eq. (33) But

$$(\underset{\sim}{a}_i)_t = (\underset{\sim}{x}_{\xi^i})_t = (\dot{\underset{\sim}{x}})_{\xi^i}$$

so that

$$(\sqrt{g})_t = \sqrt{g} \sum_{i=1}^{3} \underset{\sim}{a}^i \cdot (\dot{\underset{\sim}{x}})_{\xi^i} \qquad (120)$$

We then can write

$$C = A_t + \frac{A(\sqrt{g})_t}{\sqrt{g}} + \frac{1}{\sqrt{g}} \sum_{i=1}^{3} [\sqrt{g}\, \underset{\sim}{a}^i \cdot (\underset{\sim}{u} - \dot{\underset{\sim}{x}})A]_{\xi^i}$$

$$= \frac{1}{\sqrt{g}} \{(\sqrt{g}\, A)_t + \sum_{i=1}^{3} [(\sqrt{g}\, A)\underset{\sim}{a}^i \cdot (\underset{\sim}{u} - \dot{\underset{\sim}{x}})]_{\xi^i}\} \qquad (121)$$

which is a conservative form of the generic convective terms with regard to the quantity, $\sqrt{g}A$. By Eq. (33), the quantity

$$U^i = \underset{\sim}{a}^i \cdot (\underset{\sim}{u} - \dot{\underset{\sim}{x}}) \qquad (i = 1,2,3) \qquad (122)$$

is the contravariant velocity component in the ξ^i-direction, relative to the moving grid. Thus Eq. (121) can be written in the conservative form,

$$\sqrt{g}\, C = (\sqrt{g}\, A)_t + \sum_{i=1}^{3} (\sqrt{g}\, AU^i)_{\xi^i} \qquad (123)$$

Expanding the derivatives in Eq. (119) and using Eq. (40), we have

$$C = A_t - \dot{\underset{\sim}{x}} \cdot \sum_{i=1}^{3} \underset{\sim}{a}^i A_{\xi^i} + \sum_{i=1}^{3} \underset{\sim}{a}^i \cdot (\underset{\sim}{u}A_{\xi^i} + A\underset{\sim}{u}_{\xi^i}) \qquad (124)$$

so that the non-conservative form

$$C = A_t + \sum_{i=1}^{3} U^i A_{\xi^i} + A \sum_{i=1}^{3} \underset{\sim}{a}^i \cdot \underset{\sim}{u}_{\xi^i} \qquad (125)$$

The last summation is the divergence of the velocity, $\underset{\sim}{\nabla} \cdot \underset{\sim}{u}$. (Computationally, $\sqrt{g}$ might be included in the definition of U^i for use in the conservative form in the interest of computational efficiency, since by Eq. (33) the product $\sqrt{g}\,\underset{\sim}{a}^i$ can be evaluated directly as the cross product of the covariant base vectors.)

From Eq. (117) we have, with A taken as ξ^i,

$$\dot{\xi}^i = -\dot{\mathbf{x}} \cdot \nabla\xi^i = -\dot{\mathbf{x}} \cdot \mathbf{a}^i \qquad (126)$$

by Eq. (4). Here the time derivative of ξ^i is, of course, at a fixed position in physical space. The quantity U^i, introduced above in Eq. (122), thus could be written as

$$U^i = \mathbf{a}^i \cdot \mathbf{u} + \dot{\xi}^i = \nabla\xi^i \cdot \mathbf{u} + \dot{\xi}^i \quad (i = 1,2,3) \qquad (127)$$

Here the $\mathbf{a}^i \cdot \mathbf{u}$ are, of course, the contravariant velocity components.

C. Second derivative

The second time derivative transforms as follows:

$$\left(\frac{\partial^2\phi}{\partial t^2}\right)_{x,y} = \phi_{tt} - 2(\nabla\phi)_t \cdot \dot{\mathbf{r}} + \sum_{i=1}^{3}\sum_{j=1}^{3} \phi_{x_i x_j}\, \dot{x}_i \dot{x}_j - \nabla\phi \cdot \ddot{\mathbf{r}}$$

(128)

where the x,y subscripts on the left indicate the variables being held constant, and

$$\nabla\phi = \sum_{i=1}^{3} \mathbf{a}^i\, \phi_{\xi^i} \qquad (129)$$

$$(\nabla\phi)_t = \sum_{i=1}^{3} (\mathbf{a}^i\, \phi_{t\xi^i} + \dot{\mathbf{a}}^i\, \phi_{\xi^i}) \qquad (130)$$

$$\phi_{x_i x_j} = \sum_{k=1}^{3}\sum_{l=1}^{3} (\mathbf{a}^k)_i[(\mathbf{a}^l)_j\, \phi_{\xi^l\xi^k} + (\mathbf{a}^l_{\xi^k})_j\, \phi_{\xi^l}] \qquad (131)$$

$$(\underset{\sim}{a}^l)_{\xi^k} = \frac{1}{\sqrt{g}}(\underset{\sim}{a}_m \times \underset{\sim}{r}_{\xi^n \xi^k} - \underset{\sim}{a}_n \times \underset{\sim}{r}_{\xi^m \xi^k}) - \underset{\sim}{a}^l \sum_i^3 \sum_j^3 g^{ij} \underset{\sim}{r}_{\xi^i \xi^j} \tag{132}$$

with (l,m,n) cyclic.

Exercises

1. Obtain the covariant and contravariant base vectors for cylindrical coordinates from Eq. (3) and (4). Show that Eq. (34) holds for this system.

2. Obtain the elements of arc length, surface area, and volume for cylindrical coordinates.

3. Obtain the relations for gradient, divergence, curl, and Laplacian for cylindrical coordinates.

4. Demonstrate that the identity (21) holds for cylindrical coordinates.

5. Demonstrate that Eq. (33), (38) and (39) hold for cylindrical coordinates.

6. Repeat exercises 1 - 5 for spherical coordinates.

7. Show that the covariant base vectors may be written in terms of the contravariant base vectors by

$$\underset{\sim}{a}_i = \sqrt{g}\,(\underset{\sim}{a}^j \times \underset{\sim}{a}^k) \quad (i,j,k) \text{ cyclic}$$

   Hint: Cross $\underset{\sim}{a}^k$ into Eq. (33) and use (13), rearranging subscripts at the end. Recalling that $\sqrt{g}$ as can be expressed $(\det |g^{ij}|)^{-1}$, this gives, a relation for $(x_l)_{\xi^i}$ in terms of the derivatives $(\xi^r)_{x_s}$.

8. Show that the elements of the covariant metric tensor can be expressed in terms of the contravariant elements by

$$g_{il} = g(g^{jm}g^{kn} - g^{jn}g^{km})(i,j,k) \text{ cyclic}$$
$$(l,m,n) \text{ cyclic}$$

Hint: Follow the development of Eq. (38), but with $\underset{\sim}{a}_i \cdot \underset{\sim}{a}_1$.

9. Show that Eq. (65) is equivalent to the chain rule expression (1). Also show that the dot product of $\underset{\sim}{a}_j$ with Eq. (65) leads, after interchange of indices, to the chain rule expression (4).

10. Show that

$$(\sqrt{g})_{\xi^i} = \sqrt{g} \sum_j \sum_k g^{jk} \underset{\sim}{r}_{\xi^k \xi^i}$$

Hint: Since g = det $|g^{ij}|$ depends on ξ^i only through the g_{ij}, differential g with respect to g_{jk} and then differential g_{jk} with respect to ξ^i. Recall Eq. (38).

11. Show that $\nabla^2 \underset{\sim}{r} = 0$. Hint: Use cartesian coordinates.

12. Obtain the two-dimensional relations in Section 6 from the general expression.

13. Verify Eq. (74) for cylindrical and spherical coordinates.

14. Obtain the normal and tangential derivatives (Section 4) for cylindrical and spherical coordinates.

IV. NUMERICAL IMPLEMENTATION

1. Transformed Equations

In order to make use of a general boundary-conforming curvilinear coordinate system in the solution of partial differential equations, or of conservation equations in integral form, the equations must first be transformed to the curvilinear coordinates. Such a transformation is accomplished by means of the relations developed in the previous chapter and produces a problem for which the independent variables are time and the curvilinear coordinates. The resulting equations are of the same type as the original ones, but are more complicated in that they contain more terms and variable coefficients. The domain, on the other hand, is greatly simplified since it is transformed to a fixed rectangular region regardless of its shape and movement in physical space. This facilitates the imposition of boundary conditions and is the primary feature which makes grid generation such a valuable and important tool in the numerical solution of partial differential equations on arbitrary domains.

A numerical solution of the transformed problem can be obtained using standard techniques once the problem is discretized. Since the domain is stationary and rectangular, and since the increments of the curvilinear coordinates are arbitrary, the computation can always be done on a fixed uniform square grid. Spatial derivatives at nearly all field points in the transformed domain can therefore be represented by conventional finite-difference or finite-volume expressions, as discussed in the next section. In fact, the transformed problem has the appearance of a problem on a uniform cartesian grid and thus may be treated as such both in the formation of the difference equations and in the solution thereof.

The specific form of the transformed equations to be solved depends, of course, on which of the relations in Chapter III are used, i.e., conservative or not. As an example, consider the generic convection-diffusion equation

$$A_t + \nabla \cdot (\underline{u}A) + \nabla \cdot (\mu \nabla A) + S = 0 \tag{1}$$

Equations (III-123), (III-42), and Eq. (III-43) may be used to transform the convective terms, the gradient, and the second divergence, respectively, and thereby yield the conservative form:

$$(\sqrt{g}A)_t + \sum_{i=1}^{3} (U^i \sqrt{g}A)_{\xi^i}$$
$$+ \sum_{i=1}^{3} \sum_{j=1}^{3} [\mu \underline{a}^i \cdot (\sqrt{g}\underline{a}^j A)_{\xi_j}]_{\xi^i} + \sqrt{g}S = 0 \tag{2}$$

where now the time derivative is understood to be at a fixed point in the transformed region, and the contravariant velocity components (relative to the moving grid) are given by Eq. (III-122). Eq. (2) can also be written in the form

$$(\sqrt{g}A)_t + \sum_{i=1}^{3} [U^i \sqrt{g}A + \mu \underline{a}^i \cdot \sum_{j=1}^{3} (\sqrt{g}A\underline{a}^j)_{\xi^j}]_{\xi^i} + \sqrt{g}S = 0 \tag{3}$$

which clearly shows the conservative form. It is the product $\sqrt{g}A$, rather than the function A itself, which is conserved in this form. The derivative inside the j summation can be expanded and Eq. (III-40) invoked to obtain the simplified form:

$$(\sqrt{g}A)_t + \sum_{i=1}^{3}[\sqrt{g}(U^iA + \mu\sum_{j=1}^{3} g^{ij}A_{\xi^j})]_{\xi^i} + \sqrt{g}S = 0 \qquad (4)$$

which is still conservative in regard to the ξ^i-derivatives.

These conservative forms are in the commonly-used form

$$\underset{\sim}{B}_t + \sum_{i=1}^{3} [\underset{\sim}{F}^i(\underset{\sim}{B})]_{\xi^i} + \underset{\sim}{R} = 0$$

where the solution vector is $\underset{\sim}{B}=\sqrt{g}\underset{\sim}{A}$, the "flux" vectors $\underset{\sim}{F}^i$ are given by the brackets in (3) and (4), and the source vector is $\underset{\sim}{R} = \sqrt{g}\underset{\sim}{S}$.

The flux vectors $\underset{\sim}{F}^i$ contain metric derivatives and thus depend on time and the curvilinear coordinates through these metric elements, as well as through the solution vector $\underset{\sim}{B}$, and this must be taken into account in the construction of factored solution methods. A general formulation of split solution methods (encompassing both time splitting, e.g., approximate factorization, and spatial splitting, e.g., MacCormack method) in the curvilinear coordinates can, however, be formulated.

The non-conservative form of Eq. (1) follows using Eq. (III-125) for the convective terms, Eq. (III-65) for the gradient, and Eq. (III-67) for the divergence $\underset{\sim}{\nabla} \cdot (\mu\underset{\sim}{\nabla}A)$. The resulting equation may be written

$$A_t + \sum_{i=1}^{3}(U^i + \mu\nabla^2\xi^i)A_{\xi^i} + \sum_{i=1}^{3}\sum_{j=1}^{3} g^{ij}(\mu A_{\xi^j})_{\xi^i}$$

$$+ A\sum_{i=1}^{3} \underset{\sim}{a}^i \cdot \underset{\sim}{u}_{\xi^i} + S = 0 \qquad (5)$$

since Eq. (III-70) gives

$$\nabla^2 \xi^i = \sum_{j=1}^{3} \underset{\sim}{a}^j \cdot (\underset{\sim}{a}^i)_{\xi^j} \tag{6}$$

(The last summation in Eq. (5) is just $\underset{\sim}{\nabla} \cdot \underset{\sim}{u}$, which vanishes for incompressible flow.) Comparison of Eq. (5) with the original equation, written in the form

$$A_t + \sum_{i=1}^{3} u_i A_{x_i} + \sum_{i=1}^{3} \sum_{j=1}^{3} \delta_{ij} (\mu A_{x_j})_{x_i} + A \sum_{i=1}^{3} (u_i)_{x_i} + S = 0 \tag{7}$$

demonstrates that the equation has been complicated by the transformation only in the sense that the coefficient u_i has been replaced by the coefficient $U^i + \mu(\nabla^2 \xi^i)$, and the Kroniker delta in the double summation has been replaced by g^{ij}, thus expanding that summation from three terms to nine terms, and through the insertion of variable coefficients in the last summation. This exemplifies the fact that the use of the general curvilinear coordinate system does not introduce any significant complications into the form of the partial differential equations to be solved. When it is considered that the transformed equation (5) is to be solved on a fixed rectangular field with a uniform square grid, while the original equation (7) would have to be solved on a field with moving curved boundaries, the advantages of using the curvilinear system are clear.

These advantages are further evidenced by consideration of boundary conditions. In general, boundary conditions for the example being treated would be of the form

$$\alpha A + \beta \underset{\sim}{n} \cdot (\mu \underset{\sim}{\nabla} A) = \gamma \tag{8}$$

where $\underset{\sim}{n}$ is the unit normal to the boundary and α, β, and γ are specified. From Eq. (III-79) these conditions transform to

$$\alpha A + \beta \frac{\mu}{\sqrt{g^{ii}}} \sum_{j=1}^{3} g^{ij} A_{\xi^j} = \gamma \tag{9}$$

for a boundary on which ξ^i is constant. For comparison, the original boundary conditions (8) can be written in the form

$$\alpha A + \beta\mu \sum_{j=1}^{3} n_j A_{x_j} = \gamma \tag{10}$$

The transformed boundary conditions thus have the same form as the original conditions, but with the coefficient n_j replaced by $g^{ij}/\sqrt{g^{ii}}$. The important simplification is the fact that the boundary to which the transformed conditions are applied is fixed and flat (coincident with a curvilinear coordinate surface). This permits a discrete representation of the derivatives A_{ξ^j} along the transformed boundary without the need for interpolation. By contrast, the derivatives A_{x_j} in the original conditions cannot be discretized along the physical boundary without interpolation since the boundary is curved and may be in motion.

This discussion of a generic convection-diffusion equation and associated boundary conditions should serve to allow specific physcial equations to be transformed. References to application of these equations are given in the surveys Ref. [1] and [5]. Several examples also appear in Ref. [2].

2. Discrete Representation of Derivatives

Approximate values of the spatial derivatives of a

function which appear in the transformed equations may be found at a given point in terms of the function's value at that point and at neighboring points. As noted earlier, with the problem in the transformed space, only uniform square grids need be considered, hence the standard forms for difference representation of derivatives may be used. For example, in two dimensions the first, second, and mixed partials with respect to the curvilinear coordinates ξ and η are ordinarily represented at an interior point (i,j) by finite differences or finite-volume expressions which contain function values at no more than the nine points shown below.

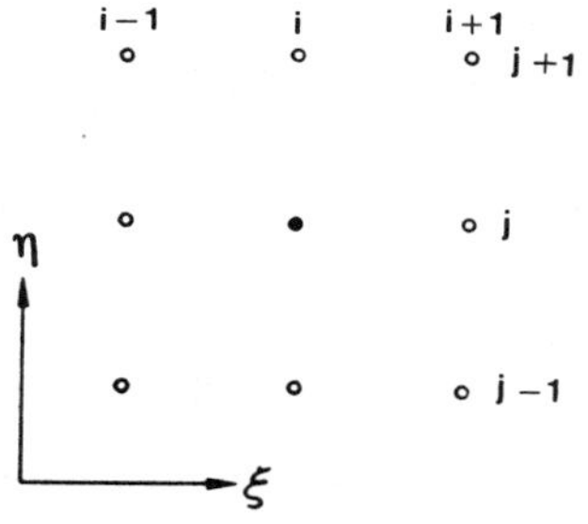

This centered, nine-point "computational molecule" is usually preferred because of the associated difference representations which are symmetry- preserving and second-order accurate. Examples of finite-difference approximations of this type are:

$$(f_\xi)_{ij} = \frac{1}{2}(f_{i+1,j} - f_{i-1,j}) \tag{11a}$$

$$(f_\eta)_{ij} = \frac{1}{2}(f_{i,j+1} - f_{i,j-1}) \tag{11b}$$

$$(f_{\xi\xi})_{ij} = f_{i+1,j} - 2f_{ij} + f_{i-1,j} \tag{12a}$$

$$(f_{\eta\eta})_{ij} = f_{i,j+1} - 2f_{ij} + f_{i,j-1} \tag{12b}$$

$$(f_{\xi\eta})_{ij} = \frac{1}{4}(f_{i+1,j+1} - f_{i+1,j-1} - f_{i-1,j+1} + f_{i-1,j-1}) \tag{13}$$

Other second-order approximations of the mixed partial $(f_{\xi\eta})_{ij}$ which use the nine-point molecule are:

$$\frac{1}{2}(f_{i+1,j+1} - f_{i+1,j} - f_{i,j+1} + 2f_{ij}$$

$$- f_{i,j-1} - f_{i-1,j} + f_{i-1,j-1}) \tag{14}$$

and

$$\frac{1}{2}(f_{i+1,j} - f_{i+1,j-1} + f_{i,j+1} - 2f_{ij}$$

$$+ f_{i,j-1} - f_{i-1,j+1} + f_{i-1,j}) \tag{15}$$

It is clear that at boundary points, where at most first partials must be represented, the computational molecule cannot be centered relative to the direction of the coordinate ξ^{α} which is constant on the boundary (see diagram below).

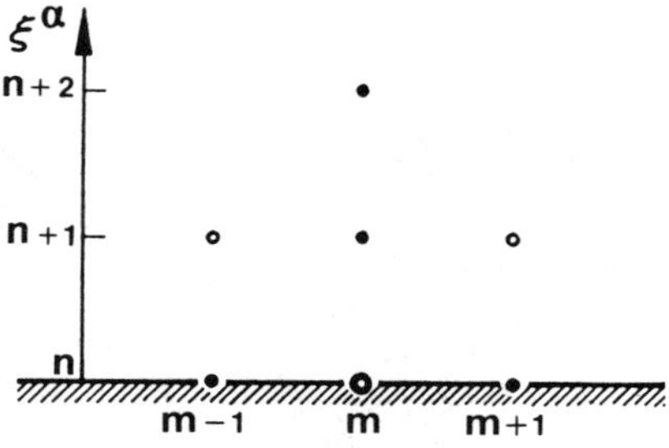

There a one-sided difference must be used to approximate $f_{\xi\alpha}$. The second-order formula appropriate for the boundary point indicated above is

$$(f_{\xi\alpha})_{mn} = \frac{1}{2}(-f_{m,n+2} + 4f_{m,n+1} - 3f_{m,n})$$

Any standard text on the subject of finite-difference methods will provide formulas of alternate order and/or based on other computational molecules.

A finite-volume approach uses function values at grid-cell centers and approximates derivatives at a cell center by line (surface in 3D) integrals about the cell boundary which are equivalent to averages over the cell. In particular, the identity

$$\nabla f_{avg} = \frac{1}{V}\int_D \nabla f \, d\tau = \frac{1}{V}\int_{\partial D} f \, \mathbf{n} \, d\sigma \tag{16}$$

is used, where V is the volume of D. Thus, if a function is assumed constant along a grid-cell face, it is a simple matter to evaluate the line integral in (16) when D is a grid cell in transformed space. In terms of the two-dimensional grid:

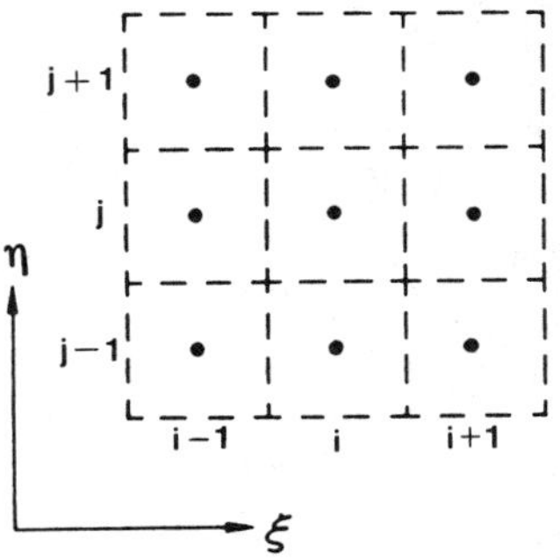

this approach gives

$$(f_\xi)_{ij} = f_{i+\frac{1}{2},j} - f_{i-\frac{1}{2},j} \tag{17a}$$

$$(f_\eta)_{ij} = f_{i,j+\frac{1}{2}} - f_{i,j-\frac{1}{2}} \tag{17b}$$

With an edge value approximated as the average of the center values of the two cells sharing that face, e.g.

$$f_{i+\frac{1}{2},j} = \frac{1}{2}(f_{i+1,j} + f_{ij}) \tag{18}$$

the values given by (17) are equivalent to ordinary central differences (cf. Eq. (11)) and hence are second-order accurate. The first partials of f may also be assumed constant along each cell edge in order to derive from (16) the following approximations of second and mixed partials at a cell center:

$$(f_{\xi\xi})_{ij} = (f_\xi)_{i+\frac{1}{2},j} - (f_\xi)_{i-\frac{1}{2},j} \tag{19a}$$

$$(f_{\xi\eta})_{ij} = (f_\xi)_{i,j+\frac{1}{2}} - (f_\xi)_{i,j-\frac{1}{2}} \tag{19b}$$

$$(f_{\eta\xi})_{ij} = (f_\eta)_{i+\frac{1}{2},j} - (f_\eta)_{i-\frac{1}{2},j} \tag{19c}$$

$$(f_{\eta\eta})_{ij} = (f_\eta)_{i,j+\frac{1}{2}} - (f_\eta)_{i,j-\frac{1}{2}} \tag{19d}$$

Now, however, the averaging scheme in (18) cannot be used to approximate edge values of the derivatives without going outside the nine-point computational molecule shown above. Instead, a second-order accurate representation can be obtained on the nine-point molecule using a forward (backward) assignment for the center value of a function and a backward (forward) assignment for the first partial on a given side. There are four possible schemes of this type. One uses

$$f_{i+\frac{1}{2},j} = f_{ij}, \quad f_{i,j+\frac{1}{2}} = f_{ij} \tag{20}$$

to evaluate $\nabla f(\xi,\eta)$ at all cell centers according to (17), and then uses

$$g_{i-\frac{1}{2},j} = g_{ij}, \quad g_{i,j-\frac{1}{2}} = g_{ij} \quad (\text{where } g = f_\xi \text{ or } f_\eta) \tag{21}$$

to evaluate the second and mixed partials given in (19). This method is equivalent to a finite-difference scheme which approximates first partials by backward differences of the function, and then approximates second and mixed partials by forward differences of the first partials. Consequently, the second derivatives which result are equal to those given in Eq. (12), while the resulting representations of the two mixed partials are unequal and only first-order accurate. If the two mixed partials are averaged, however, the second-order expression (15) is recovered. This is also true of the reverse scheme:

$$f_{i-\frac{1}{2},j} = f_{ij}, \quad f_{i,j-\frac{1}{2}} = f_{ij} \tag{22}$$

$$g_{i+\frac{1}{2},j} = g_{ij}, \quad g_{i,j+\frac{1}{2}} = g_{ij} \quad (g = f_\xi \text{ or } f_\eta) \tag{23}$$

Expressions (12) and (14) are similarly recovered from the other two possibilities (Eq. 20a, 21a, 22b, and 23b or Eq. 20b, 21b, 22a, and 23a). The symmetry-preserving form (13) can be recovered by averaging the averaged mixed partial obtained in one of the first two schemes mentioned and that obtained in one of the remaining two.

The manner in which boundary conditions are treated in a finite volume approach depends on the type of conditions imposed. When Dirichlet conditions are prescribed, it is advantageous to treat the boundary as the center line (plane in three-dimensions) of a row of cells straddling the boundary. The centers of these cells then fall on the physical boundary where the function values are known. When Neumann or mixed conditions are given, however, the boundary is best treated as coincident with cell faces.

Suppose, for example, that boundary condition (9) is to be imposed at the cell edge $\eta=j-1/2$ indicated below.

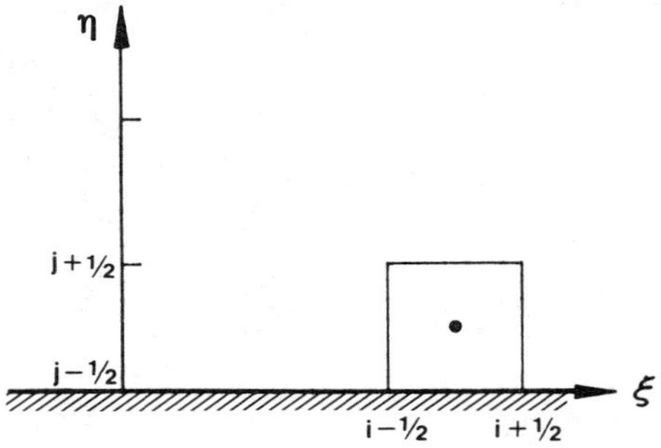

The edge value of $f_{i,j-1/2}$ cannot be approximated by the usual averaging scheme (illustrated by Eq. (18)) since there is no cell center at $\eta=j-1$. It can, however, be found in terms of neighboring cell-centered function values by using boundary condition (9) in connection with the forward/backward scheme used to approximate second derivatives at the cell centers.

Considering the scheme represented by Eq. (20) and (21), the values of f along the cell edges shown above are:

$$f_{i-\frac{1}{2},j} = f_{i-1,j}, \qquad f_{i,j+\frac{1}{2}} = f_{ij}$$

$$f_{i+\frac{1}{2},j} = f_{ij}, \qquad f_{i,j-\frac{1}{2}} = \text{undetermined} = x_i$$

It follows from Eq. (17) that the first partials of f at the cell center are

$$(f_\xi)_{ij} = f_{ij} - f_{i-1,j}, \qquad (f_\eta)_{ij} = f_{ij} - x_i$$

Eq. (21a,b) then give f_ξ and f_η along the cell edges enclosing (i,j) in terms of $f_{i-1,j}$, $f_{i-1,j+1}$, f_{ij}, $f_{i,j+1}$, $f_{i+1,j}$, x_i and x_{i+1}. In particular,

$$(f_\xi)_{i,j-\frac{1}{2}} = f_{ij} - f_{i-1,j}, \qquad (f_\eta)_{i,j-\frac{1}{2}} = f_{ij} - x_i$$

Substitution of these expressions into boundary condition (9) then determines the edge value x_i as

$$x_i = f_{i,j-\frac{1}{2}} = (\alpha - \beta\mu\sqrt{g^{22}})^{-1}$$

$$\cdot \{\gamma - \frac{\beta\mu}{\sqrt{g^{22}}}[g^{21}(f_{ij} - f_{i-1,j}) + g^{22}f_{ij}]\}$$

In this way, f, and hence f_ξ and f_η, are found on all boundary-cell edges in terms of cell-centered values of f.

The finite-difference and finite-volume techniques described thus far are appropriate for representing all derivatives with respect to the curvilinear coordinates, even those appearing in the metric quantities. In fact, as it is shown later in this chapter and in chapter V, the metric quantities should be represented numerically even when analytical expressions are available. One might have, for example,

$$(x_\xi)_{ijk} = \frac{1}{2}(x_{i+1,j,k} - x_{i-1,j,k}) \tag{24}$$

3. Special Points

Many of the expressions given in the previous section break down at so-called "special points" in the field where special attention is required in the approximation of derivatives. These points commonly arise when geometrically complicated physical domains are involved. As indicated in Chapter II, special points can occur on the domain boundary and on interfaces between subregions of a composite curvilinear coordinate system. They may be recognized in physical space as those interior points having a nonstandard number of immediate neighbors or, equivalently, those points which are vertices, or the center, of a cell with either a nonstandard number of faces or a vertex shared by a nonstandard number of other cells. (In two dimensional domains, ordinary interior points have eight immediate neighbors [refer to figure on p. 141]; standard two-dimensional interior grid cells have four sides and share each vertex with three other cells [see diagram on p. 143].) Boundary points are not special unless they are vertex-centered and have a nonstandard number of immediate neighbors (other than five in two dimensions see diagram on p. 142 for an ordinary boundary point) and then are special only when their associated boundary conditions contain spatial derivatives. Some examples of special cell-centered points and special vertex-centered points are shown below.

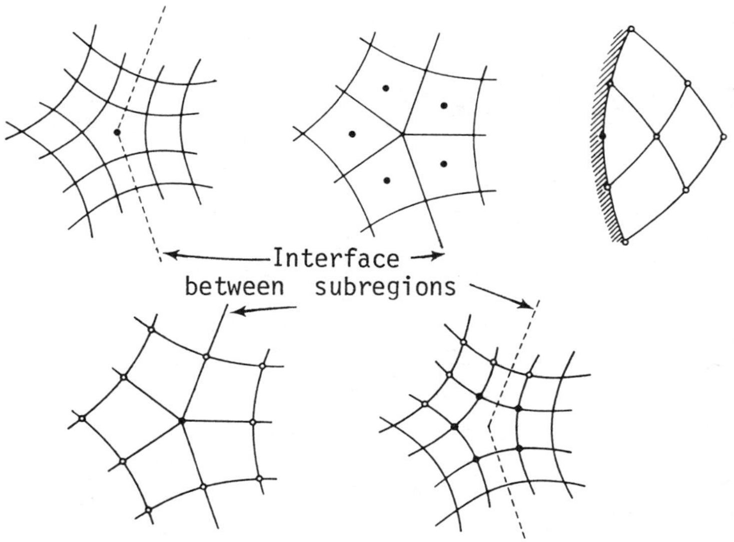

When a finite-difference formulation is used, the usual approach, as described in Section 2, can be followed at a special point P if the transformed equations and difference approximations at that point are rephrased in terms of suitable local coordinates. The local system is chosen so as to orient and label only the surrounding points to be used in the needed difference expressions. Choices appropriate to various special points are listed in Tables 1, 2, and 3.

The difficulties encountered at special points in a finite-volume approach are clearly seen by considering the image in the transformed plane. The first pair of diagrams below, for example, shows that at centers of cells having the usual number of faces but sharing a vertex with a non-standard number of cells, such difficulties amount to mere bookkeeping complications when only first partials must be approximated. Equations (17) and (18) still apply, but the indices must be defined to correctly relate the cell centers on the two sides of an interface. The following diagrams

	Diagram of special point	Characterization of special point	Image in transformed plane	Local coordinate system
I	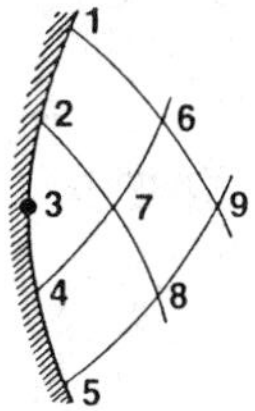	Special point 3 on smooth boundary is transformed to a corner of the computational domain.	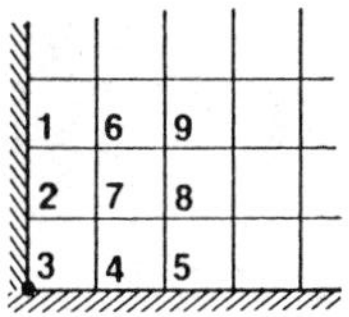	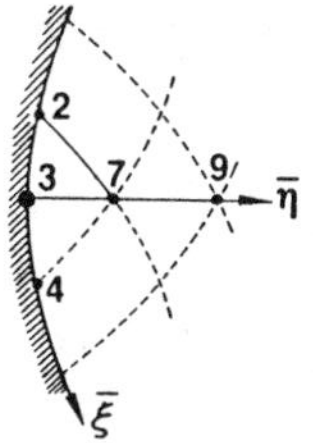
II	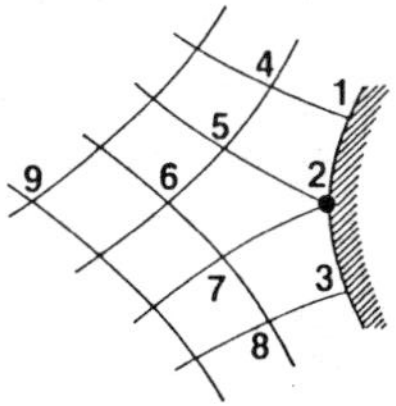	Special point 2 is on an intrusion or interior object and is transformed to a corner of an intruding or interior slab.	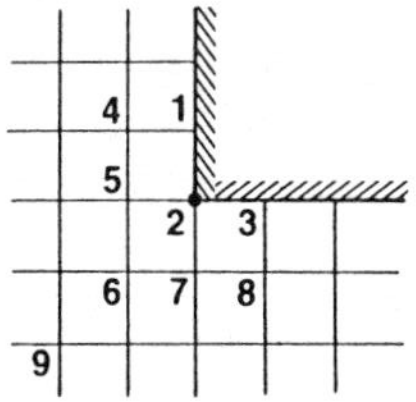	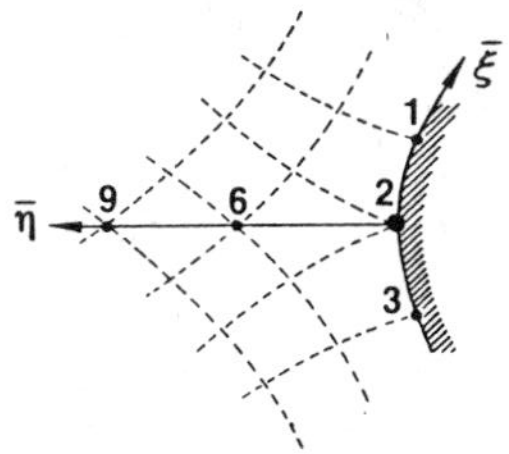
III	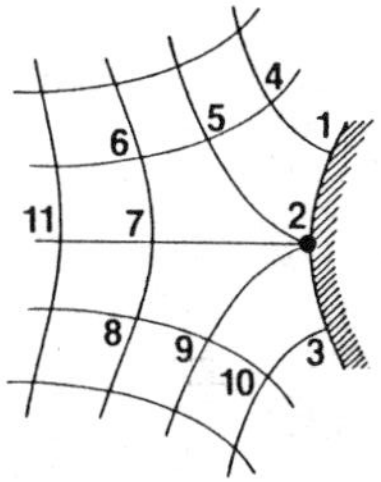	(a) Special point 2 is a branch point on an interior body, or (b) point 2 is on an interior body and is transformed to an end point of a slit.	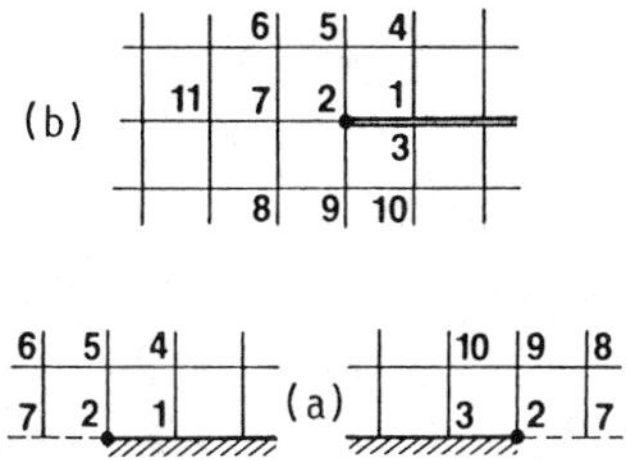	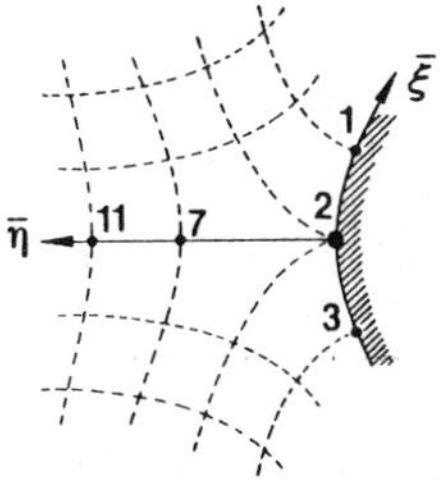

Table 1. Special boundary points.

Diagram of special point	Characterization of special point	Image in transformed plane	Local coordinate system
IV 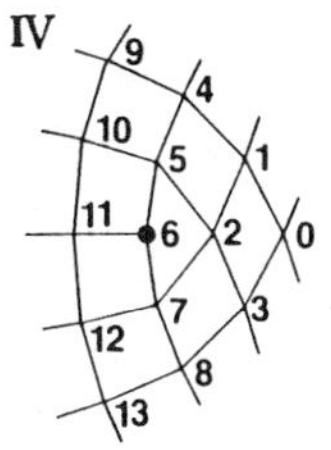	(a) Special point 6 is common to two subregions. Its image is a corner point for one and an edge point for the other. (b) Special point 6 is common to three subregions. Its image is a corner point for each.	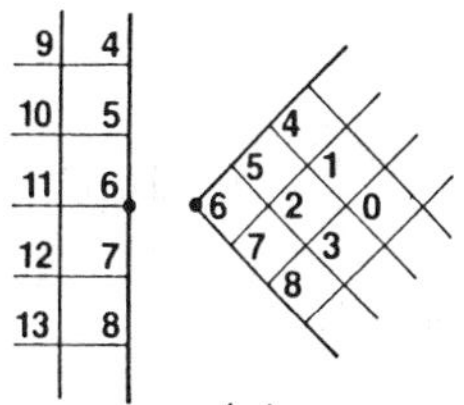 (a) 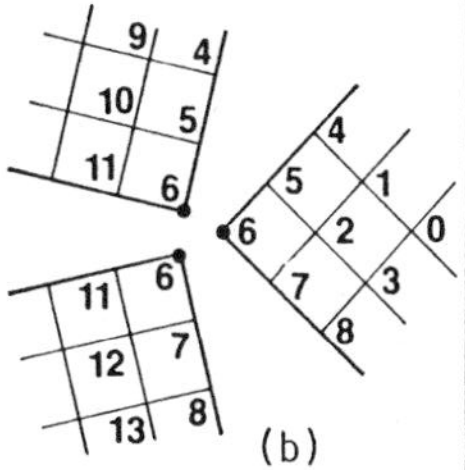 (b)	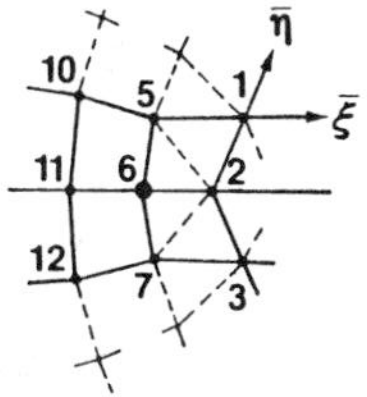
V 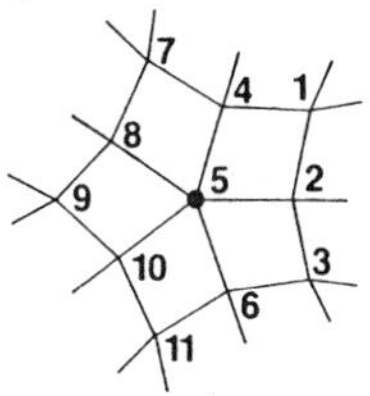	(a) Special point 5 is common to two subregions. Its image is a concave corner point for one and an edge point for the other. (b) Special point 5 is common to four subregions. Its image is a corner point for three of the segments and an edge point for the fourth. (c) Special point 5 is common to five subregions. Its image is a corner point for each. (d) Special point 5 is common to three subregions. Its image is a corner point for one segment and an edge point for the other two.	(a) 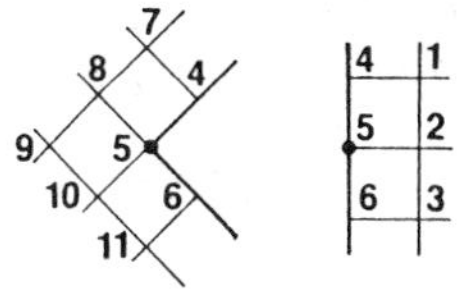 (b) 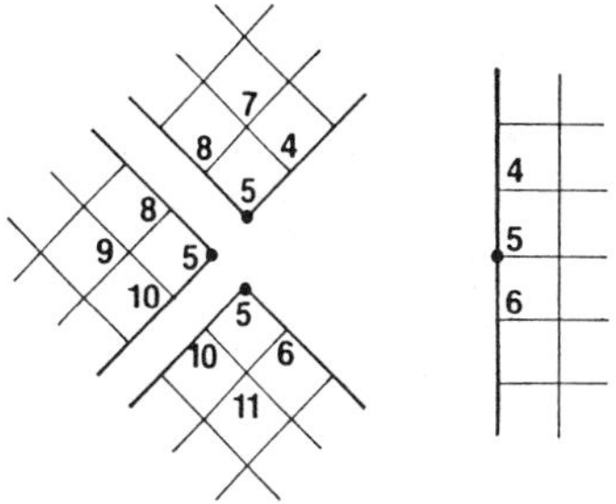 (c) Analogous to IV (b). (d) An obvious modification of V (b).	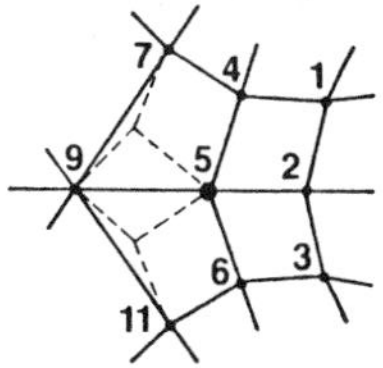 or choose points 8 and 10 instead of 7 and 11 if the skewness of segments 9-8 and 9-10 is the same or less than that of segments 9-7 and 9-11.

Table 2. Special vertex-centered interior points associated with subregions joined along grid lines.

Diagram of special point	Characterization of special point	Image in transformed plane	Local coordinate system
VI 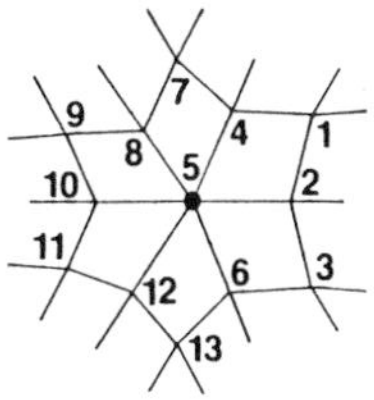	(a) Special point 5 is common to two subregions. It is a branch point in one subregion and an ordinary edge point of the other. (b) Special point 5 is common to two subregions. Its image is an endpoint of a slit in one transformed segment and is an edge point of the other. (c) Special point 5 is common to five subregions. Its image is a corner point for four of the segments and an edge point of the fifth. (d) Special point 5 is common to six subregions. Its image is a corner point for each. (e) Special point 5 is common to four subregions. Its image is a corner point for two of the segments and an edge point for the other two. (f) Special point 5 is common to three subregions and is an edge point for each.	(a) 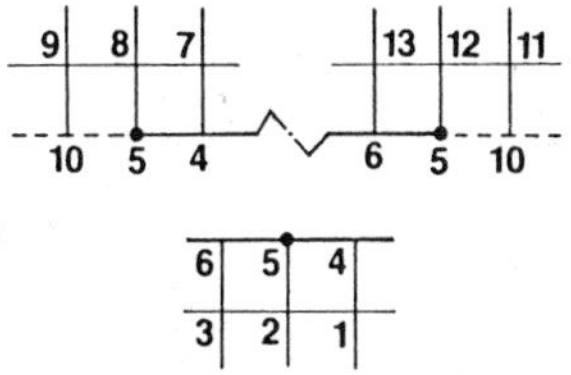 (b) 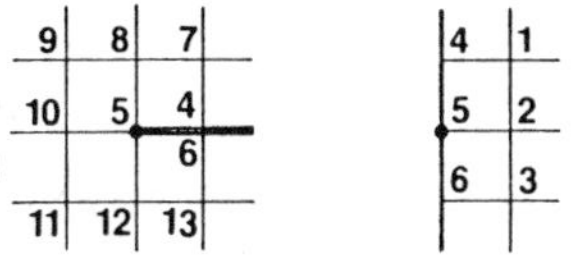 (c) Analogous to V (b). (d) Analogous to IV (b). (e) 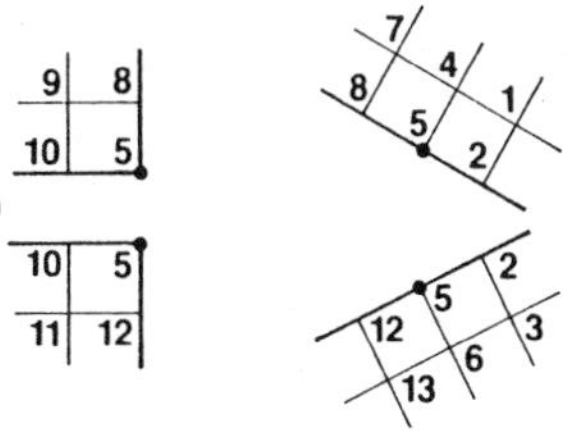 (f) An obvious modification of VI (e).	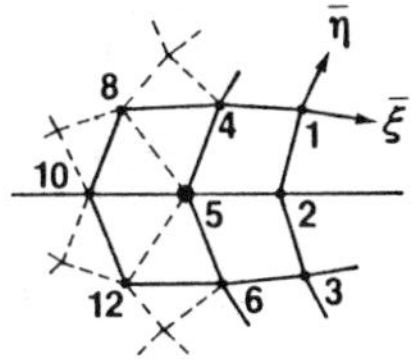 or choose points 9 and 11, or 7 and 13, instead of 8 and 12, to obtain the most reasonable system.

Table 2. continued

Diagram of special point	Characterization of special point	Image in transformed plane	Local coordinate system
VII 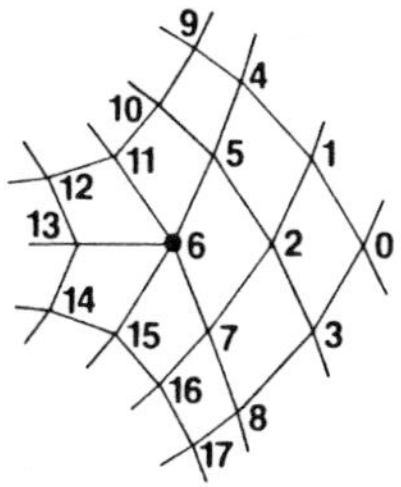	(a) Special point 6 is common to two subregions. It is a branch point in one subregion and is a corner point in the image of the other. (b) Special point 6 is common to two subregions. Its image is an endpoint of a slit in one transformed segment and is a corner point of the other. (c) Same as V (c). (d) Same as V (b). (e) Same as V (d).	(a) 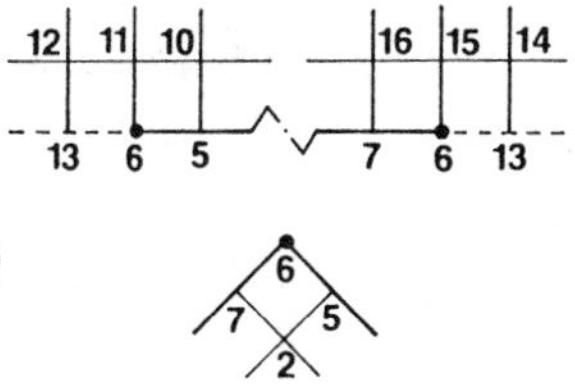 (b) 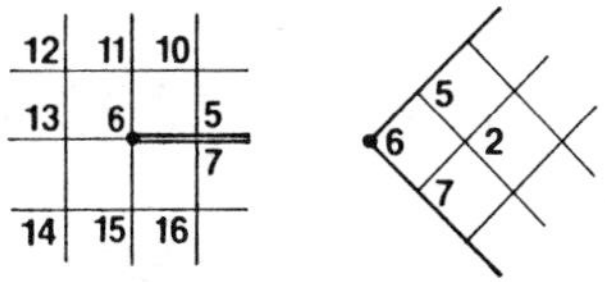 (c) Same as V (c). (d) Same as V (b). (e) Same as V (d).	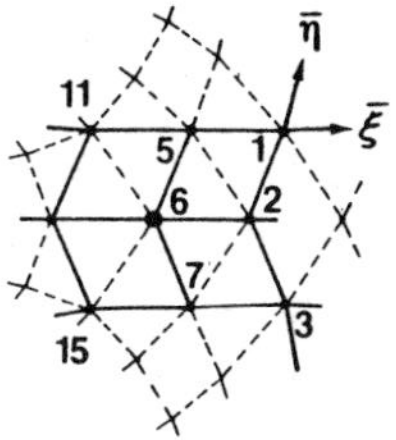 or choose points 12 and 14, or 10 and 16, instead of 11 and 15, to obtain the most reasonable system.

Table 2. continued

Diagram of special point	Characterization of associated special cell	Point in local computational molecule
VIII 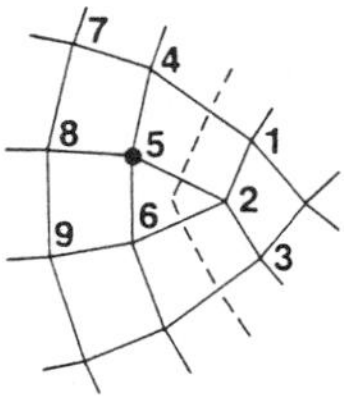	Cell center is equivalent to special point 6 in category IV.	(a) Points 1-9, or (b) points 1, 2, 4-6, 8, 9 and use the corresponding antisymmetric, 2nd-order difference for the cross derivative.
IX 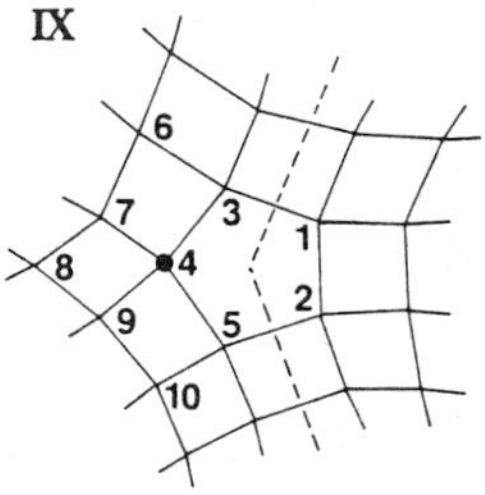	Cell center is equivalent to special point 5 in category V.	(a) Points 3-10 and 1 or 2, or (b) 3-7, 9, 10 and use corresponding antisymmetric, 2nd-order difference for cross derivative.
X 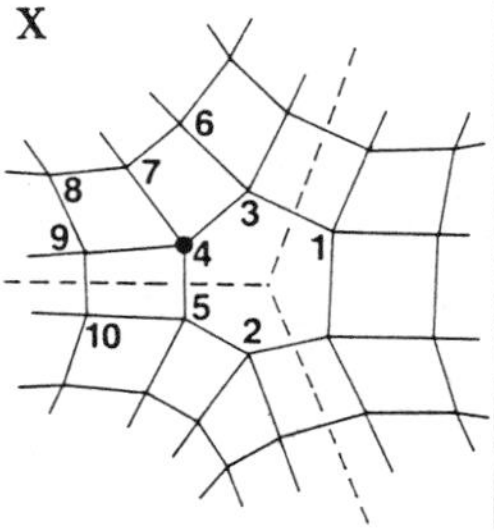	Cell center is equivalent to special point 5 in category VI.	same as IX.
XI 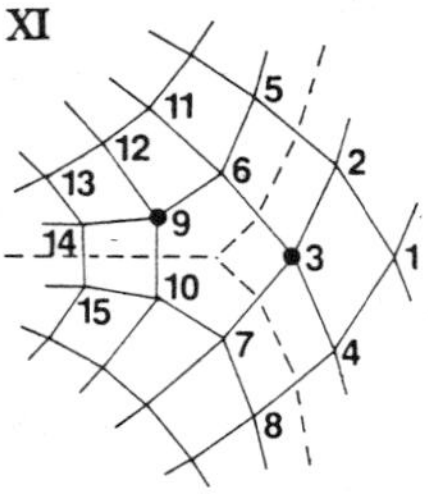	Cell center is equivalent to special point 6 in category VII.	(i) At special vertex 3: Use points 1-8 and 9 or 10. (ii) Treat special vertex 9 the same as special vertex 4 in category IX.

Table 3. Special vertex-centered interior points associated with subregions joined between grid lines.

also illustrate the breakdown at all special cell-centered points of the previously-described finite-volume schemes for approximating second and mixed partial derivatives. This is because the forward/backward orientation of the coordinate system in one segment cannot be consistently followed across the interface adjacent to, or intersecting, the special points. The second pair of diagrams displays the additional complication associated with grid cells having a nonstandard number of edges. Such a cell can occur on an interface between segments of a composite grid which are joined between grid lines. When the segments are transformed to their respective images, the separate pieces of the special grid cell cannot be joined without distorting them. It is thus unclear how to evaluate the volume and the outward normals of that transformed cell in order to use identity (16) in the transformed plane. Consequently, at special points of this type and at all special points where second derivatives must be approximated, the governing equations are best represented locally in the physical plane where such ambiguities do not exist.

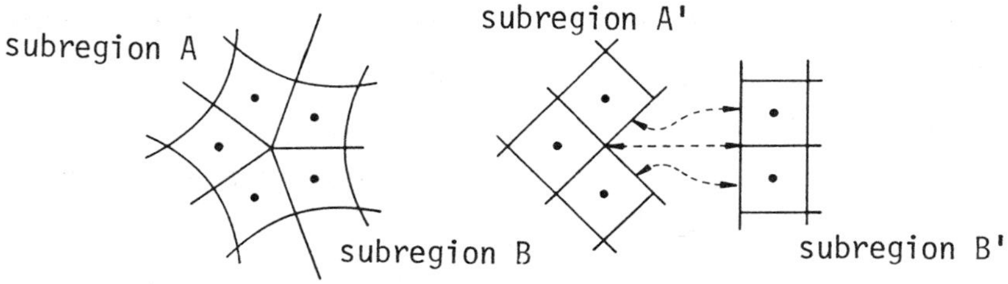

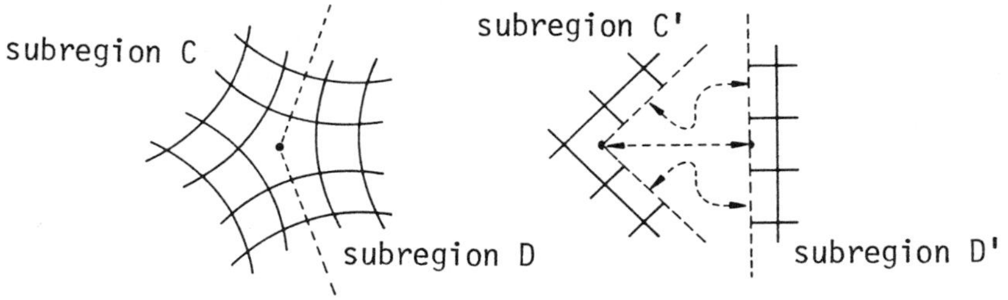

Treatment in physical space involves approximation of the original equations by means of identity (16). Thus, for a two-dimensional N-sided cell of area A with cartesian centroid P = (p_1,p_2), vertices $V^i=(v_1^i,v_2^i)$ $i=1,2,\ldots,N$, and edges s^i joining V^i and V^{i+1} $(V^{N+1}=V^1)$ along which a function f and its first partial derivatives are constant, this approach gives

$$f_x^P = A^{-1} \sum_{i=1}^{N} f^{s^i}(v_2^{i+1} - v_2^i)$$

$$f_y^P = A^{-1} \sum_{i=1}^{N} f^{s^i}(v_1^i - v_1^{i+1})$$

$$f_{xx}^P = A^{-1} \sum_{i=1}^{N} f_x^{s^i}(v_2^{i+1} - v_2^i)$$

$$f_{yy}^P = A^{-1} \sum_{i=1}^{N} f_y^{s^i}(v_1^i - v_1^{i+1})$$

where the superscripts on f and its derivatives indicate the point or face of evaluation. As in the previous section, an obvious way to approximate f^{s^i} is to average the center values of the two cells sharing edge s^i. This same averaging scheme cannot be repeated to approximate $f_x^{s^i}$, and $f_y^{s^i}$, however, without rejecting the recommended strategy of avoiding use of values at points which are not immediate neighbors of the point at which a quantity is being evaluated. Instead, we propose the averaging technique:

$$f_x^{s^i} = \frac{1}{2}(f_x^{V^i} + f_x^{V^{i+1}})$$

$$f_y^{s^i} = \frac{1}{2}(f_y^{V^i} + f_y^{V^{i+1}})$$

where the vertex values are obtained by applying identity (16) to auxiliary cells formed by joining the midpoints of the edges of each cell to the cell center. To make this more precise, let V be a vertex common to Q cells and label the cell faces emanating from V as k^i with midpoints

$$M^i=(m_1^i,m_2^i) \qquad i=1,2,\ldots,Q.$$

Then if $P^i=(p_1^i,p_2^i)$ is the center of the cell having edges k^i and k^{i+1}, and if

$$f = f^{P^i} \text{ along } M^iP^i \text{ and } P^iM^{i+1}$$

the first partial derivatives of f at V may be approximated by

$$f_x^V = A^{-1} \sum_{i=1}^{Q} f^{P^i}(m_2^{i+1} - m_2^i)$$

$$f_y^V = A^{-1} \sum_{i=1}^{Q} f^{P^i}(m_1^i - m_1^{i+1})$$

where A is the area of the 2Q-faced auxiliary cell $M^1P^1M^2P^2 \ldots M^QP^QM^1$ indicated in the following diagram.

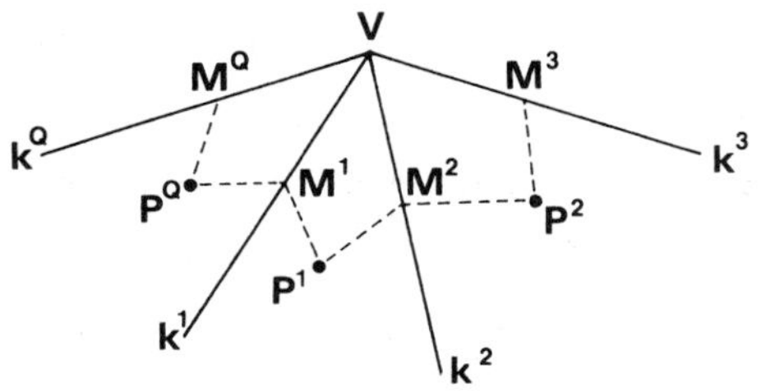

This technique is applicable to all grid cell centers; however, it is recommended for use only at points where the methods developed in section 2 break down, since the difference representations associated with those methods are simpler.

4. Metric Identities

When the transformed equations are in conservative form, it is possible for the metric coefficients to introduce spurious source terms into the equations, as has been noted in several works cited in Ref. [1] and as discussed also in Ref. [11] and [12]. This is because the metric coefficients have been included in the operand of the differential operators and if the differencing of these coefficients does not numerically satisfy identities (III-40) and (III-120), the numerical representations of derivatives of uniform physical quantities are nonvanishing.

For example, if the quantity A is constant, the conservative form for the gradient, Eq. (III-42) gives

$$0 = \sum_{i=1}^{3} (\sqrt{g}\underset{\sim}{a}^i)_{\xi^i}$$

which is precisely Eq. (III-40). Relations (III-43) - (III-45) similarly reduce to (III-40) when $\underset{\sim}{A}$ is uniform. Therefore, Eq. (III-40), or equivalently Eq. (III-21), is a metric identity which must be satisfied <u>numerically</u> in order that the conservative expressions for the gradient, divergence, curl, and Laplacian, etc., vanish when the physical variable is uniform. This consideration does not arise with the non-conservative forms since the quantity A is differentiated directly in those expressions.

Another metric identity which must be satisfied numerically arises when the grid is time-dependent. This may be seen by considering a generic conservation equation of the form

$$A_t + \nabla \cdot (\mathbf{u}A) = 0$$

The conservative relation (III-121) transforms this to

$$(\sqrt{g}A)_t + \sum_{i=1}^{3} [(\sqrt{g}A)\mathbf{a}^i \cdot (\mathbf{u} - \dot{\mathbf{x}})]_{\xi^i} = 0 \qquad (25)$$

where now the time derivative is understood to be at a fixed point in the transformed space. If A and $\mathbf{u}$ are both constants, then Eq. (25) gives

$$(\sqrt{g})_t - \sum_{i=1}^{3} (\sqrt{g}\mathbf{a}^i \cdot \dot{\mathbf{x}})_{\xi^i} = - \mathbf{u} \cdot \sum_{i=1}^{3} (\sqrt{g}\mathbf{a}^i)_{\xi^i}$$

which vanishes according to Eq. (III-40). Expansion of the left-hand summation subject to Eq. (III-40) then reveals the additional identity to be satisfied:

$$(\sqrt{g})_t - \sqrt{g} \sum_{i=1}^{3} \mathbf{a}^i \cdot (\dot{\mathbf{x}})_{\xi^i} = 0 \qquad (26)$$

which is just Eq. (III-120). This equation, therefore, is that which should be used to numerically determine updated values of the Jacobian, $\sqrt{g}$. For if $\sqrt{g}$ is instead updated directly from the new values of the cartesian coordinates, spurious source terms will appear.

The following example provides a simple illustration of differencing schemes which do, and do not, satisfy the metric identities. The conservative expression for a first derivative in two dimensions is given in Eq. (III-96) as

$$f_x = \frac{1}{\sqrt{g}} [(fy_\eta)_\xi - (fy_\xi)_\eta] \tag{27}$$

which for uniform f reduces to

$$0 = y_{\eta\xi} - y_{\xi\eta} \tag{28}$$

Suppose that f_x is to be represented at the center of the cell shown below.

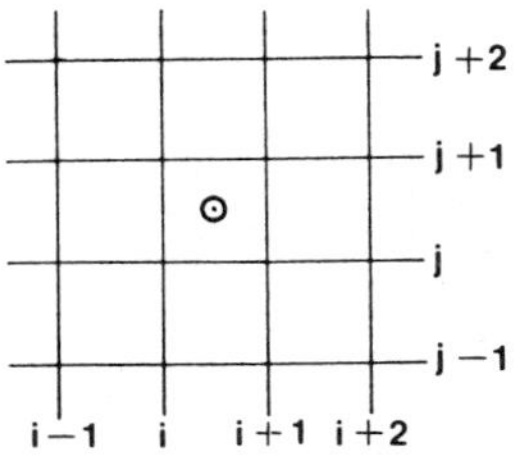

The differencing scheme should satisfy

$$(y_{\eta\xi})_{i+\frac{1}{2},j+\frac{1}{2}} = (y_{\xi\eta})_{i+\frac{1}{2},\, j+\frac{1}{2}}$$

One possible candidate is the sequence of central differences represented by

$$(y_{\eta\xi})_{i+\frac{1}{2},j+\frac{1}{2}} = (y_\eta)_{i+1,j+\frac{1}{2}} - (y_\eta)_{i,j+\frac{1}{2}} \tag{29a}$$

$$(y_\eta)_{i+1,j+\frac{1}{2}} = y_{i+1,j+1} - y_{i+1,j} \tag{29b}$$

The resulting expressions for the mixed partials are

$$(y_{\eta\xi})_{i+\frac{1}{2},j+\frac{1}{2}} = y_{i+1,j+1} - y_{i+1,j} - y_{i,j+1} + y_{ij}$$

$$(y_{\xi\eta})_{i+\frac{1}{2},j+\frac{1}{2}} = y_{i+1,j+1} - y_{i,j+1} - y_{i+1,j} + y_{ij}$$

which are indeed equal and thus satisfy identity (28). An alternate choice might be to use central differences for the second differentiation as in Eq. (29a), while approximating the required edge values of the first partials by the average of the values at the adjacent nodes, e.g.

$$(y_\eta)_{i+1,j+\frac{1}{2}} = \frac{1}{2}[(y_\eta)_{i+1,j+1} + (y_\eta)_{i+1,j}]$$

The nodal values are reasonably represented by central differences such as

$$(y_\eta)_{i+1,j+1} = \frac{1}{2}(y_{i+1,j+2} - y_{i+1,j})$$

This scheme cannot possibly satisfy (28), however, since the points used to represent $(y_{\eta\xi})_{i+\frac{1}{2},j+\frac{1}{2}}$ are:

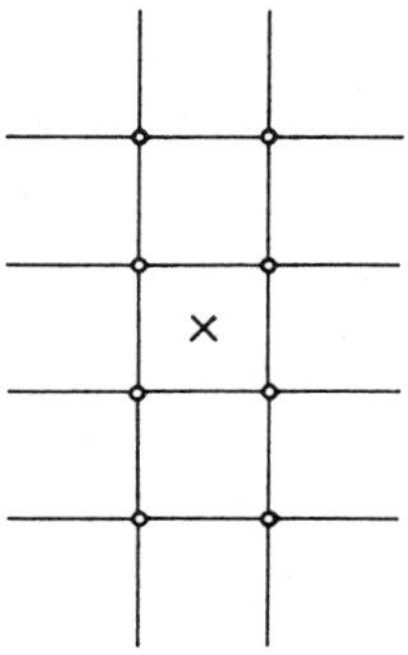

while those needed to evaluate $(y_{\xi\eta})_{i+\frac{1}{2},j+\frac{1}{2}}$ are:

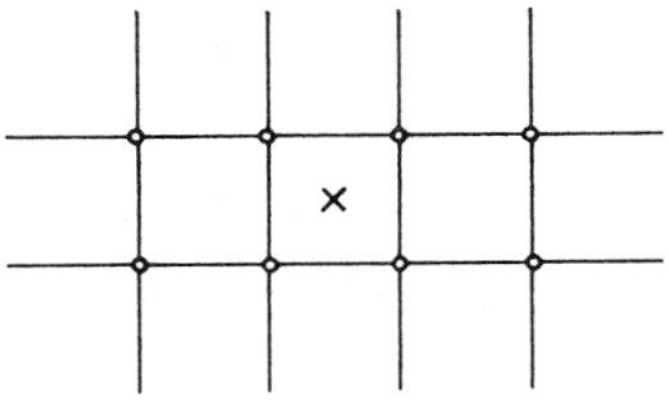

It should be noted that the representations in both of these schemes are consistent and of the same formal order of accuracy. Also, if the metric coefficients at the grid points were evaluated and stored, it would perhaps be natural to follow the second approach, using averages of the metric coefficients at the intermediate points. This, however, is not acceptable since it fails to satisfy the metric identity involved and thus would introduce spurious non-zero gradients in a uniform field.

This example suggests one basic rule that should always be followed: Never average the metric coefficients. Rather, average the coordinate values themselves, if necessary, and then calculate the metric derivatives directly. Alternatively, a coordinate system can be generated with mesh points at all of the half-integer points, as well as at the integer points used in the physical solution. The metric coefficients can then be evaluated directly by differencing between neighboring points, even at the half-integer points. For example,

$$(y_\eta)_{i+\frac{1}{2},j} = y_{i+\frac{1}{2},j+\frac{1}{2}} - y_{i+\frac{1}{2},j-\frac{1}{2}}$$

This approach was used in Ref. [13] and problems with the metric identities were thereby eliminated.

It is also possible to construct difference representations which do not involve any averaging and yet still do not satisfy the metric identities; schemes which use unsymmetric differences are an example. Fortunately, most reasonable symmetric expressions without averaging do satisfy the identities.

In the representation of the Laplacian using Eq. (III-71), $\nabla^2\xi^i$ should be calculated using Eq. (III-74), rather than using derivatives of the metric tensor elements.

Caution is required even when the coordinate transformation is known explicitly. In that case, the metric coefficients can be evaluated analytically, but the metric identities will not in general be satisfied numerically when these coefficients are differenced. This is true even in the simple case of cylindrical coordinates as the following example shows. With

$$x = \frac{\eta}{J}\cos(2\pi\frac{\xi}{I}) \qquad 0 \le \eta \le J$$

$$y = \frac{\eta}{J}\sin(2\pi\frac{\xi}{I}) \qquad 0 \le \xi \le I$$

the partials of $y(\xi,\eta)$ are

$$y_\xi = \frac{2\pi}{I}\frac{\eta}{J}\cos(2\pi\frac{\xi}{I})$$

$$y_\eta = \frac{1}{J}\sin(2\pi\frac{\xi}{I})$$

If the first partial derivative f_x is represented as in Eq. (27) and the difference in Eq. (29a) is used, but the first partials y_ξ and y_η are represented exactly, e.g.

$$(y_\eta)_{i+1,j+1/2} = \frac{1}{J}\sin(2\pi\frac{i+1}{I})$$

the bracket in Eq. (27) evaluated at $(i+\frac{1}{2},j+\frac{1}{2})$ for uniform f becomes

$$[\;] = f[\frac{1}{J}\sin(2\pi\,\frac{i+1}{I}) - \frac{1}{J}\sin(2\pi\,\frac{i}{I})$$

$$-\frac{2\pi}{I}\,\frac{j+1}{J}\cos(2\pi\frac{i+1/2}{I}) + \frac{2\pi}{I}\,\frac{j}{J}\cos(2\pi\frac{i+1/2}{I})]$$

which does not vanish identically. Thus the metric identity (28) is not satisfied when the metric derivatives, y_ξ and y_η, are evaluated analytically. But it was shown above that the difference form used here, Eq. (29a), does in fact satisfy the metric identity (28) when the metric derivatives are evaluated numerically without averaging.

The use of exact analytical expressions for the metric coefficients therefore does not necessarily increase the accuracy of the difference representations, and may actually degrade the accuracy. In Chapter V it is shown further that a detrimental contribution to the truncation error can be removed by evaluating the metric coefficients numerically rather than analytically. Accuracy in the representation of the metric coefficients thus has no inherent value in and of itself. Rather, it is the accuracy of the overall difference representation that is important.

To summarize, the metric identities can often be numerically satisfied through careful attention to the evaluation of the metric coefficients. These coefficients should be expressed as differences, not by analytical expressions. They should be evaluated directly from coordinate values wherever they are needed and should never be averaged, since the use of averaged values will almost certainly result in failure to satisfy the metric identities. Intermediate coordinate values needed to construct differences which are compatible with the metric identities can be obtained by averaging the coordinates at neighboring grid points or by using a grid with twice as many points in each direction as to be used in the actual solution.

The metric identities may become more difficult to satisfy numerically in three dimensions and in schemes involving higher-order operators or unsymmetric difference expressions, as may be needed at boundaries or near the special points discussed previously. When exact satisfaction is not achieved, the effects of the spurious source terms can be partially corrected, as discussed in Ref. [12], by subtracting off the product of the metric identities with either a uniform solution or the local solution. The former amounts to using a kind of perturbation form, while the latter is, in effect, expansion of the product derivatives involving the metric coefficients and retention of the supposedly vanishing terms, thus putting the equations into a weak conservation law form. Thus the gradient could be written, using Eq. (III-42), as

$$\underset{\sim}{\nabla} A = \frac{1}{\sqrt{g}} \sum_{i=1}^{3} (\sqrt{g}\, \underset{\sim}{a}^i A)_{\xi^i} - \frac{A_o}{\sqrt{g}} \sum_{i=1}^{3} (\sqrt{g}\underset{\sim}{a}^i)_{\xi^i}$$

where A_o is either the local value of A or a uniform value. In view of Eq. (III-40), this modification does not change the analytical expression for the gradient. Difference representations based on this form of the gradient will clearly vanish for uniform A equal to A_o. Analogous weakly-conservative expressions for all the other derivative operations of Chapter III can be inferred immediately. Subtraction of the product of the identity and the uniform free stream solution was used in Ref. [14], because of the difficulty in satisfying the metric identities exactly with flux-vector-splitting involving directional differences.

All codes should be checked for the presence of spurious source terms arising from the metric identities by running with uniform non-zero values for all of the dependent variables. If such a test run produces any changes at all, some failure to satisfy the metric identities has escaped detection (assuming the code is free from errors), and the difference representations should be modified or a change should be made to the weakly-conservative form described above.

5. Implementation Procedure

When a coordinate system has been generated, the values of the cartesian coordinates, x_i, will be available as functions of the curvilinear coordinates, ξ^i, with i = 1,2,3. Although these relations might be in the form of analytical equations in the event that the coordinate system was generated by some analytical means, a more common result is a set of values generated by a numerical solution. By definition the curvilinear coordinates take on integer values at the grid points ($\xi^i=0,1,2,\ldots N_i$, where N_i+1 is the total number of points in the ξ^i direction). Thus the values of x_1,x_2,x_3 will be available at each grid point ξ^1,ξ^2,ξ^3.

Difference expressions, such as Eq. (24), are then used at each grid point to evaluate the components of the three covariant base vectors, $\underline{a}_i$, from Eq: (III-3):

$$\begin{bmatrix} (\underset{\sim}{a}_i)_1 \\ (\underset{\sim}{a}_i)_2 \\ (\underset{\sim}{a}_i)_3 \end{bmatrix} = \begin{bmatrix} (x_1)_{\xi^i} \\ (x_2)_{\xi^i} \\ (x_3)_{\xi^i} \end{bmatrix} \quad (i = 1,2,3)$$

As discussed previously, the metric derivatives should not be averaged, but rather should always be evaluated directly from differences between grid points. Therefore, it may be necessary in some difference formulations to have coordinate values available at points between the grid points on which the solution is to be represented. In that case, the coordinate values at such points should be generated either by averaging the coordinate values between adjacent main points or by generating the coordinate system with twice as many grid points in each direction as will be used in the solution representation.

The nine elements of the covariant metric tensor can then be evaluated at each point from Eq. (III-5):

$$g_{ij} = \underset{\sim}{a}_i \cdot \underset{\sim}{a}_j \qquad (i = 1,2,3) \qquad (j = 1,2,3)$$

(Only six of these elements are distinct, of course, since the tensor is symmetric, so only six dot products actually need to be evaluated.) The Jacobian is then evaluated at each point using Eq. (III-16):

$$\sqrt{g} = \underset{\sim}{a}_1 \cdot (\underset{\sim}{a}_2 \times \underset{\sim}{a}_3)$$

Next the three components of each of the three contravariant base vectors are evaluated at each point from Eq. (III-33):

$$\mathbf{a}^i = \frac{1}{\sqrt{g}}(\mathbf{a}_j \times \mathbf{a}_k) \qquad (i = 1,2,3) \qquad (i,j,k \text{ cyclic})$$

and the nine elements of the contravariant metric tensor are evaluated at each point from Eq. (III-37):

$$g^{ij} = \mathbf{a}^i \cdot \mathbf{a}^j \qquad (i = 1,2,3) \qquad (j = 1,2,3)$$

Again only six elements are distinct.

All quantities involved in the transformed derivative operations are now available at each point. Recall that if conservative forms are to be used, the product $\sqrt{g}\mathbf{a}^i$ may be stored at each point, being evaluated from

$$\sqrt{g}\,\mathbf{a}^i = \mathbf{a}_j \times \mathbf{a}_k \qquad (i = 1,2,3) \quad (i,j,k \text{ cyclic})$$

to avoid the need for multiplication of $\mathbf{a}^i$ by $\sqrt{g}$ in all the operations.

In transforming the physical partial differential equations, the gradient, divergence, curl, and Laplacian operations will have been replaced by either the conservative expressions, Eq. (III-42)-(III-45), or by the non-conservative expressions, Eq. (III-65)-(III-71). Derivatives occurring individually will have been replaced by the expressions given by Eq. (III-50)-(III-52). Finally, derivatives occurring in boundary conditions will have been replaced by the expressions in Eq. (III-78), (III-79) or (III-82). Integrals will have been replaced by the relations given by Eq. (III-83)-(III-87). Thus, with the metric quantities evaluated at each point, as discussed above, all quantities involved in the difference representations of the transformed partial differential equations are available.

As was noted in Chapter III, the use of the conservative forms of the gradient, divergence, curl, Laplacian, etc. in the partial differential equations is equivalent to formulation of difference equations from the integral form of these equations. Hence finite-volume formulations may be set up directly from the partial differential equations by using the conservative forms for the derivative operators involved.

It should be pointed out again that the transformed partial differential equations are of the same form and type as the original equations, and are more complicated only in the sense of having variable coefficients, cross-derivatives, and more terms. The field on which these equations are solved is rectangular and the grid is fixed, uniform and square. Therefore all numerical solution algorithms that have been developed for partial differential equations on cartesian coordinate systems are applicable to these transformed equations, and all the simplifications that result from the use of uniform square grids are in order, as well.

Exercises

1. Verify Eq. (2).

2. Apply Eq. (16) to the (i,j) cell in the diagram below this equation to obtain Eq. (17). In (16) interpret the gradient as

$$\nabla = e^{\xi}\frac{\partial}{\partial\xi}+e^{\eta}\frac{\partial}{\partial\eta}$$

where e^{ξ} and e^{η} are unit vectors in the ξ and η directions, respectively, in the transformed space. The normal n is $\pm\, e^{\xi}$ or $\pm\, e^{\eta}$, as appropriate, on each of the faces of the cell. Recall $\Delta\xi=\Delta\eta=1$.

3. Following the procedure given with Eq. (20) and (21), obtain Eq. (12) from Eq. (16) applied to the (i,j) cell. Show that the two mixed partials obtained in this manner are not equal, but that their average gives Eq. (15).

4. Verify the boundary value, $f_{i,j-1/2}$ given on p.147

5. In cylindrical coordinates show that the conservative expression for ∇f does not vanish for uniform f when the metric coefficients are evaluated analytically.

V. TRUNCATION ERROR

Difference representations on curvilinear coordinate systems are constructed by first transforming derivatives with respect to cartesian coordinates into expressions involving derivatives with respect to the curvilinear coordinates (the metric coefficients). The derivatives with respect to the curvilinear coordinates are then replaced with difference expressions on the uniform grid in the transformed region. The "order" of a difference representation refers to the exponential rate of decrease of the truncation error with the point spacing. On a uniform grid this concerns simply the behavior of the error as the point spacing decreases. With a nonuniform point distribution, there is some ambiguity in the interpretation of order, in that the spacing may be decreased locally either by increasing the number of points in the field or by changing the distribution of a fixed number of points. Both of these could, of course, be done simultaneously, or the points could even be moved randomly, but to be meaningful the order of a difference representation must relate to the error behavior as the point spacing is decreased according to some pattern. This is a moot point with uniform spacing, but two senses of order on a nonuniform grid emerge: the behavior of the error (1) as the number of points in the field is increased while maintaining the same relative point distribution over the field, and (2) as the relative point distribution is changed so as to reduce the spacing locally with a fixed number of points in the field.

On curvilinear coordinate systems the definition of order of a difference representation is integrally tied to point distribution functions. The order is determined by the error behavior as the spacing varies with the points fixed in a certain distribution, either by increasing the number of points or by changing a parameter in the distribution, not simply by consideration of the points used in the difference expression as being unrelated to each other. Actually, global order is meaningful only in the first sense, since as the spacing is reduced locally with a fixed number of points in the field, the spacing somewhere else must certainly increase. This second sense of order on a nonuniform grid then is relevant only locally in regions where the spacing does in fact decrease as the point distribution is changed.

In the following sections an illustrative error analysis is given. The general development from which this is taken appears in Ref. [17], together with references to related work.

1. Order On Nonuniform Spacing

A general one-dimensional point distribution function can be written in the form

$$x(\xi) = q(\frac{\xi}{N}) \qquad 0 \leq \xi \leq N \tag{1}$$

In the following analysis, x will be considered to vary from 0 to 1. (Any other range of x can be constructed simply by multiplying the distribution functions given here by an appropriate constant.) With this form for the distribution function, the effect of increasing the number of points in a discretization of the field can be seen explicitly by defining the values of ξ at the points to be successive integers from 0 to N. In this form, N+1 is then the number of points in the discretization, so that the dependence of the error expressions on the number of points in the field will be displayed explicitly by N. This form removes the confusion that can arise in interpretation of analyses based on a fixed interval $(0 \leq \xi \leq 1)$, where variation of the number of points is represented by variation of the interval $\Delta\xi$. The form of the distribution function, i.e., the relative concentration of points in certain areas while the total number of points in the field is fixed, is varied by changing parameters in the function.

Considering the first derivative in one dimension:

$$f_x = \frac{f_\xi}{x_\xi} \tag{2}$$

with a central difference for f_ξ we have the following difference expression (with $\Delta\xi = 1$ as noted above):

$$f_x = \frac{1}{2x_\xi}\,(f_{i+1} - f_{i-1}) + T_1 \tag{3}$$

where T_1 is the truncation error. A Taylor series expansion then yields

$$T_1 = -\frac{1}{6}\,\frac{f_{\xi\xi\xi}}{x_\xi} - \frac{1}{120}\,\frac{f_{\xi\xi\xi\xi\xi}}{x_\xi} \quad \cdots \tag{4}$$

Here the metric coefficient, x_ξ, is considered to be evaluated analytically, and hence has no error. (The case of numerical evaluation of the metric coefficients is considered in a later section.)

The series in (4) cannot be truncated without further consideration since the ξ-derivatives of f are dependent on the point distribution. Thus if the point distribution is changed, either through the addition of more points or through a change in the form of the distribution function, these derivatives will change. Since the terms of the series do not contain a power of some quantity less than unity, there is no indication that the successive terms become progressively smaller.

It is thus not meaningful to give the truncation error in terms of ξ-derivatives of f. Rather, it is necessary to transform these ξ-derivatives to x-derivatives, which, of course, are not dependent on the point distribution. The first ξ-derivative follows from (2):

$$f_\xi = x_\xi f_x \tag{5}$$

Then

$$f_{\xi\xi} = x_{\xi\xi} f_x + x_\xi (f_x)_\xi = x_{\xi\xi} f_x + x_\xi^2 f_{xx} \tag{6}$$

and

$$f_{\xi\xi\xi} = x_{\xi\xi\xi} f_x + 3x_\xi x_{\xi\xi} f_{xx} + x_\xi^3 f_{xxx} \tag{7}$$

Each term in $f_{\xi\xi\xi}$ contains three ξ-differentiations. This holds true for all higher derivatives also, so that each term in $f_{\xi\xi\xi\xi\xi}$ will contain five ξ-differentiations, etc.

A. Order with fixed distribution function

From Eq. (1) we have

$$x_\xi = \frac{q'}{N}, \quad x_{\xi\xi} = \frac{q''}{N^2}, \quad x_{\xi\xi\xi} = \frac{q'''}{N^3} \tag{8}$$

Therefore if the number of points in the grid is increased while keeping the same relative point distribution, it is clear that each term in $f_{\xi\xi\xi}$ will be proportional to $1/N^3$, and each term in $f_{\xi\xi\xi\xi\xi}$ will be proportional to $1/N^5$, etc.

It then follows that the series in Eq. (4) can be truncated in this case, so that the truncation error is given by the first term, which is, using Eq. (6),

$$T_1 = -\frac{1}{6}\frac{x_{\xi\xi\xi}}{x_\xi}f_x - \frac{1}{2}x_{\xi\xi}f_{xx} - \frac{1}{6}x_\xi^2 f_{xxx} \tag{9}$$

The first two terms arise from the nonuniform spacing, while the last term is the familiar term that occurs with uniform spacing as well.

From (9) it is clear that the difference representation (3) is second-order regardless of the form of the point distribution function, in the sense that the truncation error goes to zero as $1/N^2$ as the number of points increases. This means that the error will be quartered when the number of points is doubled in the same distribution function. Thus all difference representations maintain their order on a nonuniform grid with any distribution of points in the formal sense of the truncation error decreasing as the number of points is increased while maintaining the same relative point distribution over the field.

The critical point here is that the same relative point distribution, i.e., the same distribution function, is used as the number of points in the field is increased. If this is the case, then the error will be decreased by a factor that is a power of the inverse of the number of points in the field as this number is increased. Random addition of points will, however, not maintain order. In a practical vein this means that with twice as many points the solution will exhibit one-fourth of the error (for second-order representations in the transformed plane) when the same point distribution function is used. However, if the number of points is doubled without maintaining the same relative distribution, the error reduction may not be as great as one-fourth.

From the standpoint of formal order in this sense there is no need for concern over the form of the point distribution. However, formal order in this sense relates only to the behavior of the truncation error as the number of points is increased, and the coefficients in the series may become large as the parameters in the distribution are altered to reduce the local spacing with a given number of points in the field. Thus, although the error will be reduced by the same order for all point distributions as the number of points is increased, certain distributions will have smaller error than others with a given number of points in the field, since the coefficients in the series, while independent of the number of points, are dependent on the distribution function.

B. Order with fixed number of points

An alternate sense of order for point distributions is based on expansion of the truncation error in a series in ascending powers of the spacing, x_ξ, with the number of points in the grid kept fixed and the point distribution changed to decrease the local spacing. From Eq. (9) second-order requires that

$$x_{\xi\xi\xi} \sim x_\xi^3 \quad \text{and} \quad x_{\xi\xi} \sim x_\xi^2 \tag{10}$$

This is a severe restriction that is unlikely to be satisfied. This is understandable, however, since with a fixed number of points the spacing must necessarily increase somewhere when the local spacing is decreased.

The difference between these two approaches to order should be kept clear. The first approach concerns the behavior of the truncation error as the number of points in the field increases with a fixed relative distribution of points. The series there is a power series in the inverse of the number of points in the field, and formal order is maintained for all point distributions. The coefficients in the series may, however, become large for some distribution functions as the local spacing decreases for any given number of points. The other approach concerns the behavior of the error as the local spacing decreases with a fixed number of points in the field. This second sense of order is thus more stringent, but the conditions seem to be unattainable.

2. Effect of Numerical Metric Coefficients

The above analysis has assumed the use of exact values of x_ξ, the metric coefficient. If the metric coefficient is evaluated numerically, we have, in place of Eq. (3), the difference expression

$$f_x = \frac{f_{i+1} - f_{i-1}}{x_{i+1} - x_{i-1}} + T_2 \qquad (11)$$

The Taylor expansion yields

$$T_2 = f_x - \{f_x(x_{i+1} - x_{i-1}) + \frac{1}{2}f_{xx}\,[(x_{i+1} - x_i)^2 - (x_{i-1} - x_i)^2]$$
$$+ \frac{1}{6}f_{xxx}[(x_{i+1} - x_i)^3 - (x_{i-1} - x_i)^3]\}/(x_{i+1} - x_{i-1})$$

or

$$T_2 = -\frac{1}{2}f_{xx}(x_{i+1} - 2x_i + x_{i-1})$$
$$- \frac{1}{6}f_{xxx}\frac{(x_{i+1} - x_i)^3 - (x_{i-1} - x_i)^3}{(x_{i+1} - x_{i-1})} \tag{12}$$

The coefficient of f_{xx} here is the difference representation of $x_{\xi\xi}$, while that of f_{xxx} reduces to a difference expression of x_ξ^2. We thus have T_2 given by the last two terms of T_1, and the first term of T_1 has been eliminated from the truncation error by evalutating the metric coefficient numerically rather than analytically.

Thus the use of numerical evaluation of the coordinate derivative, rather than exact analytical evaluation, eliminates the f_x term from the truncation error. Since this term is the most troublesome part of the error, being dependent on the derivative being represented, it is clear that numerical evaluation of the metric coefficients by the same difference representation used for the function whose derivative is being represented is preferable over exact analytical evaluation. It should be understood that there is no incentive, per se, for accuracy in the metric coefficients, since the object is simply to represent a discrete solution accurately, not to represent the solution on some particular coordinate system. The only reason for using any function at all to define the point distribution is to ensure a smooth distribution. There is no reason that the representations of the coordinate derivatives have to be accurate representations of the analytical derivatives of that particular distribution function.

We are thus left with truncation error of the form

$$T = -\frac{1}{2} x_{\xi\xi} f_{xx} - \frac{1}{6} x_\xi^2 f_{xxx} \quad (13)$$

when the metric coefficient is evaluated numerically. As noted above, the last term occurs even with uniform spacing. The first term is proportional to the second derivative of the solution and hence represents a numerical diffusion, which is dependent on the rate-of-change of the grid point spacing. This numerical diffusion may even be negative and hence destabilizing. Attention must therefore be paid to the variation of the spacing, and large changes in spacing from point to point cannot be tolerated, else significant truncation error will be introduced.

3. Evaluation of Distribution Functions

In Ref. [17] and Ref. [18] several distribution functions are evaluated on the basis of the size of the coefficients in the error expression. Some of this evaluation procedure is illustrated in the exercises. It appears that the following conclusions can be reached on basis of these comparisons:

(1) The exponential is not as good as the hyperbolic tangent or the hyperbolic sine. (Implementation procedures for all three of these are given in Chapter VIII.)

(2) The hyperbolic sine is the best function in the lower part of the boundary layer. Otherwise this function is not as good as the hyperbolic tangent.

(3) The error function and the hyperbolic tangent are the best functions outside the boundary layer. Between these two, the hyperbolic tangent is the

better inside, while the error function is the better outside. The error function is, however, more difficult to use.

(4) The logarithm, sine, tangent, arctangent, inverse hyperbolic tangent, quadratic, and the inverse hyperbolic sine are not suitable.

Although, as has been shown, all distribution functions maintain order in the formal sense with nonuniform spacing as the number of points in the field is increased, these comparisons of particular distribution functions show that considerable error can arise with nonuniform spacing in actual applications. If the spacing doubles from one point to the next we have, approximately, $x_{\xi\xi} = 2x_\xi - x_\xi = x_\xi$ so that the ratio of the first term in Eq. (13) to the second is inversely proportional to the spacing x_ξ. Thus for small spacing, such a rate-of-change of spacing would clearly be much too large. Obviously, all of the error terms are of less concern where the solution does not vary greatly. The important point is that the spacing not be allowed to change too rapidly in high gradient regions such as boundary layers or shocks.

4. Two-Dimensions Forms

The two-dimensional transformation of the first derivative is given by

$$f_x = (y_\eta f_\xi - y_\xi f_\eta)/\sqrt{g} \qquad (14)$$

where the Jacobian of the transformation is

$$\sqrt{g} = x_\xi y_\eta - x_\eta y_\xi \tag{15}$$

With two-point central difference representations for all derivatives the leading term of the truncation error is

$$T_x = \frac{1}{2\sqrt{g}}(y_\xi x_\eta x_{\eta\eta} - x_\xi y_\eta x_{\xi\xi})f_{xx} + \frac{1}{2\sqrt{g}}(y_\xi y_\eta)(y_{\eta\eta} - y_{\xi\xi})f_{yy}$$

$$+ \frac{1}{2\sqrt{g}}[y_\xi y_\eta(x_{\eta\eta} - x_{\xi\xi}) + x_\eta y_\xi y_{\eta\eta} - x_\xi y_\eta y_{\xi\xi}]f_{xy}$$

$$+ \text{second-order terms in the spacing} \tag{16}$$

where the coordinate derivatives are to be understood here to represent central difference expressions, e.g.,

$$x_\xi = \frac{1}{2}(x_{i+1,j} - x_{i-1,j}), \quad x_\eta = \frac{1}{2}(x_{i,j+1} - x_{i,j-1})$$

$$x_{\xi\xi} = x_{i+1,j} - 2x_{ij} + x_{i-1,j}, \quad x_{\eta\eta} = x_{i,j+1} - 2x_{ij} + x_{i,j-1}$$

These contributions to the truncation error arise from the nonuniform spacing. The familiar terms proportional to a power of the spacing occur in addition to these terms as has been noted.

Sufficient conditions can now be stated for maintaining the order of the difference representations, with a fixed number of points in each distribution. First, as in the one-dimensional case, the ratios

$$\frac{x_{\xi\xi}}{|r_\xi|^2}, \frac{y_{\xi\xi}}{|r_\xi|^2}, \frac{x_{\eta\eta}}{|r_\eta|^2}, \frac{y_{\eta\eta}}{|r_\eta|^2}$$

must be bounded as x_ξ, x_η, y_ξ, y_η approach zero. A second condition must be imposed which limits the rate at which the

Jacobian approaches zero. This condition can be met by simply requiring that $\cot\theta$ remain bounded, where θ is the angle between the ξ and η coordinate lines. The fact that this bound on the nonorthogonality imposes the correct lower bound on the Jacobian follows from the fact that $|\cot\theta| \leq M$ implies

$$g \geq \frac{1}{M^2+1} |\mathbf{r}_\xi|^2 \cdot |\mathbf{r}_\eta|^2 \tag{17}$$

With these conditions on the ratios of second to first derivatives, and the limit on the nonorthogonality satisfied, the order of the first derivative approximations is maintained in the sense that the contributions to the truncation error arising for the nonuniform spacing will be second-order terms in the grid spacing.

The truncation error terms for second derivatives that are introduced when using a curvilinear coordinate system are very lengthy and involve both second and third derivatives of the function f. However, it can be shown that the same sufficient conditions, together with the condition that

$$\frac{x_{\xi\eta}}{|\mathbf{r}_\xi|\cdot|\mathbf{r}_\eta|} \quad \text{and} \quad \frac{y_{\xi\eta}}{|\mathbf{r}_\xi|\cdot|\mathbf{r}_\eta|}$$

remain bounded, will insure that the order of the difference representations is maintained.

It was noted above that a limit on the nonorthogonality, imposed by (17), is required for maintaining the order of difference representations. The degree to which nonorthogonality affects truncation error can be stated more precisely, as follows. The truncation error for a first derivative f_x can be written

$$T_x = (y_\eta T_\xi - y_\xi T_\eta)/\sqrt{g} \tag{18}$$

where T_ξ and T_η are the truncation errors for the difference expressions of f_ξ and f_η. Now all coordinate derivatives can be expressed using direction cosines of the angles of inclination, ϕ_ξ and ϕ_η, of the ξ and η coordinate lines. After some simplification, the truncation error has the form

$$T_x = \frac{1}{\sin(\phi_\eta - \phi_\xi)} (\sin\phi_\eta \cos\phi_\xi \frac{T_\xi}{x_\xi} - \sin\phi_\xi \cos\phi_\eta \frac{T_\eta}{x_\eta}) \tag{19}$$

Therefore the truncation error, in general, varies inversely with the sine of the angle between the coordinate lines. Note that there is also a dependence on the direction of the coordinate lines. To further clarify the effect of nonorthogonality, the truncation error terms arising from nonuniform spacing are considered.

The contribution from nonorthogonality can be isolated by considering the case of skewed parallel lines with $x_\eta = x_{\eta\eta} = x_{\xi\eta} = y_{\xi\xi} = y_{\xi\eta} = 0$ as diagrammed below:

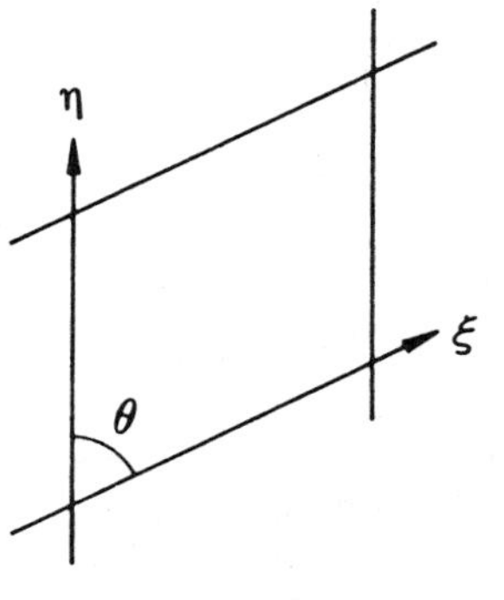

Here (16) reduces to

$$T_x = -\frac{1}{2} x_{\xi\xi} f_{xx} + \frac{1}{2}(\frac{y_\xi}{x_\xi}) y_{\eta\eta} f_{yy} - \frac{1}{2}(\frac{y_\xi}{x_\xi}) x_{\xi\xi} f_{xy}$$

Since $\cot\Theta = \frac{y_\xi}{x_\xi}$, this may be written

$$T_x = -\frac{1}{2} x_{\xi\xi} f_{xx} + \frac{1}{2} (y_{\eta\eta} f_{yy} - x_{\xi\xi} f_{xy}) \cot\theta \qquad (20)$$

This first term occurs even on an orthogonal system and corresponds to the first term in (13). The last two terms arise from the departure from orthogonality. For $\theta \leq 45°$, these terms are no greater than those from the nonuniform spacing. Reasonable departure from orthogonality is therefore of little concern when the rate-of-change of grid spacing is reasonable. Large departure from orthogonality may be more of a problem at boundaries where one-sided difference expressions are needed. Therefore grids should probably be made as nearly orthogonal at the boundaries as is practical. Note that the contribution from nonorthogonality vanishes on a skewed uniform grid.

Exercises

1. Verify Eq. (4).
2. Derive Eq. (6) and (7) by repeated differentiation of Eq. (5).
3. Verify Eq. (12).
4. Show that the coefficient of f_{xxx} in Eq. (12) can be reduced to a difference representation of x_ξ^2.
5. (a) Show that with an exponential distribution function,

$$x(\xi) = \frac{\exp(\alpha\xi/N) - 1}{\exp(\alpha) - 1} \qquad 0 \leq \xi \leq N$$

the ratio of the second term in Eq. (9) to the third term for very small spacing, s, at $\xi = 0$ is approximately equal to 1/Ns at $\xi = 0$ and to 1 at $\xi = N$. Hint: Note that $s = (x_\xi)_o$ approaches zero as α approaches infinity, and that for large α, $\alpha/(e^\alpha - 1)$ approaches $1/e^\alpha$.

(b) Show also that the average value of this ratio over the field is $[Ns \ln (1/Ns)]^{-1}$. Hint: Note that

$$\frac{1}{N}\int_o^N \frac{x_{\xi\xi}}{x_\xi^2}\, d\xi = \frac{1}{N}\left[\frac{1}{(x_\xi)_o} - \frac{1}{(x_\xi)_N}\right]$$

(c) Finally, show that the first term in Eq. (9) causes a fractional error of approximately $-1/6N^2 \ln^2 (1/Ns)$ in f_x that does not vary over the field. (Recall that this term can be eliminated by using numerical metrics, however.)

6. Show that with a hyperbolic sine distribution function,

$$x(\xi) = \frac{\sinh(\alpha\xi/N)}{\sinh \alpha} \qquad 0 \leq \xi \leq N$$

the ratio of the second term in Eq. (9) to the third term for very small spacing, s, at $\xi = 0$ vanishes at $\xi = 0$ and is approximately equal to 1 at $\xi = N$. Show also, however, that the maximum value of this ratio occurs near $\xi/N = 0.9/\ln(2/Ns)$ and is approximately equal to 1/2Ns. Finally, show that the average value of the ratio over the field is equal to $[Ns \ln(2/Ns]^{-1}$. Hint: See the preceding exercise. (Note that this distribution gives a smaller error due to the rate-of-change in the spacing than does the exponential distribution of the preceding exercise and is particularly advantageous near $\xi = 0$ where the spacing is the smallest.)

7. Show that with a hyperbolic tangent distribution function,

$$x(\xi) = 1 - \frac{\tanh[\alpha(1-\xi/N)]}{\tanh \alpha} \quad 0 \leq \xi \leq N$$

the ratio of the second term in Eq. (9) to the third term for very small spacing, s, at $\xi = 0$ is approximately equal to 1/2Ns at $\xi = 0$ and vanishes at $\xi = N$. Show also that the average of this ratio over the field is the same as for the hyperbolic sine distribution of the preceding exercise. This distribution is thus also superior to the exponential distribution.

8. With the distribution function of the form of Eq. (1), show that the truncation error in Eq. (3) is a power

series in inverse powers of N. (Hint: see Ref. [17]).

9. Verify Eq. (17).

10. Expand the differences f_ξ and f_η of Eq. (14) in Taylor series about the grid point x_{ij}. Substitute these expansions back in Eq. (14) thereby verifying Eq. (16). Certain identities will be useful, such as

$$(x_{i+1,j} - x_{i,j})^2 - (x_{i-i,j} - x_{i,j})^2 = 2x_\xi x_{\xi\xi}$$

11. Use the identity $\mathbf{r}_\xi \cdot \mathbf{r}_\eta = |\mathbf{r}_\xi|\, |\mathbf{r}_\eta| \cos\theta$ to verify the inequality in (17):

12. Use the following relations to write the truncation error in Eq. (18) in the form of Eq. (19).

$$\xi = |\mathbf{r}_\xi| \sin \phi_\xi, \quad x_\xi = |\mathbf{r}_\xi| \cos \phi_\xi$$

VI. ELLIPTIC GENERATION SYSTEMS

As noted in Chapter II, the generation of a boundary-conforming coordinate system is accomplished by the determination of the values of the curvilinear coordinates in the interior of a physical region from specified values (and/or slopes of the coordinate lines intersecting the boundary) on the boundary of the region:

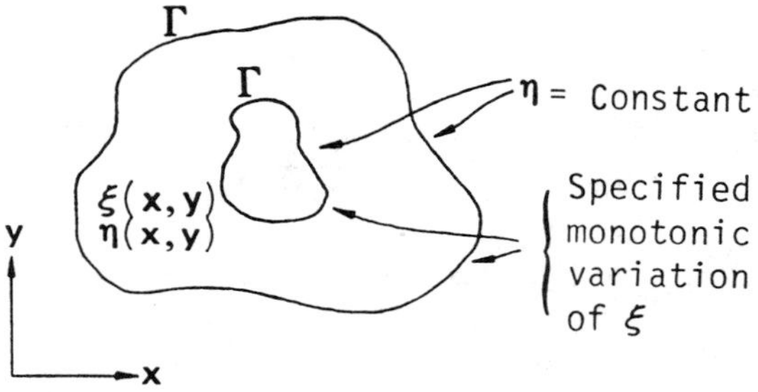

One coordinate will be constant on each segement of the physical boundary curve (surface in 3D), while the other varies monotonically along the segment (cf. Chapter II).

The equivalent problem in the transformed region is the determination of values of the physical (cartesian or other) coordinates in the interior of the transformed region from specified values and/or slopes on the boundary of this region, as discussed in Chapter II:

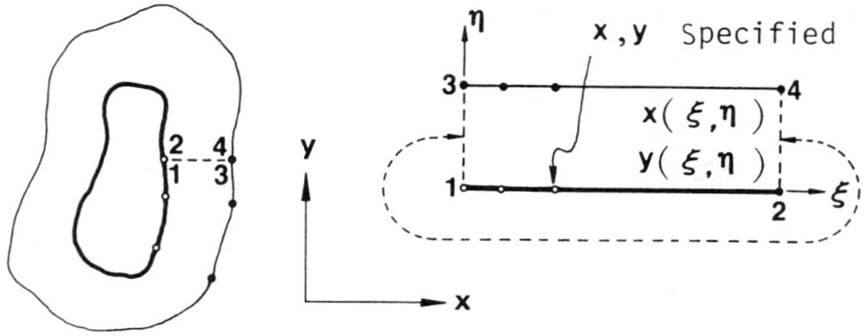

This is a more amenable problem for computation, since the boundary of the transformed region is comprised of horizontal and vertical segments, so that this region is composed of rectangular blocks which are contiguous, at least in the sense of being joined by re-entrant boundaries (branch cuts), as described in Chapter II.

The generation of field values of a function from boundary values can be done in various ways, e.g., by interpolation between the boundaries, etc., as is discussed in Chapter VIII. The solution of such a boundary-value problem, however, is a classic problem of partial differential equations, so that it is logical to take the coordinates to be solutions of a system of partial differential equations. If the coordinate points (and/or slopes) are specified on the entire closed boundary of the physical region, the equations must be elliptic, while if the specification is on only a portion of the boundary the equations would be parabolic or hyperbolic. This latter case would occur, for instance, when an inner boundary of a physical region is specified, but a surrounding outer boundary is arbitrary. The present chapter, however, treats the general case of a completely specified boundary, which requires an elliptic partial differential system. Hyperbolic and parabolic generation systems are discussed in Chapter VII .

Some general discussion of elliptic generation systems has been given in Ref. [19], and numerous references to the application thereof appear in the surveys given by Ref. [1] and [5].

1. Generation Equations

The extremum principles, i.e., that extrema of solutions cannot occur within the field, that are exhibited by some elliptic systems can serve to guarantee a one-to-one mapping between the physical and transformed regions (cf. Ref. [20] and [21]). Thus, since the variation of the curvilinear coordinate along a physical boundary segment must be monotonic, and is over the same range along facing boundary segments (cf. Chapter II), it clearly follows that extrema of the curvilinear coordinates cannot be allowed in the interior of the physical region, else overlapping of the coordinate system will occur. Note that it is the extremum principles of the partial differential system in the physical space, i.e., with the curvilinear coordinates as the dependent variables, that is relevant since it is the curvilinear coordinates, not the cartesian coordinates, that must be constant or monotonic on the boundaries. Thus it is the form of the partial differential equations in the physical space, i.e., containing derivatives with respect to the cartesian coordinates, that is important.

Another important property in regard to coordinate system generation is the inherent smoothness that prevails in the solutions of elliptic systems. Furthermore, boundary slope discontinuities are not propagated into the field. Finally, the smoothing tendencies of elliptic operators, and the extremum principles, allow grids to be generated for any configurations without overlap of grid lines. Some examples appear below:

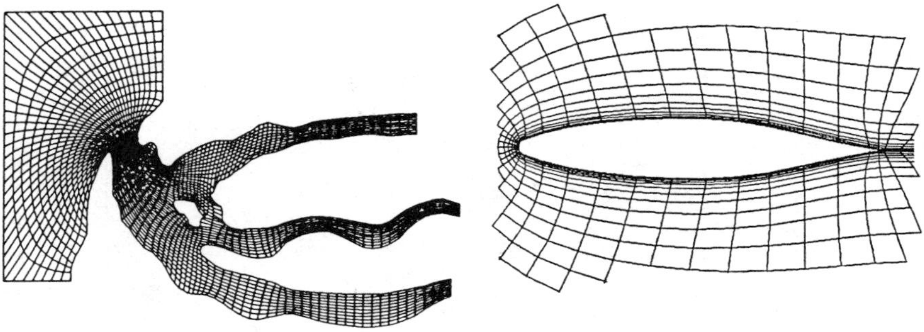

There are thus a number of advantages to using a system of elliptic partial differential equations as a means of coordinate system generation. A disadvantage, of course, is that a system of partial differential equations must be solved to generate the coordinate system.

The historical progress of the form of elliptic systems used for grid generation has been traced in Ref. [1]. Consequently, references to all earlier work will not be made here. Numerous examples of the generation and application of coordinate systems generated from elliptic partial differential equations are covered in the above reference, as well as in Ref. [2].

A. Laplace system

The most simple elliptic partial differential system, and one that does exhibit an extremum principle and considerable smoothness is the Laplace system:

$$\nabla^2 \xi^i = 0 \qquad (i = 1,2,3) \tag{1}$$

This generation system guarantees a one-to-one mapping for boundary-conforming curvilinear coordinate systems on general closed boundaries.

These equations can, in fact, be obtained from the Euler equations for the minimization of the integral

$$I = \iiint \sum_{i=1}^{3} |\nabla \xi^i|^2 \, dV \tag{2}$$

as is discussed further in Chapter XI. Since the coordinate lines are located at equal increments of the curvilinear coordinate, the quantity $|\nabla \xi^i|$ can be considered a measure of the grid point density along the coordinate line on which

ξ^i varies, i.e., ξ^i must change rapidly in physical space where grid points are clustered. Minimization of this integral thus leads to the smoothest coordinate line distribution over the field.

With this generating system the coordinate lines will tend to be equally spaced in the absence of boundary curvature because of the strong smoothing effect of the Laplacian, but will become more closely spaced over convex boundaries, and less so over concave boundaries, as illustrated below. (In this and other illustrations and applications in two dimensions, ξ^1 and ξ^2 will be denoted ξ and η, respectively, while x and y will be used for x_1 and x_2.)

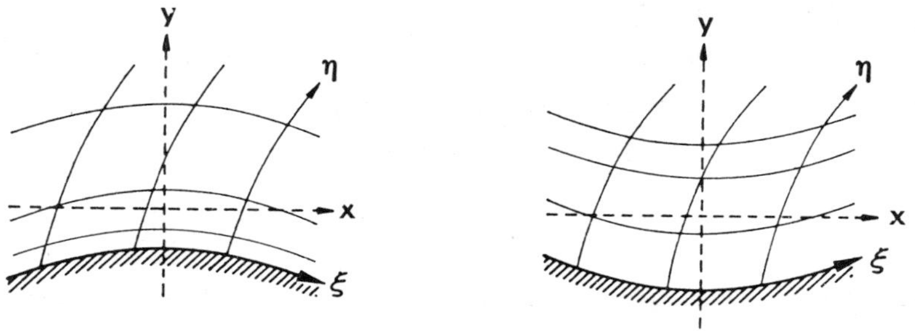

In the left figure we have $\eta_{xx} > 0$ because of the convex (to the interior) curvature of the lines of constant η (η-lines). Therefore it follows that $\eta_{yy} < 0$, and hence the spacing between the η-lines must increase with y. The η-lines thus will tend to be more closely spaced over such a convex boundary segment. For concave segments, illustrated in the right figure, we have $\eta_{xx} < 0$, so that η_{yy} must be positive, and hence the spacing of the η-lines must decrease outward from this concave boundary. Some examples of grids generated from the Laplace system are shown below. The inherent smoothness and the behavior near concave and convex boundaries are evident in these examples.

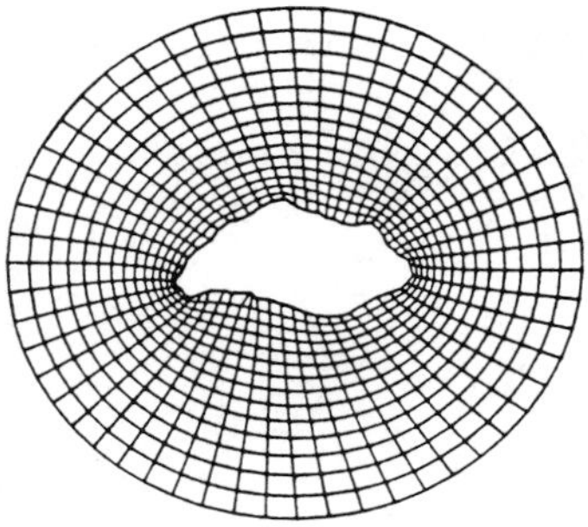

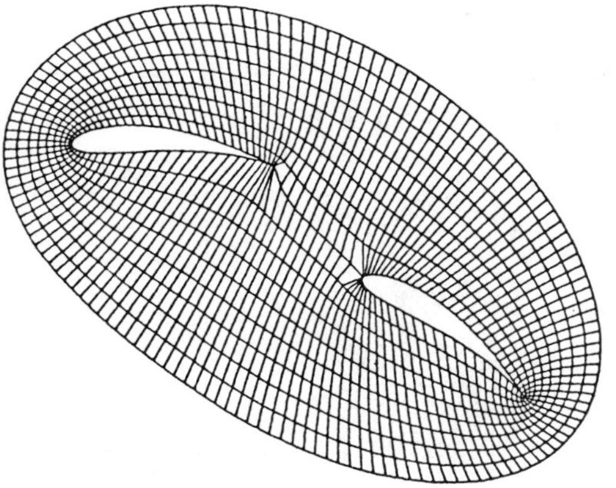

B. Poisson system

Control of the coordinate line distribution in the field can be exercised by generalizing the elliptic generating system to Poisson equations:

$$\nabla^2 \xi^i = P^i \qquad (i = 1,2,3) \tag{3}$$

in which the "control functions" P^i can be fashioned to control the spacing and orientation of the coordinate lines. The extremum principles may be weakened or lost completely with such a system, but the existence of an extremum principle is a sufficient, but not a necessary, condition for a one-to-one mapping, so that some latitude can be taken in the form of the control functions.

Considering the equation $\nabla^2 \eta = Q$ and the figures above ($P^1 = P$ and $P^2 = Q$ in the illustrations here), since a negative value of the control function would tend to make η_{yy} more negative, it follows that negative values of Q will tend to cause the coordinate line spacing in the cases shown above to increase more rapidly outward from the boundary. Generalizing, negative values of the control function Q will cause the η-lines to tend to move in the direction of decreasing η, while negative values of P in $\nabla^2\xi$ = P will cause ξ-lines to tend to move in the direction of decreasing ξ. These effects are illustrated below for an η-line boundary:

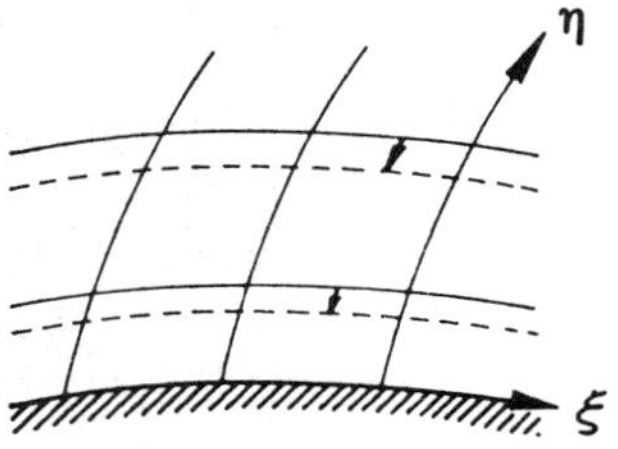

Q < 0

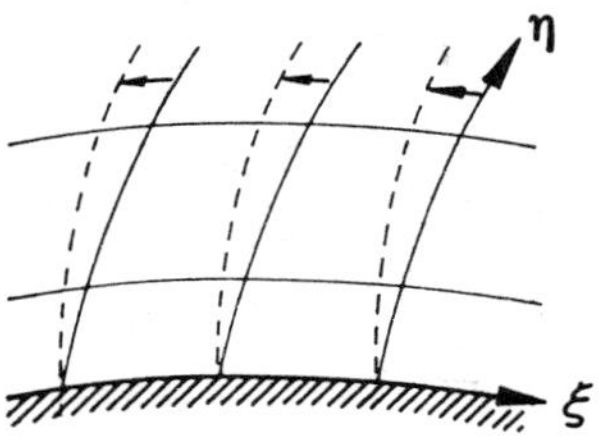

P < 0

With the boundary values fixed, the ξ-lines here cannot change the intersection with the boundary. The effect of the control function P in this case is to change the angle of intersection at the boundary, causing the ξ-lines to lean in the direction of decreasing ξ.

These effects are illustrated in the following figures:

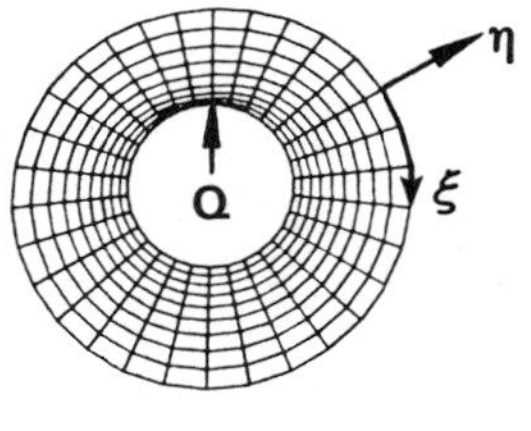

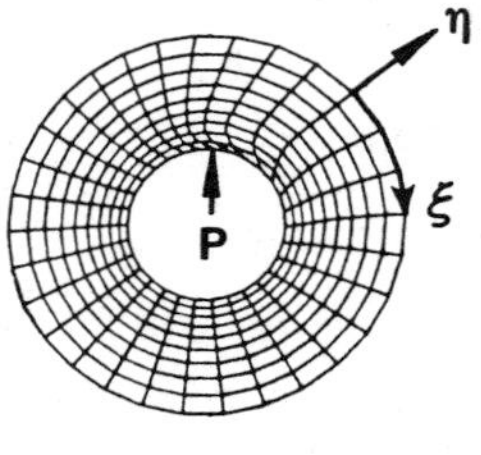

Here the ξ-lines are radial and the η-lines are circumferential. In the left illustration the control function Q is locally non-zero near a portion of the inner boundary as indicated, so the η-lines move closer to that portion of the boundary while in the right figure, P is locally non-zero, resulting in a change in intersection angle of the ξ-lines with that portion of the boundary. If the intersection angle, instead of the point location, on the boundary is specified, so that the points are free to move along the boundary, then the ξ-lines would move toward lines with lower values of ξ:

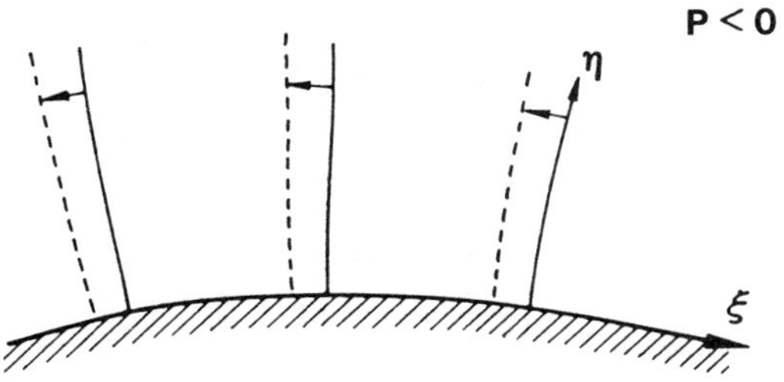

In general, a negative value of the Laplacian of one of the curvilinear coordinates causes the lines on which that coordinate is constant to move in the direction in which that coordinate decreases. Positive values of the Laplacian naturally result in the opposite effect.

C. Effect of boundary point distribution

Because of the strong smoothing tendencies that are inherent in the Laplacian operator, in the absence of the control functions, i.e., with $P^i = 0$, the coordinate lines will tend to be generally equally spaced away from the boundaries regardless of the boundary point distribution. For example, the simple case of a coordinate system comprised of horizontal and vertical lines in a rectangular physical region, (cf. the right figure below) cannot be obtained as a solution of Eq. (3) with P=Q=0 unless the boundary points are equally spaced.

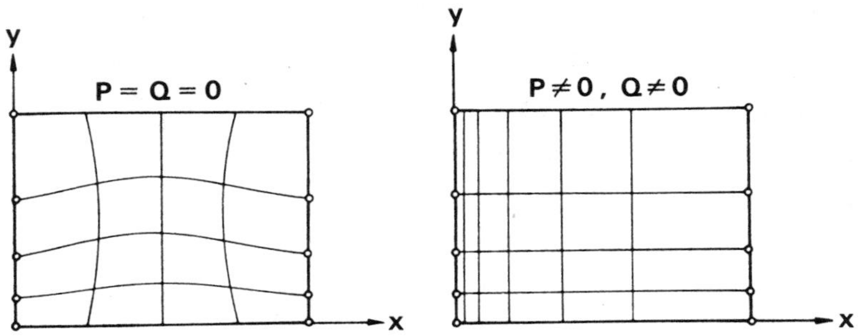

With $\xi_{yy} = \eta_{xx} = 0$, Eq. (3) reduces to

$$\xi_{xx} = P, \qquad \eta_{yy} = Q$$

and thus P and Q cannot vanish if the point distribution is not uniform on the horizontal and vertical boundaries, respectively. With P=Q=0 the lines tend to be equally-spaced away from the boundary. These effects are illustrated further in the figures below. Here the control functions are zero in the left figure.

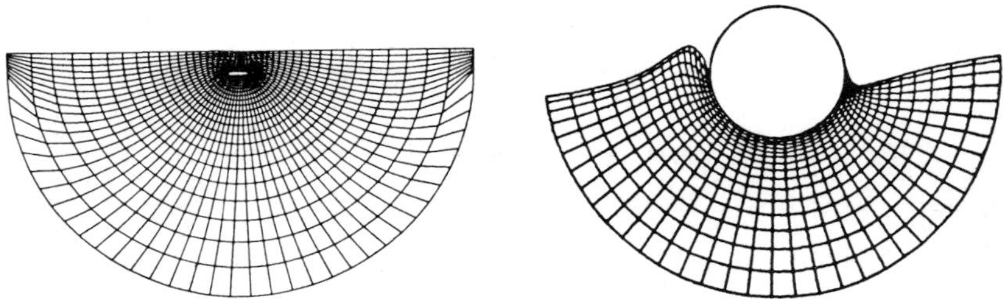

Although the spacing is not uniform on the semi-circular outer boundary in this figure, the angular spacing is essentially uniform away from the boundary. By contrast, nonzero control functions in the right figure, evaluated from the boundary point distribution, cause the field spacing to follow that on the boundary. Thus, if the coordinate lines in the interior of the region are to have the same general spacing as the point distributions on the boundaries which these lines connect, it is necessary to evaluate the control functions to be compatible with the boundary point distribution. This evaluation of the control functions from the boundary point distribution is discussed more fully in Section 2 of this chapter.

D. General Poisson-type systems

If a curvilinear coordinate system, $\bar{\xi}^i$ (i = 1,2,3), which satisfies the Laplace system

$$\nabla^2 \bar{\xi}^i = 0 \qquad (i = 1,2,3)$$

is transformed to another coordinate system, ξ^i (i = 1,2,3), then the new curvilinear coordinates, ξ^i, satisfy the inhomogeneous elliptic system (cf. Ref. [19])

$$\nabla^2 \xi^i = P^i \qquad (i = 1,2,3) \tag{4}$$

where

$$P^i = \sum_{j=1}^{3} \sum_{k=1}^{3} g^{jk} P^i_{jk} \tag{5}$$

with the P^i_{jk} defined by the transformation from $\bar{\xi}^i$ to ξ^i:

$$P^i_{jk} = \sum_{m=1}^{3} \sum_{n=1}^{3} \frac{\partial \bar{\xi}^m}{\partial \xi^j} \frac{\partial \bar{\xi}^n}{\partial \xi^k} \frac{\partial^2 \xi^i}{\partial \bar{\xi}^m \partial \bar{\xi}^n} \tag{6}$$

(It may be noted that if the subsequent transformation is one-dimensional, i.e., if $\partial \bar{\xi}^i / \partial \xi^j = \delta^i_j \, \partial f^i / \partial \xi^j$ then only the three functions P^i_{ii}, with i = 1,2,3, are nonzero.)

These results show that a grid with lines concentrated by applying a subsequent transformation (often called a "stretching" transformation) to a grid generated as the solution of the Laplace system could have been generated directly as the solution of the Poisson system (4) with appropriate "control functions", P^i_{jk}, derived from the subsequent concentrating transformation according to Eq. (6). Therefore, it is appropriate to adopt this Poisson system (4) as the generation system, but with the control functions specified directly rather than through a subsequent transformation.

Thus an appropriate generation system can be defined by Eqs. (4) and (5):

$$\nabla^2 \xi^i = \sum_{j=1}^{3} \sum_{k=1}^{3} g^{jk} P^i_{jk} \qquad (i = 1,2,3) \tag{7}$$

with the control functions, P^i_{jk}, considered to be specified. The basis of the generation system (7) is that it produces a coordinate system that corresponds to the subsequent application of a stretching transformation to a coordinate system generated for maximum smoothness. From Eq. (6), the three control functions P^i_{ii} $(i = 1,2,3)$ correspond to one-dimensional stretching in each coordinate direction and thus are the most important of the control functions. In applications, in fact, the other control functions have been taken to be zero, i.e., $P^i_{jk}=\delta^i_j \delta^i_k P_i$, so that the generation system becomes

$$\nabla^2 \xi^i = g^{ii} P_i \qquad (i = 1,2,3) \tag{8}$$

It may be noted that, using Eq. (III-37), Eq. (7) can be written as

$$\nabla^2 \xi^i = \sum_{j=1}^{3} \sum_{k=1}^{3} P^i_{jk} (\underset{\sim}{\nabla} \xi^j \cdot \underset{\sim}{\nabla} \xi^k) = 0 \tag{9}$$

Actual computation is to be done in the rectangular transformed field, as discussed in Chapter II, where the curvilinear coordinates, ξ^i, are the independent variables, with the cartesian coordinates, x_i, as dependent variables. The transformation of Eq. (9) is obtained using Eq. (III-71). Thus we have

$$\nabla^2 \underset{\sim}{r} = \sum_{i=1}^{3} \sum_{j=1}^{3} g^{ij} \underset{\sim}{r}_{\xi^i \xi^j} + \sum_{k=1}^{3} (\nabla^2 \xi^k) \underset{\sim}{r}_{\xi^k} \tag{10}$$

But $\nabla^2 \underset{\sim}{r}=0$ and then using Eq. (7), we have

$$\sum_{i=1}^{3} \sum_{j=1}^{3} g^{ij} (\underset{\sim}{r}_{\xi^i \xi^j} + \sum_{k=1}^{3} P_{ij}^k \underset{\sim}{r}_{\xi^k}) = 0 \tag{11}$$

This then is the quasi-linear elliptic partial differential equation which is to be solved to generate the coordinate system. (In computation, the Jacobian squared, g, can be omitted from the evaluation of the metric coefficients, g^{ij}, in this equation since it would cancel anyway, cf. Eq. III-38.) As noted above, the more common form in actual use has been that with only three control functions, Eq. (8), which in the transformed region is

$$\sum_{i=1}^{3} \sum_{j=1}^{3} g^{ij} \underset{\sim}{r}_{\xi^i \xi^j} + \sum_{k=1}^{3} g^{kk} P_k \underset{\sim}{r}_{\xi^k} = 0 \tag{12}$$

Most of the following discussion therefore will center on the use of this last equation as the generation system. This form becomes particularly simple in one dimension, since then we have

$$x_{\xi\xi} + P x_\xi = 0 \tag{13}$$

which can be integrated to give, with $0 \le \xi \le I$,

$$x(\xi) = x_0 + (x_I - x_0) \frac{\int_0^{\xi} \exp[-\int_0^{\xi'} P(\xi'')d\xi'']d\xi'}{\int_0^{I} \exp[-\int_0^{\xi'} P(\xi'')d\xi'']d\xi'}$$

The one-dimensional control function corresponding to a distribution $x(\xi)$ thus is given by

$$P(\xi) = -\frac{x_{\xi\xi}}{x_\xi} \tag{14}$$

In two dimensions, Eq. (11) reduces to the following form, using the two-dimensional relations given in Section 8 of Chapter III (with $\xi^1 = \xi$ and $\xi^2 = \eta$)

$$D^{(3)}\underset{\sim}{r} = 0 \tag{15}$$

where

$$D^{(3)} = g_{22}\frac{\partial^2}{\partial\xi\partial\xi} - 2g_{12}\frac{\partial^2}{\partial\xi\partial\eta} + g_{11}\frac{\partial^2}{\partial\eta\partial\eta} + S\frac{\partial}{\partial\xi} + T\frac{\partial}{\partial\eta} \tag{16}$$

$$S = g_{22}P^1_{11} - 2g_{12}P^1_{12} + g_{11}P^1_{22} \tag{17a}$$

$$T = g_{22}P^2_{11} - 2g_{12}P^2_{12} + g_{11}P^2_{22} \tag{17b}$$

with $\underset{\sim}{r} = \underset{\sim}{i}x + \underset{\sim}{j}y$ and

$$g_{11} = x_\xi^2 + y_\xi^2 \tag{18a}$$

$$g_{22} = x_\eta^2 + y_\eta^2 \tag{18b}$$

$$g_{12} = x_\xi x_\eta + y_\xi y_\eta \tag{18c}$$

This corresponds to the following system in the physical space, from (7),

$$\nabla^2\xi = \frac{S}{g} \tag{19a}$$

$$\nabla^2\eta = \frac{T}{g} \tag{19b}$$

where $g = (x_\xi y_\eta - x_\eta y_\xi)^2$.

The two-dimensional form of the simpler generation system (12) with only two control functions is

$$g_{22}(\underset{\sim}{r}_{\xi\xi} + P\underset{\sim}{r}_{\xi}) + g_{11}(\underset{\sim}{r}_{\eta\eta} + Q\underset{\sim}{r}_{\eta}) - 2g_{12}\, \underset{\sim}{r}_{\xi\eta} = 0 \qquad (20)$$

for which the system in the physical space is, from (8),

$$\nabla^2\xi = \frac{g_{22}}{g} P \qquad (21a)$$

$$\nabla^2\eta = \frac{g_{11}}{g} Q \qquad (21b)$$

This generation system has been widely used, and a number of applications are noted in Ref. [1] and [5]. Several examples appear in Ref. [2].

Substitution of (3) in (10) gives the transformation of the original Poisson system (3) as

$$\sum_{i=1}^{3} \sum_{j=1}^{3} g^{ij}\, \underset{\sim}{r}_{\xi^i \xi^j} + \sum_{k=1}^{3} P^k\, \underset{\sim}{r}_{\xi^k} = 0 \qquad (22)$$

This generation system has also been widely used, cf. Ref. [1] and [2], and the two-dimensional form is

$$g_{22}\, \underset{\sim}{r}_{\xi\xi} + g_{11}\, \underset{\sim}{r}_{\eta\eta} - 2g_{12}\, \underset{\sim}{r}_{\xi\eta} + g(P\underset{\sim}{r}_{\xi} + Q\underset{\sim}{r}_{\eta}) = 0 \quad (23)$$

corresponding in the physical space to

$$\nabla^2\xi = P \qquad (24a)$$

$$\nabla^2\eta = Q \qquad (24b)$$

This system has also been widely used (cf. Ref. [1] and [5]), and its use predates that of Eq. (21). In general, however, the form of (12), corresponding to the system (8), is probably preferable over that of (22), which corresponds to (3), because of the simple form to which the former reduces in one dimension, and because the control functions in (8) are orders of magnitude smaller than those in (3) for similar effects.

E. Other systems

Other elliptic systems of the general form (4) have been considered, such as with $P^i = gP_i$, where the P_i are the specified control functions, and with $P^i = -(\nabla D \cdot \nabla \xi^i)/D$, where D is the control function. The latter form puts Eq. (4) in the form of a diffusion equation with the control function in the role of a variable diffusivity:

$$\nabla \cdot (D \nabla \xi^i) = 0 \tag{25}$$

This system also corresponds to the Euler equations for maximization of the smoothness, but now with the coefficient, D, serving as a weight function, i.e., multiplying the integrand in Eq. (2), so that the smoothness is emphasized where D is large. Both of these systems have actually been implemented only in two dimensions, although the formulations are general. Specific references to these and other related systems are given in Ref. [1] and [5].

Another elliptic system for the generation of an orthogonal grid has been constructed by combining the orthogonality conditions, $g_{ij} = 0$ $(i \neq j)$ with a specified distribution of the Jacobian over the field, $\sqrt{g} = f(\mathbf{r})$. (This system is discussed further in Chapter IX.) Some two-dimensional applications appear in Ref. [2], as noted in Ref. [5].

The second-order systems allow the specification of either the point distribution on the boundary (Dirichlet problem):

or the coordinate line slope at the boundary (Neumann problem):

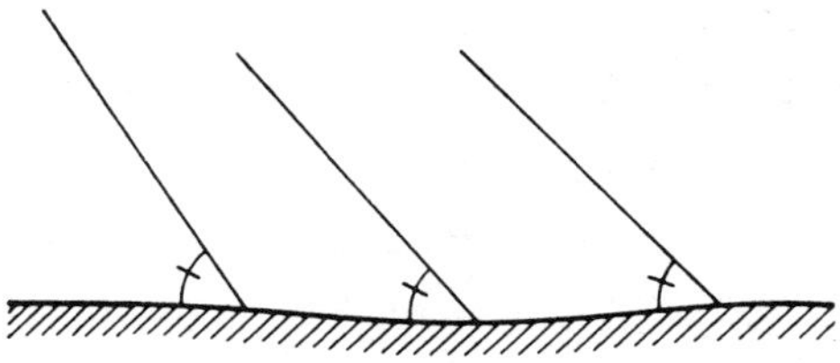

but not both. Thus it is not possible with such systems to generate grids which are orthogonal at the boundary with specified point distribution thereon. (This assumes that the control functions are specified. It is possible to adjust the control functions to achieve orthogonality at the boundary as is discussed in Section 2.)

A fourth-order elliptic system can be formulated by replacing the Laplacian operator, ∇^2, with the biharmonic operator, ∇^4. The analogous form to (4) then is

$$\nabla^4\xi^i = P^i \quad (i = 1,2,3) \tag{26}$$

which can be implemented as a system of two second-order equations:

$$\nabla^2\xi^i = R^i \tag{27a}$$

$$\nabla^2 R^i = P^i \tag{27b}$$

From (III-71) and (22) above, the transformed system is

$$\sum_{i=1}^{3} \sum_{j=1}^{3} g^{ij} R^{l}_{\xi^i \xi^j} + \sum_{k=1}^{3} R^{k} R^{l}_{\xi^k} = P^{l} \quad (l = 1,2,3) \tag{28a}$$

$$\sum_{i=1}^{3} \sum_{j=1}^{3} g^{ij} \underset{\sim}{r}_{\xi^i \xi^j} + \sum_{k=1}^{3} R^{k} \underset{\sim}{r}_{\xi^k} = 0 \tag{28b}$$

This generation system, being of higher order, allows more boundary conditions, so that the coordinate line intersection angles, as well as the point locations, can be specified on the boundary. It is therefore possible with this system to generate a coordinate system which is orthogonal at the boundary with the point distribution on the boundary specified, and for which the first coordinate surface off the boundary is at a specified distance from the boundary:

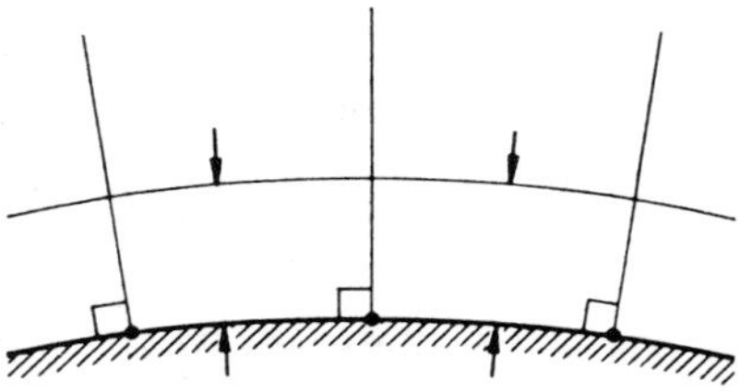

This allows segmented grids to be patched together with slope continuity as discussed in Chapter II.

In the above discussions, generation systems have been formulated based on linear differential operators in the physical space, e.g., the Laplacian with respect to the cartesian coordinates, resulting in quasi-linear equations in the transformed space where the computation is actually performed. It is also possible to formulate the generation system using linear differential operators in the transformed space, e.g., the Laplacian with respect to the curvilinear coordinates:

$$\sum_{i=1}^{3} r_{\xi^i \xi^i} = P^i \tag{29}$$

The use of some such generation systems is noted in Ref. [1], and such a biharmonic system is noted in Ref. [5]. Although this certainly produces simpler equations to be solved, since the computation is done in the transformed space, such systems transform to quasi-linear equations in the physical space, and hence the extremum principles are lost in the physical space. This means that there is a possiblity of coordinate lines overlapping in general configurations. Therefore it is generally best to formulate the generation system using linear operators in the physical space.

As noted above, other variations of elliptic systems of the type discussed here are noted in Ref. [1] and [5]. Elliptic generation systems may also be produced from the Euler equations resulting from the application of variational principles to produce adpative grids, as is discussed in Chapter XI. Still another system, based on the successive generation of curved surfaces in the three-dimensional region, is given in Section 3B of this chapter. Finally, quasiconformal mapping (Ref. [22] and [23]) is another example of an elliptic generation system.

2. Control Functions

For the elliptic generation system given by Eq. (12), the control functions that will produce a specified line distribution for a rectangular region, and for an annular region, are given as Eq. (14) and in Exercise 8, respectively. These functions could be used in other regions, of course, with the same general effect. In such extended use, the former would be more appropriate for simply-connected regions, while the latter would be appropriate for multiply-connected regions. Use of the rectangular function in a multiply-connected region produces a stronger concentration than was intended because of the concentration over convex boundaries that is inherent in Poisson-type generation systems (cf. Section 1A).

With generation systems of the Poisson type, negative values of the control function P^i in Eq. (4), or P_i in Eq. (8) (since $g^{ii} > 0$), will cause the ξ^i coordinate lines to concentrate in the direction of decreasing ξ^i (cf. Section 1B). Several approaches to the determination of these control functions are discussed below.

A. Attraction to coordinate lines/points

This effect can be utilized to achieve attraction of coordinate lines to other coordinate lines and/or points by taking the form of the control functions to be, in 2D, (again with $\xi^1=\xi$, $\xi^2=\eta$, $P^1=P$, $P^2=Q$)

$$P(\xi,\eta) = -\sum_{i=1}^{N} a_i \operatorname{sign}(\xi - \xi_i) \exp(-c_i|\xi-\xi_i|) \qquad (30)$$

$$- \sum_{i=1}^{M} b_i \operatorname{sign}(\xi - \xi_i) \exp\{-d_i[(\xi-\xi_i)^2+(\eta-\eta_i)^2]^{1/2}\}$$

and an analogous form for $Q(\xi,\eta)$ with ξ and η interchanged. (Here the subscripts identify particular ξ-lines and are not to be confused with the superscripts used to refer to the curvilinear coordinates in general.) In this form, the control functions are functions only of the curvilinear coordinates.

In the P function, the effect of the amplitude a_i is to attract ξ-lines toward the ξ_i-line:

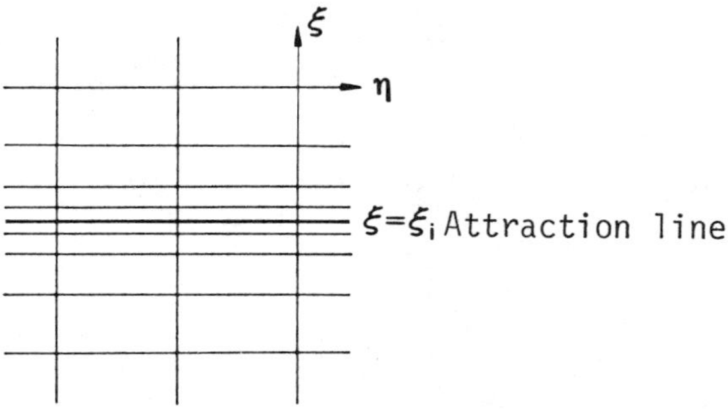

while the effect of the amplitude b_i is to attract ξ-lines toward the single point (ξ_i, η_i):

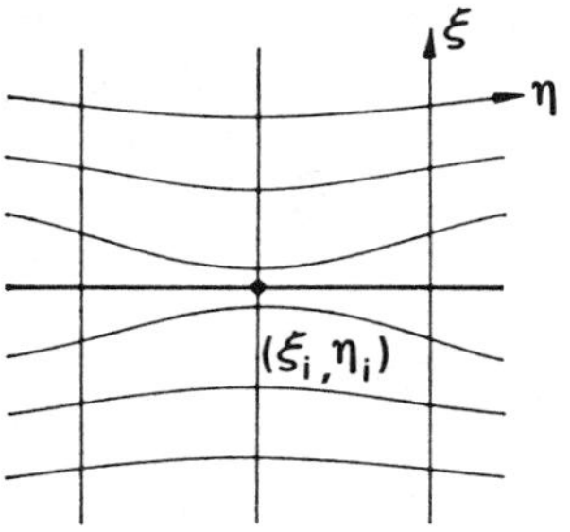

Note that this attraction to a point is actually attraction of ξ-lines to a point on another ξ-line, and as such acts normal to the ξ-line through the point. There is no attraction of η-lines to this point via the P function. In each case the effect of the attraction decays with distance in ξ-η space from the attraction site according to the decay factors, c_i and d_i. This decay depends only on the ξ-distance from the ξ_i-line in the first term, so that entire ξ-lines are attracted to the entire ξ_i line. In the second term, however, the decay depends on both the ξ and η distances from the attraction point (ξ_i, η_i), so that the effect is limited to portions of the ξ-lines. With the inclusion of the sign changing function, the attraction occurs on both sides of the ξ-line, or the (ξ_i, η_i) point, as the case may be. Without this function, attraction occurs only on the side toward increasing ξ, with repulsion occuring on the other side. A negative amplitude simply reverses all of these effects, i.e., attraction becomes repulsion, and vice versa. The effect of the Q function on η-lines follows analogously.

In the case of a boundary that is an η-line, positive amplitudes in the Q function will cause η-lines off the boundary to move closer to the boundary, assuming that η increases off the boundary. The effect of the P function will be to alter the angle at which the ξ-lines intersect the boundary, if the points on the boundary are fixed, with the ξ-lines tending to lean in the direction of decreasing ξ. These effects have been noted in figures above, and further examples are given below:

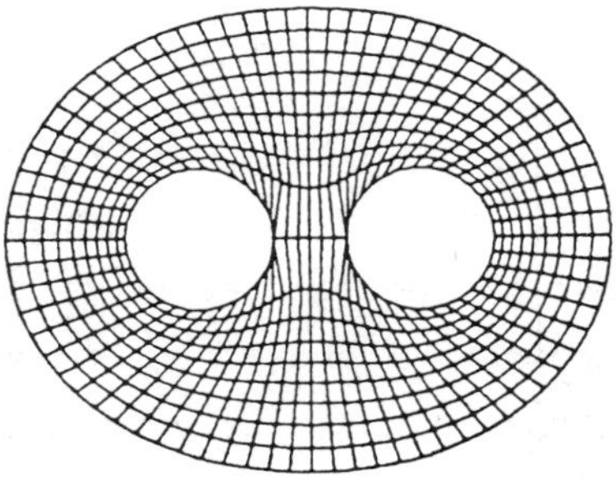

No control function

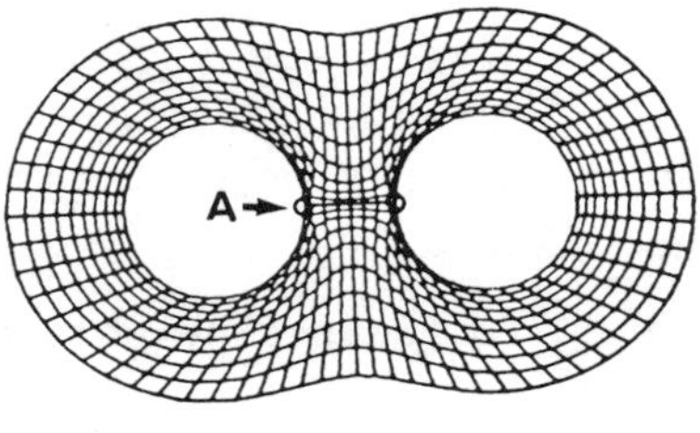

A: Attraction to point

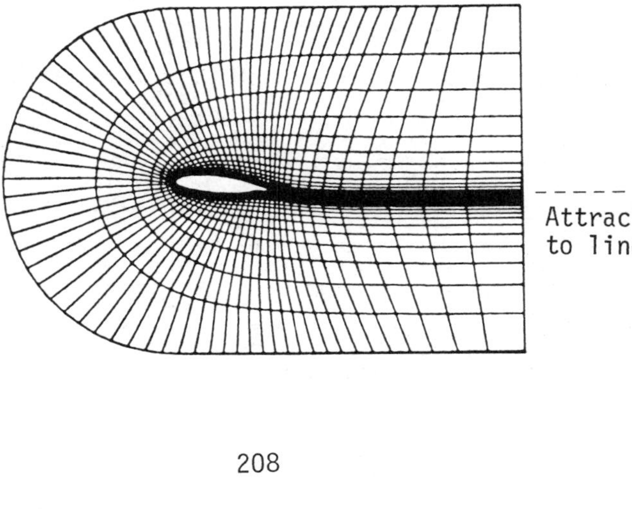

Attraction to line

The first two figures here show the result of attraction to the two circled points, in comparison with the case with no control function. The last figure illustrates strong attraction to the coordinate line coincident with the inner boundary and the branch cut in this C-type system. If the boundary is such that η decreases off the boundary, then the amplitudes in the Q function must be negative to achieve attraction to the boundary. In any case, the amplitudes a_i cause the effects to occur all along the boundary (as in the last figure above), while the effects of the amplitudes b_i occur only near selected points on the boundary (second figure above).

In configurations involving branch cuts, the attraction lines and/or points in this type of evaluation of the control function strictly should be considered to exist on all sheets. In the O-type configuration shown on p. 29, where the two sides of the cut are on opposite sides of the transformed region, the control function P for attraction to the ξ_i-line must be constructed as follows: In the figure below,

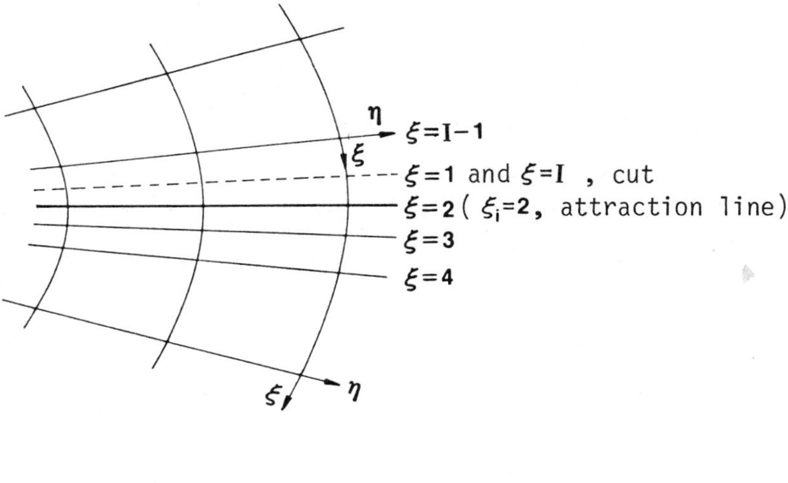

when the attraction line is the $\xi_i=2$ line, the $\xi=I-1$ line experiences a counterclockwise attraction to this line at a distance of $(I-1)-2$. However, the $\xi_i=2$ attraction line also appears as a $I+(\xi_i-1)=I+(2-1)=I+1$ attraction line on the next sheet as the cut is crossed. Therefore, the $\xi=I-1$ line also experiences a clockwise attraction to this $I+1$ line at a distance of $(I+1)-(I-1)=2$, and this attraction is, of course, stronger than the first mentioned. In fact, since the attraction line is repeated on all sheets there strictly must be a summation over all sheets in Eq. (30), i.e., a summation over k, with ξ_i replaced by $\xi_i+k\Delta\xi$ where $\Delta\xi$ is the jump in ξ at the cut ($\Delta\xi=I-1$ in the above figure). Thus ξ_i in Eq. (30) would be replaced by the $\xi_i+k\Delta\xi$, and the right side would be summed from $k=-\infty$ to $+\infty$. However, because of the exponential decay, the terms decrease rapidly as k increases, so that only the term with the smallest distance in the k summation really needs to be included, i.e., only the term giving clockwise attraction at a distance of 2 from the attraction line for the $\xi=I-1$ line in the above figure. Since there is no jump in η across the cut in this configuration, the evaluation of Q is affected by this cut only through the replacement of ξ_i as above in the term for the point attraction, with summation over k of only this part of the right side. Again only the term with the smallest distance need actually be included.

For the C-type configuration on p. 30, with the two sides of the cut on the same side of the transformed region, η is reflected in the cut, and the construction of the control function Q is as follows. With reference to the figure below,

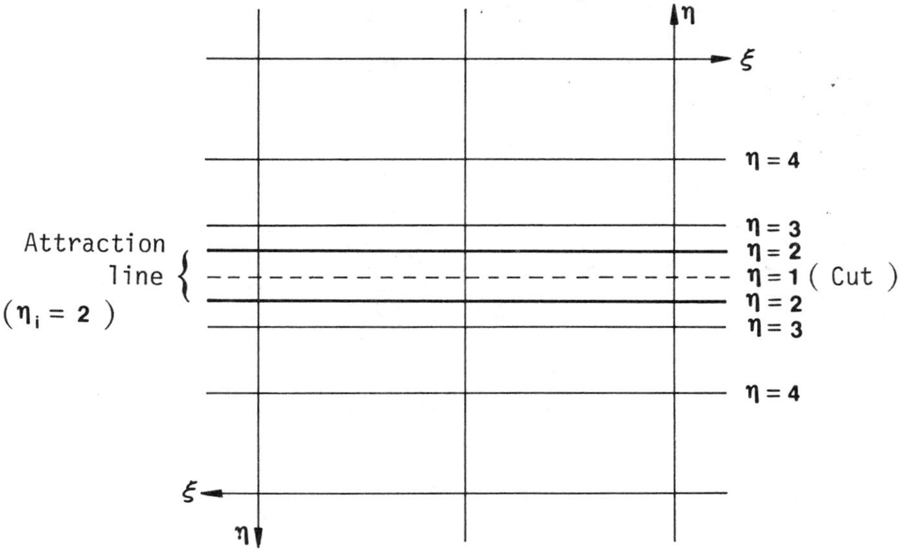

the attraction line, $\eta_i=2$, is located on both sides of the cut in this configuration. Now the $\eta=3$ line above the cut experiences a downward attraction toward the $\eta_i=2$ attraction line at a distance of 3-2=1. Strictly speaking, this line above the cut should also experience a downward attraction toward the portion of the $\eta_i=2$ attraction line below the cut as it appears on the next sheet (and, in fact, on all other sheets), i.e., at $\eta_{cut}-(\eta_i-\eta_{cut})$, where η_{cut} is the value of η on the cut ($\eta_{cut}=1$ here). This attraction line on the next sheet is at a distance $\eta-[\eta_{cut}-(\eta_i-\eta_{cut})]$ from the η-line of interest, i.e., at 3-[1-(2-1)]=3 from the $\eta=3$ line above the cut. This attraction line on the next sheet is therefore farther away and hence its effect can perhaps be neglected. However, for lines between the attraction line and the cut, the effect of the attraction line on the next sheet should be considered. In any case it is necessary to take into account the attraction lines appearing on the next sheet, those on all other sheets being too far away to be of consequence. Here the evaluation of the control function P is affected by the cut only through the point attraction part, with η_i replaced as above.

The third type of cut, illustrated on p. 40, for which the two sides of the cut face across a void of the transformed region, is treated by replacing η_i with $\eta_i-\Delta\eta$ in both the control functions, where $\Delta\eta-1$ is the number of η-lines in the void. There is no additional summation in this case.

The case on p.52, where the coordinate species changes sign at the cut, requires individual attention at each cut. For example, the contribution to the control functions in region A at a point (ξ,η) from an attraction site (ξ_i,η_i) in region B would be evaluated using distances of $(\xi-\xi_{max})+(\eta_{max}-\eta_i)$ and $(\eta-\xi_i)$ in place of $\xi-\xi_i$ and $\eta-\eta_i$, respectively.

B. Attraction to lines/points in space

If the attraction line and/or points are in the field, rather than on a boundary, then the above attraction is not to a fixed line or point in space, since the attraction line or points are themselves determined by the solution of the generation system and hence are free to move. It is, of course, also possible to take the control functions to be funtions of x and y instead of ξ and η, and thus achieve attraction to fixed lines and/or points in the physical field. This case becomes somewhat more complicated, since it must be ensured that coordinate lines are not attracted parallel to themselves.

With the attraction discussed in the previous section, η-lines are attracted to other η-lines, and ξ-lines are attracted to other ξ-lines. It is unreasonable, of course, to attempt to attract η-lines to ξ-lines, since that would have the effect of collapsing the coordinate system. When, however, the attraction is to be to certain fixed lines in the physical region, defined by curves $y = f(x)$, care must be exercised to avoid attempting to attract coordinate lines to specified curves that cut the coordinate lines at large angles. Thus, in the figure below,

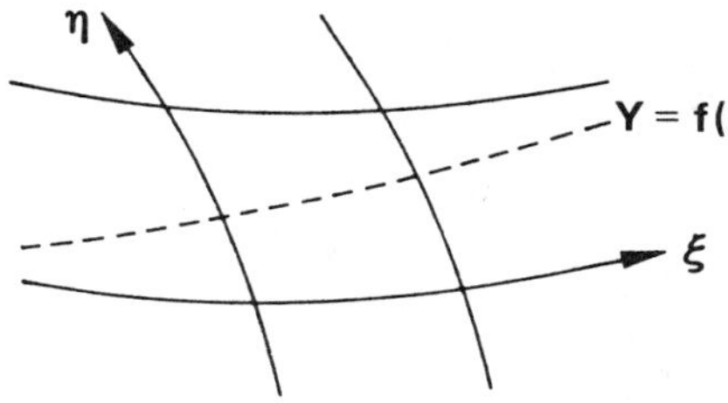

it is unreasonable to attract ξ-lines to the curve $y=f(x)$, while it is natural to attract the η-lines to this curve.

However, in the general situation, the specified line $y=f(x)$ will not necessarily be aligned with either a ξ or η line along its entire length. Since it is unreasonable to attract a line tangentially to itself, some provision is necessary to decrease the attraction to zero as the angle between the coordinate line and the given line $y = f(x)$ approaches 90°. This can be accomplished by multiplying the attraction function by the cosine of the angle between the coordinate line and the line $y=f(x)$. It is also necessary to change the sign on the attraction function on either side of the line $y=f(x)$. This can be done by multiplying by the sine of the angle between the line $y=f(x)$ and the vector to the point on the coordinate line.

These two purposes can be accomplished as follows. Let a general point on the ξ-line be located by the vector $\underline{R}(x,y)$, and let the attraction line $y = f(x)$ be specified by the collection of points $S(x_i,y_i)$, $i=1,2,\ldots N$. Let the unit tangent to the attraction line be $\underset{\sim}{t}(x_i, y_i)$, and the unit tangent to a ξ-line be $\underset{\sim}{\tau}^{(\xi)}$. Then, with $\underset{\sim}{k}$ the unit vector normal to the two-dimensional plane, and with reference to the following figure,

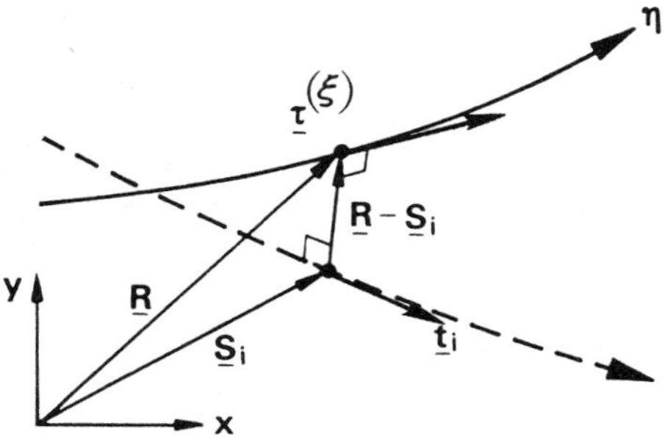

the control functions, P(x,y) and Q(x,y), may logically be taken as

$$P(x,y) = -\sum_{i=1}^{N} a_i(\underset{\sim}{t}_i \cdot \underset{\sim}{\tau}^{(\xi)}) \frac{[\underset{\sim}{t}_i x(\underset{\sim}{R}-\underset{\sim}{S}_i)]\cdot\underset{\sim}{k}}{|\underset{\sim}{R}-\underset{\sim}{S}_i|} \exp(-d_i|\underset{\sim}{R}-\underset{\sim}{S}_i|) \quad (31)$$

The equation for Q simply has ξ replaced by η in the above. These functions depend on x and y through both $\underset{\sim}{R}$ and $\underset{\sim}{\tau}^{(\xi)}$ or $\underset{\sim}{\tau}^{(\eta)}$, and thus must be recalculated at each point as the iterative solution proceeds. This form of coordinate control will therefore be more expensive to implement than that based on attraction to other coordinate lines.

There is no real distinction between "line" and "point" attraction with this type of attraction. "Line" attraction here is simply attraction to a group of points that form a line, y=f(x). If line attraction is specified then the tangent to the line y=f(x) is computed from the adjacent points on the line. If point attraction is specified, then the "tangent" must be input for each point. The unit tangents to the coordinate lines are computed from Eq. (III-3):

$$\underset{\sim}{t}^{(\xi)} = \frac{1}{\sqrt{g_{22}}}(\underset{\sim}{i}x_\eta + \underset{\sim}{j}y_\eta)$$

$$\underset{\sim}{t}^{(\eta)} = \frac{1}{\sqrt{g_{11}}}(\underset{\sim}{i}x_\xi + \underset{\sim}{j}y_\xi)$$

The presence of branch cuts introduces no complication with this type of attraction since the distances involved are in terms of the cartesian coordinates, rather than the curvilinear coordinates. This form of attraction makes the control functions dependent on both the curvilinear and cartesian coordinates, and thus attraction to space lines and/or points involves more complicated equations in the transformed region than does attraction to other coordinate lines and/or points, since for the former, coefficients of the first derivatives are functions of the dependent variables. Attraction to lines and/or points in space has not been widely used, and the use of Eq. (31) has not been fully tested.

C. Evaluation along a coordinate line

As has been noted above, if it is desired that the spacing of the coordinate lines in the field generally follow that of the points on the boundary, the control functions must be evaluated so as to correspond to this boundary point distribution. This can be accomplished as follows. (The developments in this and the next two sections are generalizations of that given in Ref. [12], and other works cited therein and in Ref. [5].)

The projection of Eq. (12) along a coordinate line on which ξ^1 varies is found by forming the dot product of this equation with the base vector $\underset{\sim}{a}_1 = \underset{\sim}{r}_{\xi^1}$, which is tangent to the line.

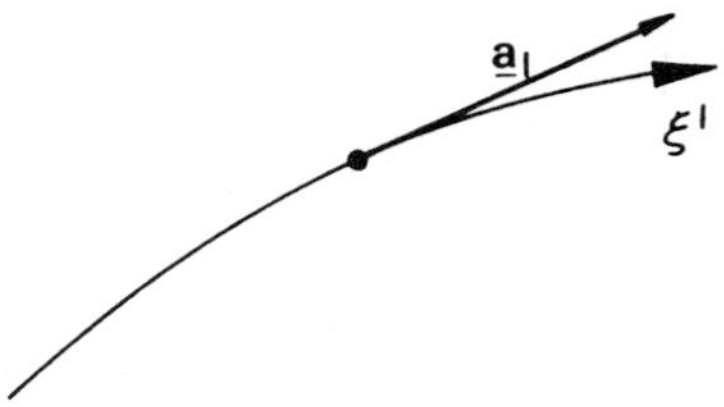

Thus we have

$$\sum_{i=1}^{3} \sum_{j=1}^{3} g^{ij} \underset{\sim}{r}_{\xi^l} \cdot \underset{\sim}{r}_{\xi^i \xi^j} + \sum_{k=1}^{3} g^{kk} P_k \underset{\sim}{r}_{\xi^l} \cdot \underset{\sim}{r}_{\xi^k} = 0 \qquad (32)$$

Now assume for the moment that the two coordinate lines crossing the coordinate line of interest do so orthogonally. Then on this line we have

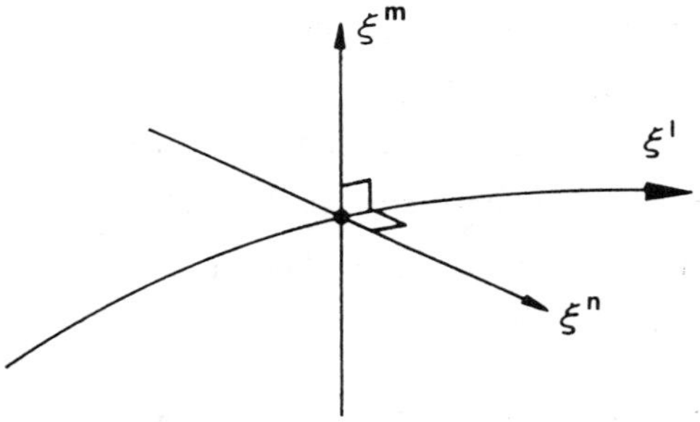

and

$$\underset{\sim}{r}_{\xi^l} \cdot \underset{\sim}{r}_{\xi^k} = \underset{\sim}{a}_l \cdot \underset{\sim}{a}_k = g_{lk} = \delta_{lk} g_{ll}$$

which leads to an explicit equation for P_l on the coordinate line of interest:

$$P_l = \frac{-1}{g^{ll} g_{ll}} \underset{\sim}{r}_{\xi^l} \cdot \sum_{i=1}^{3} \sum_{j=1}^{3} g^{ij} \underset{\sim}{r}_{\xi^i \xi^j} \qquad (33)$$

If it is further assumed for the moment that the two coordinate lines crossing the coordinate line of interest are also orthogonal to each other, i.e., complete orthogonality on the line of interest,

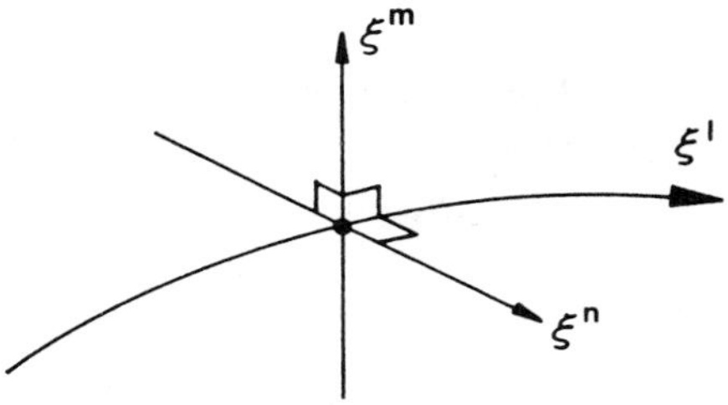

we have on this line $g^{ij} = \delta_{ij} g^{ii}$ and $g_{ij} = \delta_{ij} g_{ii}$ Also, from Eq. (III-38),

$$g^{11} = \frac{1}{g}(g_{mm}g_{nn} - g_{mn}^2) = \frac{1}{g} g_{mm}g_{nn} \qquad (1,m,n\ )\ \text{cyclic}$$

since $m \neq n$. But also by Eq. (III-16) $g = g_{11}g_{mm}g_{nn}$ so that $g^{11}g_{11}=1$. Then Eq. (33) becomes

$$P_1 = -\sum_{i=1}^{3} \frac{1}{g_{ii}} \underset{\sim}{r}_{\xi^1} \cdot \underset{\sim}{r}_{\xi^i \xi^i} \quad (1 = 1,2,3) \tag{34}$$

which can also be written, using Eq. (III-3), as

$$P_1 = -\sum_{i=1}^{3} \frac{1}{g_{ii}} \underset{\sim}{a}_1 \cdot (\underset{\sim}{a}_i)_{\xi^i} \tag{35}$$

By Eq. (III-7) the derivative of arc length along the coordinate line on which ξ^1 varies is

$$s_{\xi^1} = |\underset{\sim}{r}_{\xi^1}| \tag{36}$$

Then

$$s_{\xi^1\xi^1} = |\underset{\sim}{r}_{\xi^1}|_{\xi^1} = \frac{\underset{\sim}{r}_{\xi^1} \cdot \underset{\sim}{r}_{\xi^1\xi^1}}{|\underset{\sim}{r}_{\xi^1}|} \tag{37}$$

so that the logarithmic derivative of arc length along this coordinate line is given by

$$S_1 = \frac{s_{\xi^1\xi^1}}{s_{\xi^1}} = \frac{\underset{\sim}{r}_{\xi^1} \cdot \underset{\sim}{r}_{\xi^1\xi^1}}{|\underset{\sim}{r}_{\xi^1}|^2} = \frac{\underset{\sim}{r}_{\xi^1} \cdot \underset{\sim}{r}_{\xi^1\xi^1}}{g_{11}} \tag{38}$$

which is exactly the i=1 term in the summation in Eq. (34).

The unit tangent to a coordinate line on which ξ^m varies is

$$\underset{\sim}{T}^m = \frac{\underset{\sim}{r}_{\xi^m}}{|\underset{\sim}{r}_{\xi^m}|} = \frac{\underset{\sim}{a}_m}{|\underset{\sim}{a}_m|} \tag{39}$$

and the derivative of this unit tangent with respect to arc length is a vector that is normal to this line, the magnitude of which is the curvature, K, of the line. The unit vector in this normal direction is the principal normal, $\underset{\sim}{N}$, to the line.

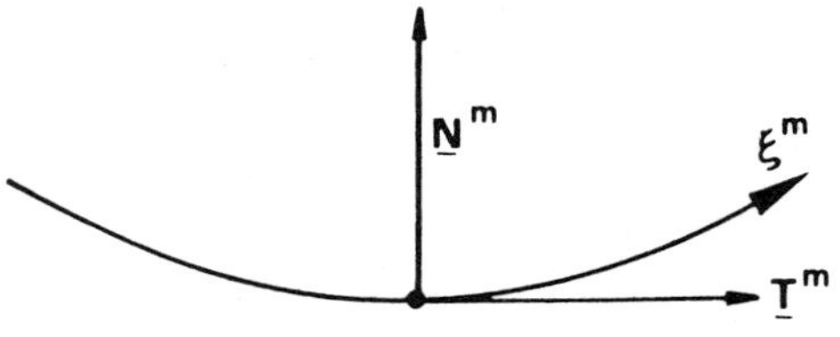

Thus, using Eq. (36),

$$K^m \underset{\sim}{N}^m = (\underset{\sim}{T}^m)_s = (\underset{\sim}{T}^m)_{\xi^m}\, \xi^m_s = \frac{(\underset{\sim}{T}^m)_{\xi^m}}{|\underset{\sim}{r}_{\xi^m}|} \tag{40}$$

Then

$$K^m \underset{\sim}{N}^m = \frac{1}{|\underset{\sim}{r}_{\xi^m}|} \left(\frac{\underset{\sim}{r}_{\xi^m}}{|\underset{\sim}{r}_{\xi^m}|} \right)_{\xi^m}$$

$$= \frac{|\underset{\sim}{r}_{\xi^m}|^2 \, \underset{\sim}{r}_{\xi^m \xi^m} - (\underset{\sim}{r}_{\xi^m} \cdot \underset{\sim}{r}_{\xi^m \xi^m}) \underset{\sim}{r}_{\xi^m}}{|\underset{\sim}{r}_{\xi^m}|^4} \tag{41}$$

so that the curvature is

$$K^m = \frac{1}{|\underset{\sim}{r}_{\xi^m}|^3} \left[|\underset{\sim}{r}_{\xi^m}|^2 \, |\underset{\sim}{r}_{\xi^m \xi^m}|^2 - (\underset{\sim}{r}_{\xi^m} \cdot \underset{\sim}{r}_{\xi^m \xi^m})^2 \right]^{1/2} \tag{42}$$

The component of $K^m \underset{\sim}{N}^m$ along the coordinate line on which ξ^1 varies

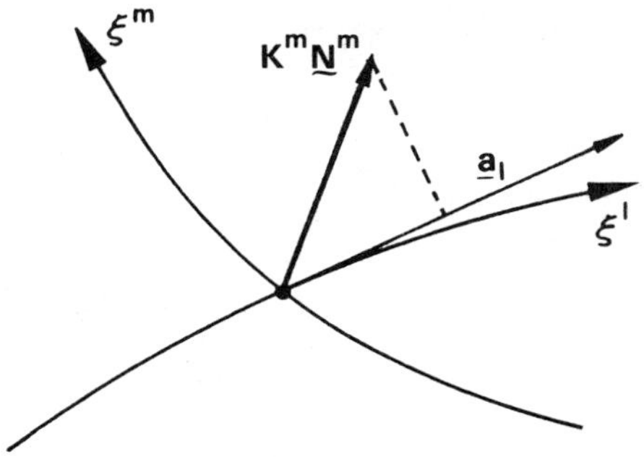

is given by

$$(K^m \underset{\sim}{N}^m)^{(1)} = K^m \underset{\sim}{N}^m \cdot \frac{\underset{\sim}{r}_{\xi^1}}{|\underset{\sim}{r}_{\xi^1}|} = \frac{\underset{\sim}{r}_{\xi^1} \cdot \underset{\sim}{r}_{\xi^m \xi^m}}{|\underset{\sim}{r}_{\xi^1}| \, |\underset{\sim}{r}_{\xi^m}|^2} \tag{43}$$

since $\underset{\sim}{r}_{\xi^l} \cdot \underset{\sim}{r}_{\xi^m} = 0$ for $m \neq l$. Then the two terms of the summation in Eq. (34) for which $m \neq l$ can be written as

$$\frac{\underset{\sim}{r}_{\xi^l} \cdot \underset{\sim}{r}_{\xi^m \xi^m}}{g_{mm}} = |\underset{\sim}{r}_{\xi^l}| (K^m \underset{\sim}{N}^m)^{(l)} = \sqrt{g_{ll}}\, (K^m \underset{\sim}{N}^m)^{(l)} \qquad (m \neq l) \qquad (44)$$

Thus Eq. (34) can be written

$$P_l = - S_l - \sqrt{g_{ll}}\, [(K^m \underset{\sim}{N}^m)^{(l)} + (K^n \underset{\sim}{N}^n)^{(l)}] \qquad (l=1,2,3) \qquad (45)$$

where (l,m,n) are cyclic, and using Eq. (III-3), we have

$$S_l = \frac{s_{\xi^l \xi^l}}{s_{\xi^l}} = \frac{\underset{\sim}{a}_l \cdot (\underset{\sim}{a}_l)_{\xi^l}}{g_{ll}} = \frac{(g_{ll})_{\xi^l}}{2g_{ll}} \qquad (46)$$

and

$$(K^m \underset{\sim}{N}^m)^{(l)} = \frac{\underset{\sim}{a}_l \cdot (\underset{\sim}{a}_m)_{\xi^m}}{\sqrt{g_{ll}} g_{mm}} \qquad (47)$$

with an analogous equation for $(K^n \underset{\sim}{N}^n)^{(l)}$. The arc length in the expressions (45) for the control function P_l along the coordinate line on which ξ^l varies can be determined entirely from the grid point distribution on the line using Eq. (46). The other two terms in P_l, however, involve derivatives off this line and therefore must either be determined by specifications of the components of the curvature, $K\underset{\sim}{N}$, of the crossing lines along the line of interest, or by interpolation between values evaluated on coordinate surfaces intersecting the ends of this line.

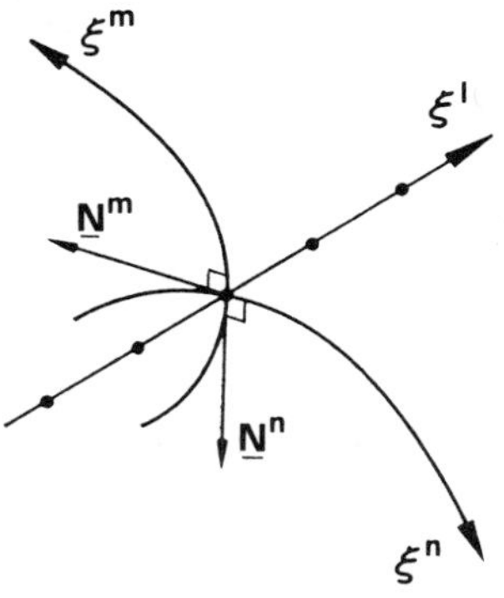

If it is assumed that the curvatures of these crossing lines vanish on the coordinate line of interest, then the last two terms in Eq. (45) are zero, and the control function becomes simply

$$P_l = - S_l \tag{48}$$

and then can be evaluated entirely from the specified point distribution on the coordinate line of interest.

The neglect of the curvature terms, however, is ill-advised since the elliptic system already has a strong tendency to concentrate lines over a convex boundary, as has been discussed earlier in this chapter. Therefore neglect of the curvature terms will result in control functions which will produce a stronger concentration than intended over convex boundaries (and weaker over concave). When interpolation from the end points is used to determine the curvature term, the entire term ($K\underline{N}$) should be interpolated, since individual interpolation of the vectors $\underset{\sim}{a}_l$ and $(\underset{\sim}{a}_m)_{\xi^m}$ can give an inappropriate value for the dot product.

It should be noted that the assumptions of orthogonality, and perhaps vanishing curvature, that were made in the course of the development of these expressions for the control functions on a coordinate line are not actually enforced on the resulting coordinate system, but merely served to allow some reasonable relations for these control functions corresponding to a specified point distribution on a coordinate line to be developed. This should not be considered a source of error since the control functions are arbitrary in the generation system (12).

D. Evaluation on a coordinate surface

In a similar fashion, expressions for the control functions on a coordinate surface on which ξ^l is constant can be obtained from the projections of Eq. (12) along the two coordinate lines lying on the surface, i.e., the lines on which ξ^m and ξ^n vary, (l,m,n) being cyclic.

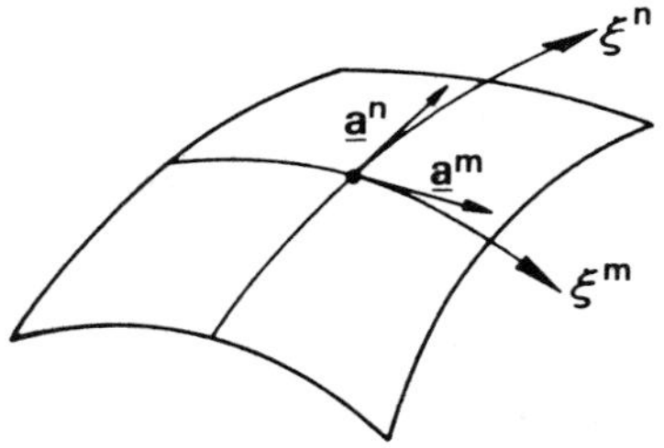

These projections are given by Eq. (32) with l replaced by m and n, respectively. If it is assumed for the moment that the coordinate line crossing the coordinate surface of interest is orthogonal to the surface

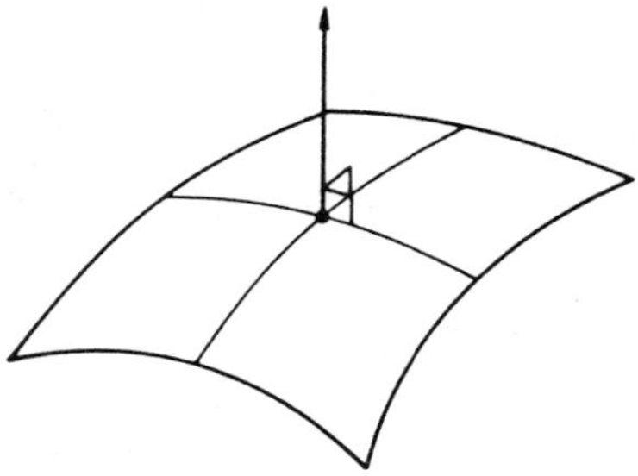

then $\underset{\sim}{r}_{\xi^1} \cdot \underset{\sim}{r}_{\xi^m} = \underset{\sim}{r}_{\xi^1} \cdot \underset{\sim}{r}_{\xi^n} = 0$, so that P_1 is removed from both of these two equations to yield the equation

$$\sum_{i=1}^{3} \sum_{j=1}^{3} g^{ij} \underset{\sim}{r}_{\xi^m} \cdot \underset{\sim}{r}_{\xi^i\xi^j} + g^{mm} g_{mm} P_m + g^{nn} g_{mn} P_n = 0$$

and an analogous equation with m and n interchanged. Solution of these two equations for P_m and P_n then yields

$$P_m = \frac{-(g_{nn} \underset{\sim}{r}_{\xi^m} - g_{mn} \underset{\sim}{r}_{\xi^n}) \cdot \sum_{i=1}^{3} \sum_{j=1}^{3} g^{ij} \underset{\sim}{r}_{\xi^i\xi^j}}{g^{mm}(g_{mm}g_{nn} - g_{mn}^2)} \qquad (49)$$

with an analogous equation for P_n with m and n interchanged. Since $g_{1m} = g_{1n} = 0$, we have by Eq. (III-38), $g^{1m} = g^{1n} = 0$. Therefore only the five terms, 11, mm, nn, mn, nm, are non-zero in the summation. Also from Eq. (III-38) we have

$$g^{mm} = \frac{g_{nn}}{g_{mm}g_{nn} - g_{mn}^2}$$

since here $g = \det|g_{ij}| = g_{11}(g_{mm}g_{nn} - g_{mn}^2)$. An analogous equation for g^{nn} is obtained by interchanging m and n.

Then Eq. (49) can be rewritten as

$$P_m = -(\underset{\sim}{r}_{\xi^m} - \frac{g_{mn}}{g_{nn}} \underset{\sim}{r}_{\xi^n}) \cdot (\sum_{i=1}^{3} g^{ii} \underset{\sim}{r}_{\xi^i\xi^i} + 2g^{mn} \underset{\sim}{r}_{\xi^m\xi^n}) \qquad (50)$$

and an analogous equation for P_n with m and n interchanged. But, again using Eq. (III-38), we have

$$g^{11} = \frac{1}{g_{11}} \quad \text{and} \quad g^{mn} = \frac{-g_{mn}}{g_{mm}g_{nn} - g_{mn}^2}$$

Therefore

$$P_m = -\frac{1}{g_{11}}\left(\underline{r}_{\xi^m} - \frac{g_{mn}}{g_{nn}}\underline{r}_{\xi^n}\right) \cdot \underline{r}_{\xi^1\xi^1}$$

$$- \left(\frac{\underline{r}_{\xi^m} - \frac{g_{mn}}{g_{nn}}\underline{r}_{\xi^n}}{g_{mm}g_{nn} - g_{mn}^2}\right)$$

$$\cdot \left(g_{nn}\,\underline{r}_{\xi^m\xi^m} + g_{mm}\,\underline{r}_{\xi^n\xi^n} - 2g_{mn}\,\underline{r}_{\xi^m\xi^n}\right) \qquad (51)$$

and the analogous equation with m and n interchanged. This can also be written as

$$P_m = -\frac{1}{g_{11}}\left(\underline{a}_m - \frac{g_{mn}}{g_{nn}}\underline{a}_n\right) \cdot (\underline{a}_1)_{\xi^1} \qquad (52)$$

$$- \left(\frac{\underline{a}_m - \frac{g_{mn}}{g_{nn}}\underline{a}_n}{g_{mm}g_{nn} - g_{mn}^2}\right) \cdot \left[g_{nn}(\underline{a}_m)_{\xi^m} + g_{mm}(\underline{a}_n)_{\xi^n} - 2g_{mn}(\underline{a}_m)_{\xi^n}\right]$$

and the analogous equation for P_n.

All of the terms, except the first, in the above equations can be evaluated completely from the point distribution on the coordinate surface of interest.

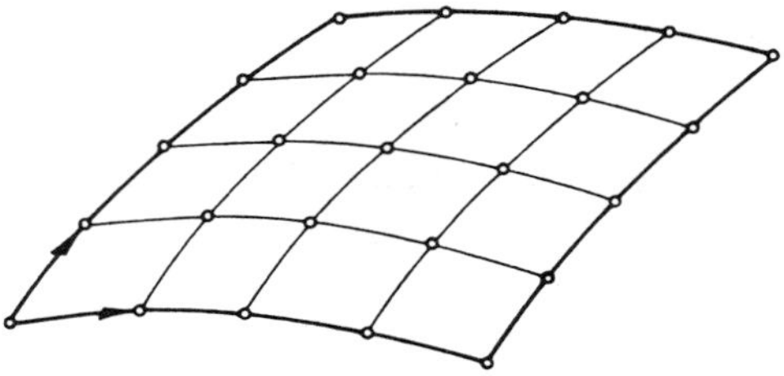

From Eq. (47) the first term in (52) can be written

$$- [\sqrt{g_{mm}}\, (K^1\underset{\sim}{N}^1)^{(m)} - \frac{g_{mn}}{\sqrt{g_{nn}}} (K^1\underset{\sim}{N}^1)^{(n)}] \tag{53}$$

where $(K^1\underset{\sim}{N}^1)^{(m)}$ and $(K^1\underset{\sim}{N}^1)^{(n)}$ are the components of the curvature $K\underset{\sim}{N}$ for the coordinate line crossing the coordinate surface of interest along the two coordinate lines on the surface.

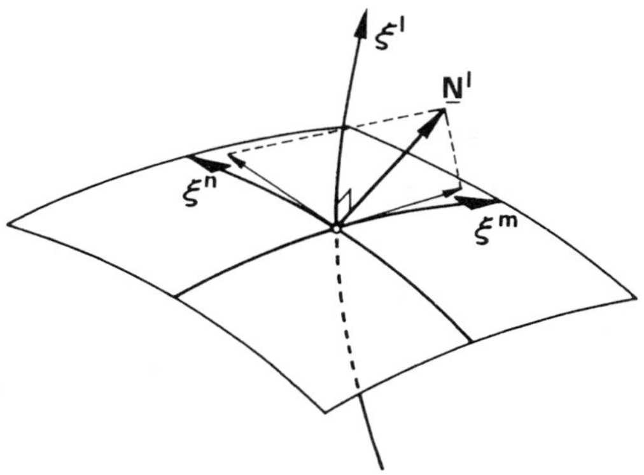

These quantities must be either specified on the surface or interpolated from values evaluated on its intersections with the other coordinate surfaces. If it is further assumed that the curvature of the crossing line vanishes at the surface, then this first term in Eq. (52) vanishes also.

As was noted for the control functions on a line, the curvature terms should not be neglected, however, else the concentration will be stronger than intended over convex boundaries and weaker over concave. Also, it is the entire term $K\underset{\sim}{N}$ which should be interpolated, not the individual vectors involved, else the dot product can have inappropriate values.

E. Evaluation from boundary point distribution

Using the relations developed in the previous two sections for the control functions on a coordinate line and on a coordinate surface, an interpolation procedure can be formulated for evaluation of the control functions in the entire field. If the point distribution is specified on all the boundary surfaces of a three-dimensional field, the control functions can be evaluated on these boundaries using the relations in Section D, and then the control functions in the entire three-dimensional field can be interpolated from these values on the bounding surfaces using transfinite interpolation (discussed in connection with algebraic grid generation in Chapter VIII.)

To be definite, consider a general three-dimensional region bounded by six curved sides:

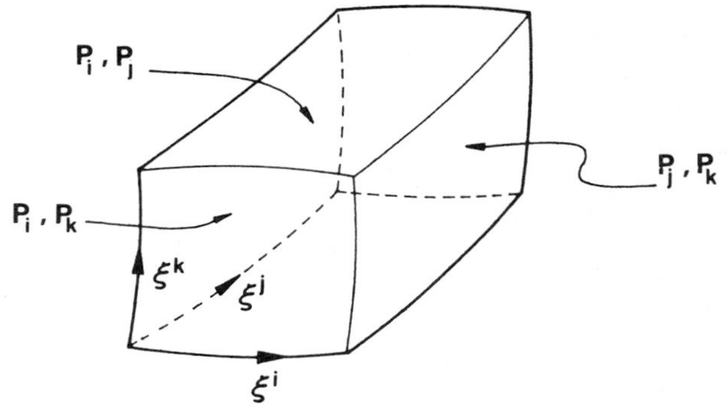

with curvilinear coordinates as shown, which transforms to a rectangular block. From Eq. (52) the two control functions, P_j and P_k, can be evaluated from the specified boundary-point distribution on the two faces on which ξ^i is constant, i.e., the left and right faces in the figure. Similar evaluations yield two control functions on each face, with the result that the control function P_k will be known on the four faces on which ξ^k varies, i.e., the front, back, left, and right faces in the figure. Thus, in general, interpolation for the control function P_k in the interior of the region is done from the boundary values on the four faces on which ξ^k varies:

$$P_k(\xi^i,\xi^j,\xi^k) =$$

$$\frac{(\xi^i - 1)}{(I^i - 1)} P_k(I^i,\xi^j,\xi^k) - \frac{(\xi^i - 1)}{(I^i - 1)} \frac{(\xi^j - 1)}{(I^j - 1)} P_k(I^i,I^j,\xi^k)$$

$$+ \frac{(I^i - \xi^i)}{(I^i - 1)} P_k(1,\xi^j,\xi^k) - \frac{(\xi^i - 1)}{(I^i - 1)} \frac{(I^j - \xi^j)}{(I^j - 1)} P_k(I^i,1,\xi^k)$$

$$+ \frac{(\xi^j - 1)}{(I^j - 1)} P_k(\xi^i,I^j,\xi^k) - \frac{(I^i - \xi^i)}{(I^i - 1)} \frac{(\xi^j - 1)}{(I^j - 1)} P_k(1,I^j,\xi^k)$$

$$+ \frac{(I^j - \xi^j)}{(I^j - 1)} P_k(\xi^i,1,\xi^k) - \frac{(I^i - \xi^i)}{(I^i - 1)} \frac{(I^j - \xi^j)}{(I^j - 1)} P_k(1,1,\xi^k) \quad (54)$$

Here I^i is the maximum value of ξ^i, etc., i.e., $1 \leq \xi^i \leq I^i, i=1,2,3$. In an analogous manner all three control functions can be determined in the interior of the region.

It may be desirable in some cases to generate a two-dimensional coordinate system on a curved surface, as discussed in Section 3, rather than specifing the point distribution on the surface. The two control functions needed on the surface for this purpose can be determined by interpolation from values evaluated on the four edges of the surface:

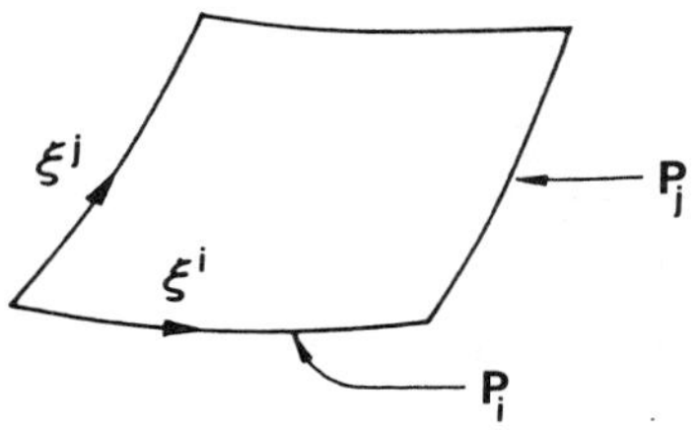

Eq. (45) allows the control function P_i to be evaluated on the edges on which ξ^i varies, i.e., the top and bottom edges in the figure. This control function on the surface can then be evaluated by interpolation between these two edges:

$$P_i(\xi^i,\xi^j,\xi^k) = \frac{(\xi^j - 1)}{(I^j - 1)} P_i(\xi^i,I^j,\xi^k) + \frac{(I^j - \xi^j)}{(I^j - 1)} P_i(\xi^i,1,\xi^k) \qquad (55)$$

Both of the necessary control functions on the surface can thus be determined from the specified boundary point distributions on the edges of the surface.

F. Iterative determination

As noted above, a second-order elliptic generation system allows either the point locations on the boundary or the coordinate line slope at the boundary to be specified, but not both. It is possible, however, to iteratively adjust the control functions in the generation system of the Poisson type discussed above until not only a specified line slope but also the spacing of the first coordinate surface off the boundary is achieved, with the point locations on the boundary specified.

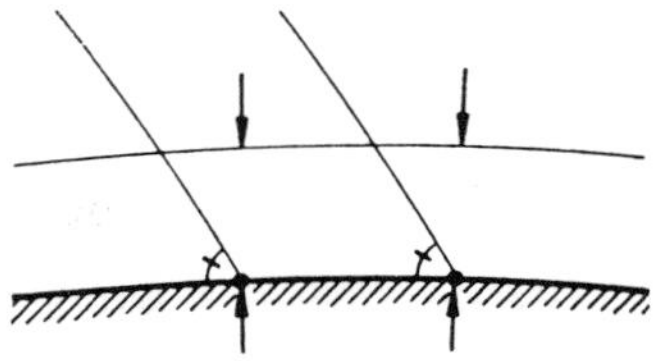

In three dimensions the specification of the coordinate line slope at the boundary requires the specification of two quantities, e.g., the direction cosines of the line with two tangents to the boundary.

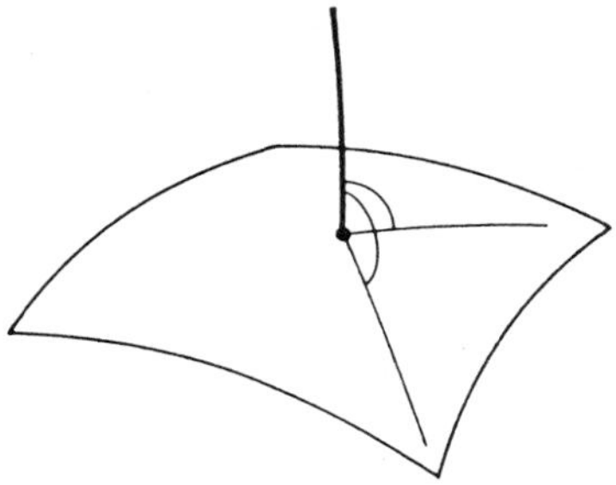

The specification of the spacing of the first coordinate surface off the boundary requires one more quantity,

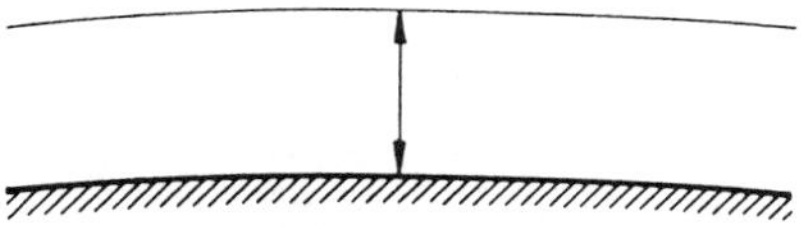

and therefore the three control functions in the system (12) are exactly sufficient to allow these three specified quantities to be achieved, while the one boundary condition allowed by the second-order system provides for the point locations on the boundary to be specified.

The capability for achieving a specified coordinate line slope at the boundary makes it possible to generate a grid which is orthogonal at the boundary, with a specified point distribution on the boundary, and also a specified spacing of the first coordinate surface off the boundary. This feature is important in the patching together of segmented grids, with slope continuity, as discussed in Chapter II, for embedded systems.

An iterative procedure can be constructed for the determination of the control functions in two dimensions as follows (cf. Ref. [25]): Consider the generation system given by Eq. (20). On a boundary segment that is a line of constant η we have $\mathbf{r}_\xi$ and $\mathbf{r}_{\xi\xi}$ known from the specified boundary point distribution

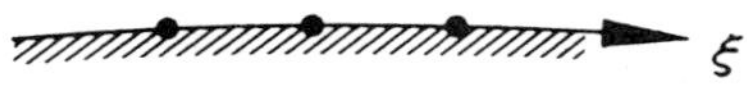

also $|\mathbf{r}_\eta|$, the spacing off this boundary, is specified

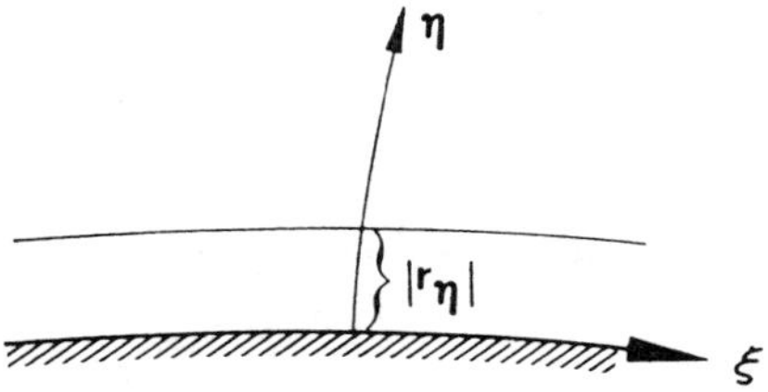

as is the condition of orthogonality at the boundary, i.e., $\mathbf{r}_\xi \cdot \mathbf{r}_\eta = 0$,

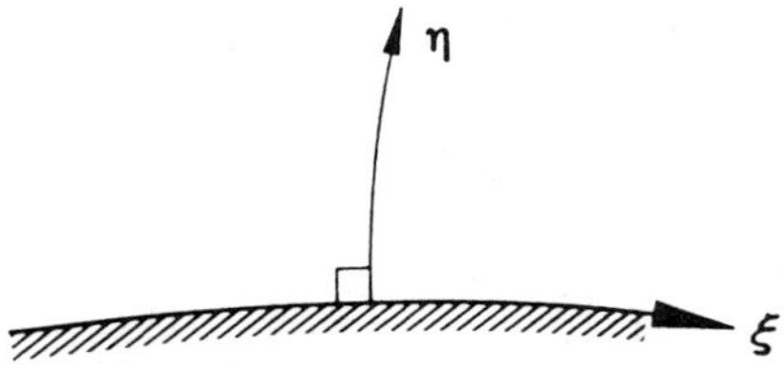

But specification of $|\mathbf{r}_\eta| = \sqrt{x_\eta^2 + y_\eta^2}$, together with the condition $\mathbf{r}_\xi \cdot \mathbf{r}_\eta = x_\xi x_\eta + y_\xi y_\eta = 0$ provides two equations for the determination of x_η and y_η in terms of the already known values of the x_ξ and y_ξ. Therefore $\mathbf{r}_\eta$ is known on the boundary.

Because of the orthogonality at the boundary, Eq.

(20) (Eq. (23) is used instead in Ref. [25]) reduces to the following equation on the boundary:

$$|\underset{\sim}{r}_\eta|^2(\underset{\sim}{r}_{\xi\xi} + P\underset{\sim}{r}_\xi) + |\underset{\sim}{r}_\xi|^2(\underset{\sim}{r}_{\eta\eta} + Q\underset{\sim}{r}_\eta) = 0$$

Dotting $\underset{\sim}{r}_\xi$ and $\underset{\sim}{r}_\eta$ into this equation, and again using the condition of orthogonality, yields the following two equations for the control functions on the boundary:

$$P = -\frac{\underset{\sim}{r}_\xi \cdot \underset{\sim}{r}_{\xi\xi}}{|\underset{\sim}{r}_\xi|^2} - \frac{\underset{\sim}{r}_\xi \cdot \underset{\sim}{r}_{\eta\eta}}{|\underset{\sim}{r}_\eta|^2} \tag{56a}$$

$$Q = -\frac{\underset{\sim}{r}_\eta \cdot \underset{\sim}{r}_{\eta\eta}}{|\underset{\sim}{r}_\eta|^2} - \frac{\underset{\sim}{r}_\eta \cdot \underset{\sim}{r}_{\xi\xi}}{|\underset{\sim}{r}_\xi|^2} \tag{56b}$$

All of the quantities in these equations are known on the boundary except $\underset{\sim}{r}_{\eta\eta}$. (On a boundary that is a line of constant ξ, the same equations for the control functions result, but now with $\underset{\sim}{r}_{\xi\xi}$ the unknown quantity.)

The iterative solution thus proceeds as follows:

(1). Assume values for the control function on the boundary.

(2). Solve Eq. (20) to generate the grid in the field.

(3). Evaluate $\underset{\sim}{r}_{\eta\eta}$ on η-line boundaries, and $\underset{\sim}{r}_{\xi\xi}$ on ξ-line boundaries, from the result of Step (2), using one-sided difference representations. Then evaluate the control functions on the boundary from Eq. (56). Evaluate the control functions in the field by interpolation from the boundary values.

Steps (2) and (3) are then repeated until convergence.

This type of iterative solution has been implemented in the GRAPE code of Ref. [24] - [26], some results of which are shown below:

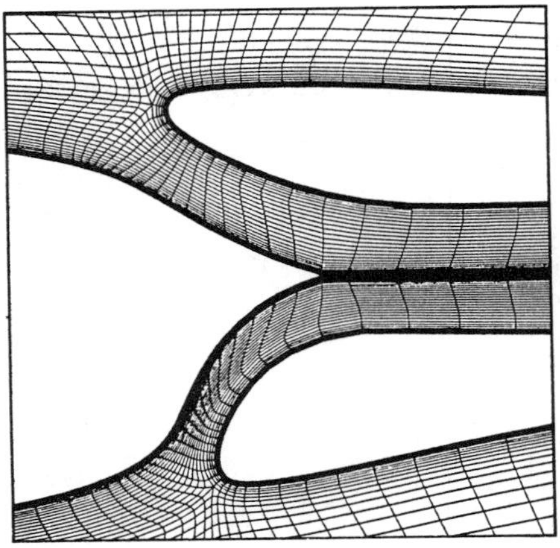

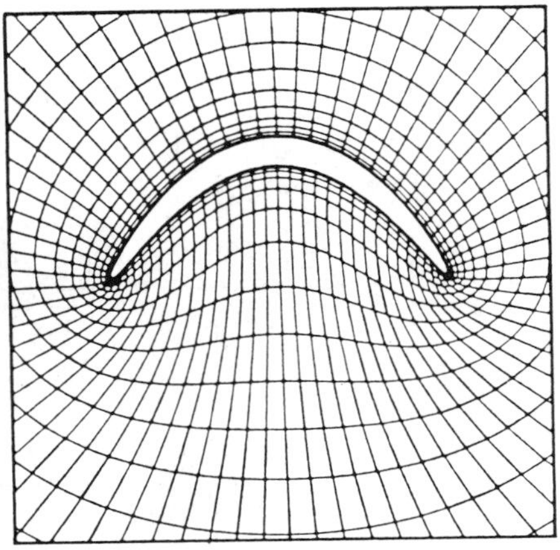

These grids are orthogonal at the boundary, and the spacing of the first coordinate surface (line in 2D) off the boundary is specified at each boundary point, the locations of which are specified.

An iterative solution procedure for the determination of the three control functions for the general three-dimensional case can be constructed as follows. Eq. (52) gives the two control functions, P_m and P_n, for a coordinate surface on which ξ^l is constant (l,m,n cyclic) for the case where the coordinate line crossing the surface is normal to the surface. Taking the projection of the generation equation (12) on the coordinate line along which ξ^l varies, we have on this same surface,

$$\sum_{i=1}^{3} \sum_{j=1}^{3} g^{ij} \underset{\sim}{r}_{\xi^l} \cdot \underset{\sim}{r}_{\xi^i \xi^j} + g^{ll} g_{ll} P_l = 0$$

since $g_{1m} = g_{1n} = 0$ on the surface. Using the relations for the metric components obtained for this situation in Section D, this equation reduces to

$$P_1 = -\frac{1}{g_{11}} \underset{\sim}{r}_{\xi^1} \cdot \underset{\sim}{r}_{\xi^1\xi^1}$$

$$- \frac{\underset{\sim}{r}_{\xi^1}}{g_{mm}g_{nn} - g_{mn}^2} \cdot (g_{nn}\, \underset{\sim}{r}_{\xi^m\xi^m} + g_{mm}\, \underset{\sim}{r}_{\xi^n\xi^n} - 2g_{mn}\, \underset{\sim}{r}_{\xi^m\xi^n}) \tag{57}$$

Since the coordinate line intersecting the surface is to be normal to the surface, we may write

$$\underset{\sim}{r}_{\xi^1} = \underset{\sim}{a}_1 = \sqrt{g_{11}}\, \frac{\underset{\sim}{a}_m \times \underset{\sim}{a}_n}{|\underset{\sim}{a}_m \times \underset{\sim}{a}_n|}$$

$$= \frac{\sqrt{g_{11}}}{\sqrt{g_{mm}g_{nn} - g_{mn}^2}}\, \underset{\sim}{a}_m \times \underset{\sim}{a}_n \tag{58}$$

since

$$|\underset{\sim}{a}_m \times \underset{\sim}{a}_n|^2 = g_{mm}g_{nn} - g_{mn}^2$$

using the identity (III-9). Eq. (57) can then be written

$$P_1 = -\frac{(g_{11})_{\xi^1}}{2g_{11}} - \frac{\sqrt{g_{11}}}{(g_{mm}g_{nn} - g_{mn}^2)^{3/2}} (\underset{\sim}{a}_m \times \underset{\sim}{a}_n)$$

$$\cdot\ [g_{nn}(\underset{\sim}{a}_m)_{\xi^m} + g_{mm}(\underset{\sim}{a}_n)_{\xi^n} - 2g_{mn}(\underset{\sim}{a}_m)_{\xi^n}] \tag{59}$$

With the spacing along the coordinate line intersecting the surface specified at the surface, we have $|\underset{\sim}{r}_{\xi^1}| = \sqrt{g_{11}}$ known on the surface. Since all the quantities subscripted m or n in Eq. (52) and (59) can be evaluated completely from the specified point distribution on the surface, we then have all quantities in these equations for the three control functions on the surface known except for $(g_{11})_{\xi^1}$ and $(\underset{\sim}{a}_1)_{\xi^1}$. These two quantities are not independent, and using Eq. (58), we have

$$(g_{11})_{\xi^1} = 2\underset{\sim}{a}_1 \cdot (\underset{\sim}{a}_1)_{\xi^1} = \frac{2\sqrt{g_{11}}}{\sqrt{g_{mm}g_{nn} - g_{mn}^2}} (\underset{\sim}{a}_m \times \underset{\sim}{a}_n) \cdot (\underset{\sim}{a}_1)_{\xi^1} \tag{60}$$

Recall also that $(\underset{\sim}{a}_1) = \underset{\sim}{r}_{\xi^1\xi^1}$.

Therefore, with the control functions in the field determined from the values on the boundary by interpolation, as discussed in the preceeding section, Eq. (52) and (59) can be applied to determine the new boundary values of the control functions in terms of the new values of $(\underset{\sim}{a}_1)_{\xi^1}$ in an iterative solution. Upon convergence, the coordinate system then will have the coordinate lines intersecting the boundary normally at fixed locations and with the specified spacing on these lines off the boundary.

A similar iterative determination of the two control functions for use in generating a coordinate system on a surface can be set up using only Eq. (52), and the analogous equation for P_n, with the first term either omitted, amounting to the assumption of vanishing curvature of the crossing line at the surface or with this term considered as specified on the surface, either directly or by interpolation from the edges of the surface. These equations are applied on the edges of the surface (the edges assumed to be on coordinate lines) to provide two equations for the two control functions, P_m and P_n, on these edges. Here the two dimensional surface coordinate system is to be orthogonal on the bounding edges of the surface with the spacing off the edges, and the point distribution thereon, specified on these edges.

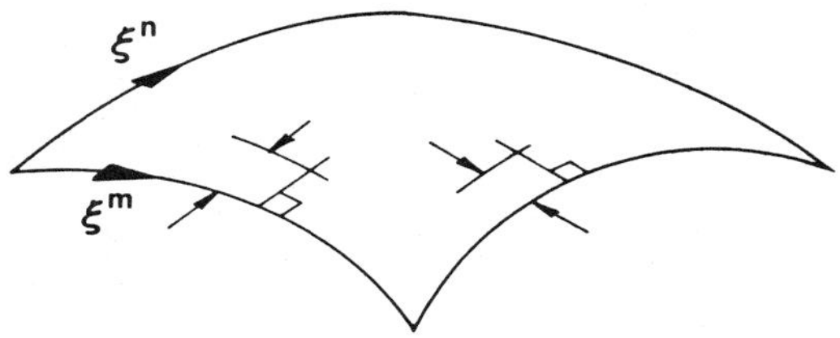

Since the coordinate system is to be orthogonal on the edges, $g_{mn}=0$ there so that the last term in the bracket in Eq. (52) vanishes. Eq. (52) then reduces to the following expression

$$P_m = -\frac{1}{g_{mm}} \underset{\sim}{a}_m \cdot (\underset{\sim}{a}_m)_{\xi^m} - \frac{1}{g_{nn}} \underset{\sim}{a}_m \cdot (\underset{\sim}{a}_n)_{\xi^n} \tag{61}$$

and the analogous equation for P_n is

$$P_n = -\frac{1}{g_{nn}} \underset{\sim}{a}_n \cdot (\underset{\sim}{a}_n)_{\xi^n} - \frac{1}{g_{mm}} \underset{\sim}{a}_n \cdot (\underset{\sim}{a}_m)_{\xi^m} \tag{62}$$

These equations can also be written

$$P_m = -\frac{(g_{mm})_{\xi^m}}{2g_{mm}} - \frac{1}{g_{nn}} \underset{\sim}{a}_m \cdot (\underset{\sim}{a}_n)_{\xi^n} \tag{63}$$

and

$$P_n = -\frac{(g_{nn})_{\xi^n}}{2g_{nn}} - \frac{1}{g_{mm}} \underset{\sim}{a}_n \cdot (\underset{\sim}{a}_m)_{\xi^m} \tag{64}$$

If the point distribution is specified along an edge on which ξ^m varies, then g_{mm}, $\underset{\sim}{a}_m$, and $(\underset{\sim}{a}_m)_{\xi^m}$ can all be calculated on this edge. The specification of the spacing from this edge to the first coordinate line off the edge determines g_{nn} on this edge. Also, because of the orthogonality on the edge, we have

$$\underset{\sim}{a}_n = \sqrt{\frac{g_{nn}}{g_{mm}}}\ \underset{\sim}{N} \times \underset{\sim}{a}_m \tag{65}$$

where $\underset{\sim}{N}$ is the unit normal to the surface. Note that $\underset{\sim}{N}$ will vary along the edge if the surface is curved. Since the surface normal, $\underset{\sim}{N}$, will be known, all quantities in Eq. (63) and (64) are known except $(g_{nn})_{\xi^n}$ and $(\underset{\sim}{a}_n)_{\xi^n}$. These two quantities are not independent and, in fact,

$$\begin{aligned}(g_{nn})_{\xi^n} &= 2\underset{\sim}{a}_n \cdot (\underset{\sim}{a}_n)_{\xi^n} \\ &= 2\sqrt{\frac{g_{nn}}{g_{mm}}}(\underset{\sim}{N} \times \underset{\sim}{a}_m) \cdot (\underset{\sim}{a}_n)_{\xi^n}\end{aligned} \tag{66}$$

On edges along which ξ^n varies, Eq. (65) and (66) are replaced by

$$\underset{\sim}{a}_m = \sqrt{\frac{g_{mm}}{g_{nn}}}\ \underset{\sim}{a}_n \times \underset{\sim}{N} \tag{67}$$

and

$$(g_{mm})_{\xi^m} = 2\sqrt{\frac{g_{mm}}{g_{nn}}}(\underset{\sim}{a}_n \times \underset{\sim}{N}) \cdot (\underset{\sim}{a}_m)_{\xi^m} \tag{68}$$

and it is $(g_{mm})_{\xi^m}$ and $(a_m)_{\xi^m}$ that are not known.

The iterative solution then proceeds as described above, with the new control functions being determined from Eq. (63) and (64), together with Eq. (65) and (66) on edges along which ξ^m varies, or with Eq. (67) and (68) on edges along which ξ^n varies.

3. Surface Grid Generation Systems

The grid generation systems discussed in the preceeding sections of this chapter have been for the generation of curvilinear coordinate systems in general three-dimensional regions. Two-dimensional forms of these systems serve to generate curvilinear coordinate systems in general two-dimensional regions in a plane. It is also of interest, however, to generate two-dimensional curvilinear coordinate systems on general curved surfaces.

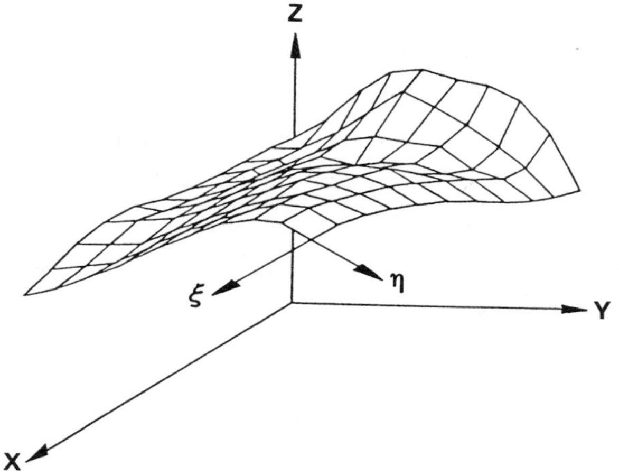

Here the surface is specified, and the problem is to generate a two-dimensional grid on that surface, the third curvilinear coordinate being constant on the surface. The configurations of the transformed region will be the same as described in Chapter II for two-dimensional systems in general, i.e., composed of contiguous rectangular blocks in a plane, with point locations and/or coordinate line slopes specified on the boundaries. These boundaries now correspond to bounding curves on the curved surface of the physical region. The problem is thus essentially the same as that discussed above for two-dimensional plane regions, except that the curvature of the surface must now enter the partial differential equations which comprise the grid generation system.

As for general regions, algebraic generation systems based on interpolation can be constructed, and such systems are discussed in Chapter VIII. The problem can also be considered as an elliptic boundary-value problem on the surface with the same general features discussed above being exhibited by the elliptic generation system.

A. Surface grid generation

An elliptic generation system for surface grids can be devised from the formulae of Gauss and Beltrami, cf. Ref. [27]. Some related, but less general, developments are noted in Ref. [9] and [5]. The starting point is the set formed by the formulae of Gauss for a surface, which for a surface, ξ^ν = constant, ($\nu = 1,2,$ or 3) are given by Eq. (34) of Appendix A:

$$\underset{\sim}{r}_{\xi^\alpha \xi^\beta} = \sum_\delta \Gamma^\delta_{\alpha\beta}\, \underset{\sim}{r}_{\xi^\delta} + b_{\alpha\beta}\, \underset{\sim}{n}^{(\nu)} \tag{69}$$

where the variation of the indices α, β and δ is over the two coordinate indices different from ν. (Greek coordinate indices are used here to set apart the coordinates generated on a surface from those generated in a three-dimensional region in general).

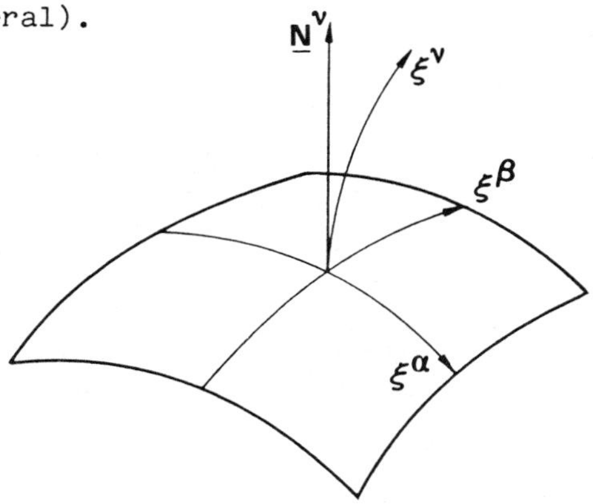

The unit normal $\underline{n}^{(\nu)}$, the coefficients $b_{\alpha\beta}$, and the surface Christoffels (T) have all been defined in Eq. (15), (20), and (33) of Appendix A, respectively. The indices α,β each assume the two values different from ν. For each ν, with (α,β,ν) taken in cyclic order, we have

$$\underline{r}_{\xi^\alpha\xi^\alpha} = \sum_\delta T^\delta_{\alpha\alpha}\,\underline{r}_{\xi^\delta} + \underline{n}^{(\nu)}b_{\alpha\alpha} \tag{70a}$$

$$\underline{r}_{\xi^\alpha\xi^\beta} = \sum_\delta T^\delta_{\alpha\beta}\,\underline{r}_{\xi^\delta} + \underline{n}^{(\nu)}b_{\alpha\beta} \tag{70b}$$

$$\underline{r}_{\xi^\beta\xi^\beta} = \sum_\delta T^\delta_{\beta\beta}\,\underline{r}_{\xi^\delta} + \underline{n}^{(\nu)}b_{\beta\beta} \tag{70c}$$

with δ assuming the two values, α,β.

A surface grid generation system that is analogous in form to that based on Poisson-type equations in a plane given earlier can be constructed by multiplying Eq. (70a,b,c), respectively, by $G_\nu g^{\alpha\alpha}$, $2G_\nu g^{\alpha\beta}$, $G_\nu g^{\beta\beta}$ and adding. This given, after some algebra,

$$L^{(\nu)}\underline{r} + G_\nu[(\Delta_2^{(\nu)}\xi^\alpha)\underline{r}_{\xi^\alpha} + (\Delta_2^{(\nu)}\xi^\beta)\underline{r}_{\xi^\beta}] = \underline{n}^{(\nu)}R^{(\nu)} \quad (71)$$

where

$$G_\nu = g_{\alpha\alpha}\, g_{\beta\beta} - g_{\alpha\beta}^2 = gg^{\nu\nu} \quad (72)$$

$$L^{(\nu)} = g_{\beta\beta}\frac{\partial^2}{\partial\xi^\alpha\xi^\alpha} - 2g_{\alpha\beta}\frac{\partial^2}{\partial\xi^\alpha\xi^\beta} + g_{\alpha\alpha}\frac{\partial^2}{\partial\xi^\beta\xi^\beta} \quad (73)$$

$$R^{(\nu)} = G_\nu(k_I^{(\nu)} + k_{II}^{(\nu)}) \quad (74)$$

$$k_I^{(\nu)} + k_{II}^{(\nu)} = g^{\alpha\alpha}b_{\alpha\alpha} + 2g^{\alpha\beta}b_{\alpha\beta} + g^{\beta\beta}b_{\beta\beta} \quad (75)$$

The quantities $k_I^{(\nu)}$ and $k_{II}^{(\nu)}$ are the local principal curvatures of the surface ξ^ν=constant. It must be noted here the $R^{(\nu)}$ as defined in (74) is based on the intrinsic values of $b_{\alpha\beta}$. That is, the $b_{\alpha\beta}$ are solely determined by the data and coordinates as available in the surface. If, however, it is desired to use Eq. (71) for generating a series of surfaces in a three-dimensional space, as in the following section, from the data of a given surface, then it is desirable to have an extrinsic form for $b_{\alpha\beta}$. To obtain the extrinsic form, we use Eq. (29) of Appendix A, i.e.,

$$\underline{r}_{\xi^\alpha\xi^\beta} = \sum_{k=1}^{3} \Gamma_{\alpha\beta}^k\, \underline{r}_{\xi^k} \quad (76)$$

Equating the right hand sides of Eq. (69) and (76), taking the dot product with $\underline{n}^{(\nu)}$ on both sides, and noting that $\underline{n}^{(\nu)}\cdot\underline{r}_{\xi^\alpha}=0$, we get

$$b_{\alpha\beta} = \Gamma_{\alpha\beta}^\nu\, \lambda^{(\nu)} \quad (77a)$$

where

$$\lambda^{(\nu)} = \underset{\sim}{n}^{(\nu)} \cdot \underset{\sim}{r}_{\xi^\nu} \tag{77b}$$

Thus

$$R^{(\nu)} = G_\nu(g^{\alpha\alpha}\Gamma^\nu_{\alpha\alpha} + 2g^{\alpha\beta}\Gamma^\nu_{\alpha\beta} + g^{\beta\beta}\Gamma^\nu_{\beta\beta})\lambda^{(\nu)} \tag{78}$$

The operator Δ_2 is called the Beltrami second-order differential operator, and in general is defined as

$$\Delta_2^{(\nu)} = \frac{1}{\sqrt{G_\nu}}\left[\frac{\partial}{\partial\xi^\alpha}\left\{\frac{1}{\sqrt{G_\nu}}\left(g_{\beta\beta}\frac{\partial}{\partial\xi^\alpha} - g_{\alpha\beta}\frac{\partial}{\partial\xi^\beta}\right)\right\} + \frac{\partial}{\partial\xi^\beta}\left\{\frac{1}{\sqrt{G_\nu}}\left(g_{\alpha\alpha}\frac{\partial}{\partial\xi^\beta} - g_{\alpha\beta}\frac{\partial}{\partial\xi^\alpha}\right)\right\}\right] \tag{79}$$

Thus

$$\Delta_2^{(\nu)}\xi^\alpha = \frac{1}{\sqrt{G_\nu}}\left[\frac{\partial}{\partial\xi^\alpha}\left(\frac{g_{\beta\beta}}{\sqrt{G_\nu}}\right) - \frac{\partial}{\partial\xi^\beta}\left(\frac{g_{\alpha\beta}}{\sqrt{G_\nu}}\right)\right] \tag{80a}$$

$$\Delta_2^{(\nu)}\xi^\beta = \frac{1}{\sqrt{G_\nu}}\left[\frac{\partial}{\partial\xi^\beta}\left(\frac{g_{\alpha\alpha}}{\sqrt{G_\nu}}\right) - \frac{\partial}{\partial\xi^\alpha}\left(\frac{g_{\alpha\beta}}{\sqrt{G_\nu}}\right)\right] \tag{80b}$$

The generation system is now formed by taking, in analogy with the system (7),

$$\Delta_2^{(\nu)}\xi^\alpha = \sum_\mu \sum_\sigma g^{\mu\sigma} P^\alpha_{\mu\sigma} \tag{81a}$$

$$\Delta_2^{(\nu)}\xi^\beta = \sum_\mu \sum_\sigma g^{\mu\sigma} P^\beta_{\mu\sigma} \tag{81b}$$

where μ and σ each assume the two values α, β in the summation. Here the $P^\delta_{\mu\sigma}$ are the symmetric control functions. Thus the equations for the generation of surface grids are (with $\underset{\sim}{r} = \underset{\sim}{i}x + \underset{\sim}{j}y + \underset{\sim}{k}z$)

$$D^{(\nu)}\underset{\sim}{r} = \underset{\sim}{n}^{(\nu)}R^{(\nu)} \tag{82}$$

where

$$D^{(\nu)} = g_{\beta\beta}\frac{\partial^2}{\partial\xi^\alpha\partial\xi^\alpha} - 2g_{\alpha\beta}\frac{\partial^2}{\partial\xi^\alpha\partial\xi^\beta} + g_{\alpha\alpha}\frac{\partial^2}{\partial\xi^\beta\partial\xi^\beta} + S\frac{\partial}{\partial\xi^\alpha} + T\frac{\partial}{\partial\xi^\beta} \tag{83}$$

$$S = g_{\beta\beta}\,P^\alpha_{\alpha\alpha} - 2g_{\alpha\beta}\,P^\alpha_{\alpha\beta} + g_{\alpha\alpha}\,P^\alpha_{\beta\beta} \tag{84}$$

$$T = g_{\beta\beta}\,P^\beta_{\alpha\alpha} - 2g_{\alpha\beta}\,P^\beta_{\alpha\beta} + g_{\alpha\alpha}\,P^\beta_{\beta\beta} \tag{85}$$

The left-hand side of Eq. (82) here corresponds exactly to that of Eq. (15) for the plane. However, here we have in place of (18) the relations

$$g_{\alpha\alpha} = x^2_{\xi^\alpha} + y^2_{\xi^\alpha} + z^2_{\xi^\alpha} \tag{86a}$$

$$g_{\beta\beta} = x^2_{\xi^\beta} + y^2_{\xi^\beta} + z^2_{\xi^\beta} \tag{86b}$$

$$g_{\alpha\beta} = x_{\xi^\alpha}x_{\xi^\beta} + y_{\xi^\alpha}y_{\xi^\beta} + z_{\xi^\alpha}z_{\xi^\beta} \tag{86c}$$

The effect of the surface curvature enters through the inhomogenous term, in particular through $R^{(\nu)}$, which is, in fact, equal to twice the product of $\sqrt{G_\nu}$ and the mean curvature of the surface. Here, as for the plane, the control functions, $P^\delta_{\alpha\beta}$, are considered to be specified. This system corresponds to the following system in the physical space, from (81),

$$\Delta_2^{(\nu)}\xi^\alpha = \frac{S}{G_\nu} \tag{87a}$$

$$\Delta_2^{(\nu)}\xi^\beta = \frac{T}{G_\nu} \tag{87b}$$

Thus the Beltrami operator on the general surface replaces the Laplacian operator in the plane. If the surface is a plane, the Beltrami operator reduces to the Laplacian.

If only the two control functions $P^\alpha_{\alpha\alpha}$ and $P^\beta_{\beta\beta}$ are included, the surface grid generation system reduces to the more practical system

$$g_{\beta\beta}(\underline{r}_{\xi^\alpha\xi^\alpha} + P\underline{r}_{\xi^\alpha}) + g_{\alpha\alpha}(\underline{r}_{\xi^\beta\xi^\beta} + Q\underline{r}_{\xi^\beta}) - 2g_{\alpha\beta}\underline{r}_{\xi^\alpha\xi^\beta}$$
$$= \underline{n}^{(\nu)}R^{(\nu)} \tag{88}$$

corresponding to the plane system given by Eq. (20). In the physical plane this system is

$$\Delta_2^{(\nu)}\xi^\alpha = \frac{g_{\beta\beta}}{G_\nu} P \tag{89a}$$

$$\Delta_2^{(\nu)}\xi^\beta = \frac{g_{\alpha\alpha}}{G_\nu} Q \tag{89b}$$

Clearly, we could also replace the system (89) with the simplier system

$$\Delta_2^{(\nu)}\xi^\alpha = P \tag{90a}$$

$$\Delta_2^{(\nu)}\xi^\beta = Q \tag{90b}$$

in analogy with the sytem (24) in the plane, to obtain the surface grid generation system

$$g_{\beta\beta}\, \underset{\sim}{r}_{\xi^\alpha\xi^\alpha} + g_{\alpha\alpha}\, \underset{\sim}{r}_{\xi^\beta\xi^\beta} - 2g_{\alpha\beta}\, \underset{\sim}{r}_{\xi^\alpha\xi^\beta} + G_\nu(P\underset{\sim}{r}_{\xi^\alpha} + Q\underset{\sim}{r}_{\xi^\beta})$$

$$= \underset{\sim}{n}^{(\nu)}\underset{\sim}{R}^{(\nu)} \tag{91}$$

which is analogous to the plane system (23).

Equation (71) is the basic equation for the generation of curvilinear coordinates in a given surface. From (74) the function $R^{(\nu)}$ depends on the principal curvatures $k_I^{(\nu)}$ and $k_{II}^{(\nu)}$. The sum $k_I^{(\nu)} + k_{II}^{(\nu)}$ is twice the mean curvature of the surface, and its value is invariant to the coordinates introduced in the given surface. If the equation of the surface in the form $x_3 = f(x_1,x_2)$ is available, then from elementary differential geometry

$$k_I^{(\nu)}+k_{II}^{(\nu)} = [(1 + q^2)r - 2pqs + (1 + p^2)t]/(1 + p^2 + q^2)^{3/2} \tag{92}$$

where

$$p = f_{x_1},\ q = f_{x_2},\ r = f_{x_1x_1},\ s = f_{x_1x_2},\ t = f_{x_2x_2}.$$

For arbitrary surfaces it is always possible to use a numerical method, e.g., the least square method, to fit an equation in the form $x_3=f(x_1,x_2)$ or $F(x_1,x_2,x_3)=0$ and to obtain the needed partial derivatives to find $k_I^{(\nu)}+k_{II}^{(\nu)}$ as a function of x_1,x_2,x_3. (Surface grids have been obtained for simply and double connected regions in a surface using the above method.)

It may be desired in some applications to generate a new coordinate system based on an already existing coordinate system in a given surface. In the formulation of this problem Eq. (71) can have the form of $R^{(\nu)}$ given in (74), (75) or (78). Let the surface on which the new grid is to be generated be specified parametrically by

$$\underset{\sim}{r} = \underset{\sim}{r}(u,v) \tag{93}$$

(For example, the parameters (u,v) might be latitude and longitude on a spherical surface.) If the specified cartesian coordinates on the surface form a finite set of discrete points, a smooth interpolation scheme is needed to recover the differentiable functions in (93). To attain the desired smoothness in the parametric representation (93), it is generally preferable to divide the given surface into a suitable number of patches such that each patch is representable by a bicubic spline with suitable blending functions. Having once established the smooth parametric functions (93), it is now possible to introduce any other desired coordinate system, say (ξ^α, ξ^β) on the surface.

For example, a surface coordinate system ξ^α, ξ^β of the configuration

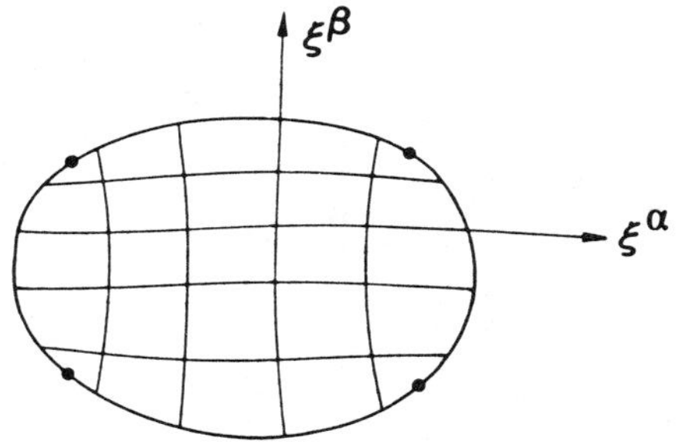

might be generated on a surface defined by the parametric coordinates (u,v) in a latitude-longitude configuration:

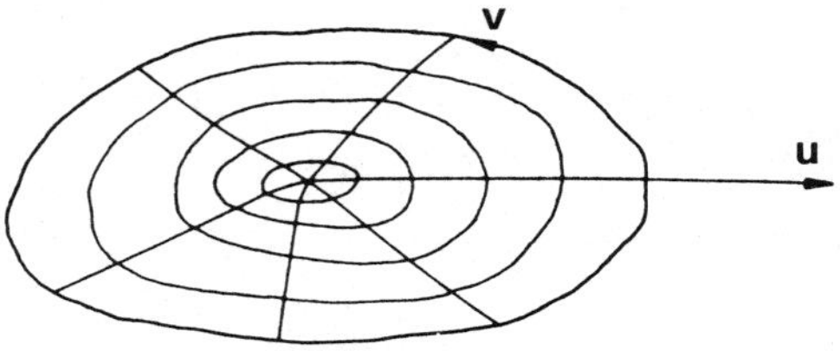

Alternatively, a surface may be defined in terms of cross-sections, in which case one of the parametric coordinates (u,v) runs around the section and the other connects the sections:

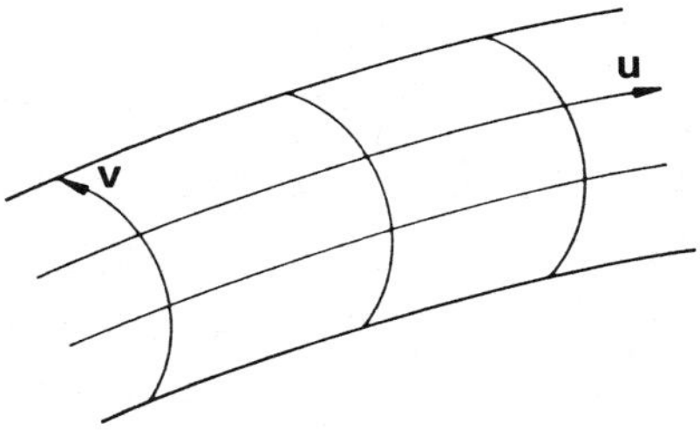

The fundamental equations for the generation of (ξ^α, ξ^β) on the surface ξ^ν = constant can be obtained from Eq. (75) in the form

$$L^{(\nu)}\underset{\sim}{r} + P'\underset{\sim}{r}_{\xi^\alpha} + Q'\underset{\sim}{r}_{\xi^\beta} = \underset{\sim}{n}^{(\nu)}R^{(\nu)} \tag{94}$$

Here we have taken

$$\Delta_2^{(\nu)}\xi^\alpha = \frac{P'}{G_\nu} \tag{95a}$$

$$\Delta_2^{(\nu)}\xi^\beta = \frac{Q'}{G_\nu} \tag{95b}$$

Now using the chain rule of differentiation, we can write $\underset{\sim}{r}_{\xi^\beta}$, $\underset{\sim}{r}_{\xi^\alpha}$, $\underset{\sim}{r}_{\xi^\alpha\xi^\beta}$, etc., in terms of $\underset{\sim}{r}_u$, $\underset{\sim}{r}_v$, $\underset{\sim}{r}_{uu}$, etc. Thus, for example,

$$\underset{\sim}{r}_{\xi\alpha} = \underset{\sim}{r}_u u_{\xi\alpha} + \underset{\sim}{r}_v v_{\xi\alpha} \tag{96}$$

$$\underset{\sim}{r}_{\xi\alpha\xi\beta} = \underset{\sim}{r}_u u_{\xi\alpha\xi\beta} + \underset{\sim}{r}_v u_{\xi\alpha\xi\beta} + (\underset{\sim}{r}_{uu} u_{\xi\beta} + \underset{\sim}{r}_{uv} v_{\xi\beta}) u_{\xi\alpha}$$
$$+ (\underset{\sim}{r}_{uv} u_{\xi\beta} + \underset{\sim}{r}_{vv} v_{\xi\beta}) v_{\xi\alpha} \tag{97}$$

$$g_{\alpha\beta} = \underset{\sim}{r}_{\xi\alpha} \cdot \underset{\sim}{r}_{\xi\beta} = (\underset{\sim}{r}_u \cdot \underset{\sim}{r}_u) u_{\xi\alpha} u_{\xi\beta} + (\underset{\sim}{r}_v \cdot \underset{\sim}{r}_v) v_{\xi\alpha} v_{\xi\beta}$$
$$+ (\underset{\sim}{r}_u \cdot \underset{\sim}{r}_v)(u_{\xi\alpha} v_{\xi\beta} + u_{\xi\beta} v_{\xi\alpha})$$
$$= \overline{g}_{\alpha\alpha} u_{\xi\alpha} u_{\xi\beta} + \overline{g}_{\beta\beta} v_{\xi\alpha} v_{\xi\alpha} + \overline{g}_{\alpha\beta}(u_{\xi\alpha} v_{\xi\beta} + u_{\xi\beta} v_{\xi\alpha}) \tag{98}$$

with the $\overline{g}$ quantities as defined below. Substituting these derivatives in (94), and also in the expressions for $\underset{\sim}{n}^{(\nu)}$ and $\underset{\sim}{R}^{(\nu)}$, we get

$$\underset{\sim}{r}_u(L^{(\nu)}u + P'u_{\xi\alpha} + Q'u_{\xi\beta}) + \underset{\sim}{r}_v(L^{(\nu)}v + P'v_{\xi\alpha} + Q'v_{\xi\beta})$$
$$+ J_\nu^2\, \overline{L}^{(\nu)}\underset{\sim}{r} = \overline{\underset{\sim}{n}}^{(\nu)}\, \overline{R}^{(\nu)}\, J_\nu^2 \tag{99}$$

where

$$\overline{L}^{(\nu)} = \overline{g}_{\beta\beta}\frac{\partial^2}{\partial u \partial u} - 2\overline{g}_{\alpha\beta}\frac{\partial^2}{\partial u \partial v} + \overline{g}_{\alpha\alpha}\frac{\partial^2}{\partial v \partial v} \tag{100}$$

$$J_\nu = u_{\xi\alpha} v_{\xi\beta} - u_{\xi\beta} v_{\xi\alpha} \tag{101}$$

$$\overline{G}_\nu = \overline{g}_{\beta\beta}\overline{g}_{\alpha\alpha} - \overline{g}_{\alpha\beta}^2 \tag{102}$$

$$\bar{g}_{\beta\beta} = \underset{\sim}{r}_v \cdot \underset{\sim}{r}_v, \quad \bar{g}_{\alpha\beta} = \underset{\sim}{r}_u \cdot \underset{\sim}{r}_v, \quad \bar{g}_{\alpha\alpha} = \underset{\sim}{r}_u \cdot \underset{\sim}{r}_u \tag{103}$$

To isolate the differential equations for u and v as dependent variables from (99), we take the dot product of Eq. (99) with $\underset{\sim}{r}_u$, and then with $\underset{\sim}{r}_v$, and use the conditions

$$\underset{\sim}{r}_u \cdot \underset{\sim}{n}^{(\nu)} = 0, \quad \underset{\sim}{r}_v \cdot \underset{\sim}{n}^{(\nu)} = 0$$

Writing

$$P = P'/\bar{J}_\nu^2, \quad Q = Q'/\bar{J}_\nu^2, \quad \bar{J}_\nu = \sqrt{\bar{G}_\nu}$$

the required equations are

$$a u_{\xi^\alpha \xi^\alpha} - 2b u_{\xi^\beta \xi^\alpha} + c u_{\xi^\beta \xi^\beta} + J_\nu^2 (P u_{\xi^\alpha} + Q u_{\xi^\beta}) = J_\nu^2 \Delta_2 u \tag{104a}$$

$$a v_{\xi^\alpha \xi^\alpha} - 2b v_{\xi^\beta \xi^\alpha} + c v_{\xi^\beta \xi^\beta} + J_\nu^2 (P v_{\xi^\alpha} + Q v_{\xi^\beta}) = J_\nu^2 \Delta_2 v \tag{104b}$$

where

$$a = (\bar{g}_{\beta\beta} v_{\xi^\beta}^2 + 2\bar{g}_{\alpha\beta} u_{\xi^\beta} v_{\xi^\beta} + \bar{g}_{\alpha\alpha} u_{\xi^\beta}^2)/\bar{J}_\nu^2 \tag{105a}$$

$$b = [\bar{g}_{\beta\beta} v_{\xi^\beta} v_{\xi^\alpha} + \bar{g}_{\alpha\beta}(u_{\xi^\beta} v_{\xi^\alpha} + u_{\xi^\beta} v_{\xi^\alpha}) + \bar{g}_{\alpha\alpha} u_{\xi^\beta} u_{\xi^\alpha}]/\bar{J}_\nu^2 \tag{105b}$$

$$c = (\bar{g}_{\beta\beta} u_{\xi^\alpha}^2 + 2\bar{g}_{\alpha\beta} u_{\xi^\alpha} v_{\xi^\alpha} + \bar{g}_{\alpha\alpha} v_{\xi^\alpha}^2)/\bar{J}_\nu^2 \tag{105c}$$

$$\Delta_2 u = [\frac{\partial}{\partial u}(\frac{\bar{g}_{\beta\beta}}{\bar{J}_\nu}) - \frac{\partial}{\partial v}(\frac{\bar{g}_{\alpha\beta}}{\bar{J}_\nu})]/\bar{J}_\nu \tag{106a}$$

$$\Delta_2 v = [\frac{\partial}{\partial v}(\frac{\bar{g}_{\alpha\alpha}}{\bar{J}_\nu}) - \frac{\partial}{\partial u})(\frac{\bar{g}_{\alpha\beta}}{\bar{J}_\nu})]/\bar{J}_\nu \tag{106b}$$

Note that the metric quantities with an overbar relate to the surface definition in terms of the parametric coordinates and therefore can be calculated directly from the surface specification, Eq. (92).

Clearly we could redefine the control functions so that (104) is replaced by the following system, which is analogous in form to the plane system (20):

$$a(u_{\xi^\alpha\xi^\alpha} + Pu_{\xi^\alpha}) + c(u_{\xi^\beta\xi^\beta} + Qu_{\xi^\beta}) - 2bu_{\xi^\alpha\xi^\beta} = J_\nu^2\,\Delta_2 u \quad (107a)$$

$$a(v_{\xi^\alpha\xi^\alpha} + Pv_{\xi^\alpha}) + c(v_{\xi^\beta\xi^\beta} + Qv_{\xi^\beta}) - 2bv_{\xi^\alpha\xi^\beta} = J_\nu^2\,\Delta_2 v \quad (107b)$$

B. Three-dimensional grids

As mentioned earlier, the system of Eq. (71), or Eq. (82), is also capable of generating three-dimensional grids. This capability in the set of equations is incorporated through $\bar{R}^{(\nu)}$ as defined in (78).

The strategy of the method is to generate a series of surfaces on each of which two curvilinear coordinates vary while the third remains fixed. The variation along the third coordinate is specified as a surface derivative condition, which in turn depends on the given boundary data.

A study of Eq. (82) - (84) shows immediately that for the solution of Eqs. (82) we need to specify the values of $\underset{\sim}{r}$ and $\underset{\sim}{r}_{\xi^\nu}$ on certain curves ξ^ν=constant. To fix ideas, let us consider the problem of coordinate generation between two given surfaces $\xi^\beta = \xi^\beta_B$ and $\xi^\beta = \xi^\beta_\infty$ as shown below:

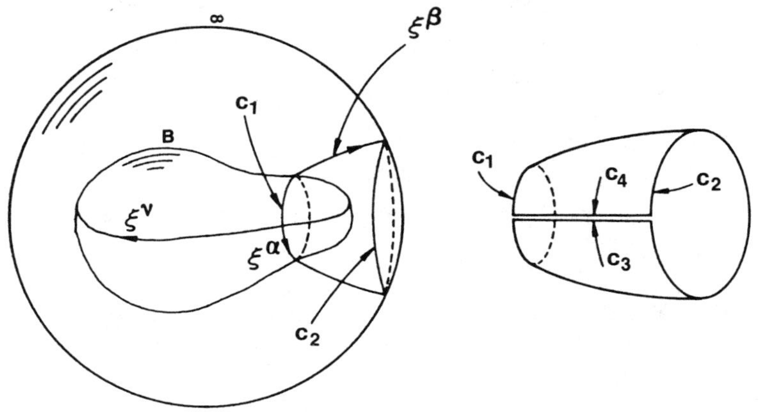

The coordinates on these surfaces are ξ^{α} and ξ^{ν}. To start solving these equations we need the values of $\underset{\sim}{r}$ and $\underset{\sim}{r}_{\xi^{\nu}}$ on the surfaces $\xi^{\beta} = \xi^{\beta}_{B}$ and $\xi^{\beta} = \xi^{\beta}_{\infty}$. These values of $\underset{\sim}{r}$ are the input conditions for the solution of Eqs. (82), and are either prescribed analytically or numerically. On the other hand, the values of $\underset{\sim}{r}_{\xi^{\nu}}$ at B and ∞ are available easily, based on the values of $\underset{\sim}{r}$, simply by numerical differentiation. The values of $r_{\xi^{\nu}}$ in the field for each surface to be generated are then obtained by interpolation between the available values of $(\underset{\sim}{r}_{\xi^{\nu}})_{B}$ and $(\underset{\sim}{r}_{\xi^{\nu}})_{\infty}$. A simple formulae which has been used with success is

$$\underset{\sim}{r}_{\xi^{\nu}} = f_1(\xi^{\beta})(\underset{\sim}{r}_{\xi^{\nu}})_{B} + f_2(\xi^{\beta})(\underset{\sim}{r}_{\xi^{\nu}})_{\infty} \tag{108}$$

where

$$f_1(\xi^{\beta}_{B}) = 1, \quad f_2(\xi^{\beta}_{B}) = 0$$

$$f_1(\xi^{\beta}_{\infty}) = 0, \quad f_2(\xi^{\beta}_{\infty}) = 1$$

4. Implementation

The setup of the transformed region configuration is done as described in Chapter II. This includes the placing of the cartesian coordinates of the selected points on the boundary of the physical region into $\underline{r}_{ijk}$ for each block and the setting of the interface correspondence between points on the surrounding layer for each block and points inside the same, or another, block via input to an image-point array as described in Section 6 of Chapter II.

A. Difference equations

Implementation of an elliptic generation system then is accomplished by devising an algorithm for the numerical solution of the partial differential equations comprising the generation system. Recall that the use of the surrounding layer for each block, as described in Section 6 of Chapter II, allows the same difference representations that are used in the interior to be used on the interfaces. The usual approach is to replace all derivatives in the partial differential equations by second-order central difference expressions, as given in Chapter IV, and then to solve the resulting system of algebraic difference equations by iteration. As noted above, most generation systems of interest are quasilinear, so that the difference equations are nonlinear.

A number of different algorithms have been used for the soltuion of these equations, including point and line SOR, ADI, and multi-grid iteration (cf. Ref. [1] and [5]). For general configurations, point SOR is certainly the most convenient to code and has been found to be rapid and dependable, using over-relaxation, for a wide variety of configurations. The optimum acceleration parameters and the convergence rate decrease as the control functions increase in magnitude. Some consideration has been given to the calculation of a field of locally-optimum acceleration parameters (cf. Ref. [1]), but the predicted values generally tend to be too high, and the desired increases in convergence rate were not obtained.

Since the system is nonlinear, convergence depends on the initial guess in iterative solutions. The algebraic grid generation procedures discussed in Chapter VIII can serve to generate this initial guess, and transfinite interpolation generally produces a more reliable initial guess than does unidirectional interpolation because of the reduced skewness in the former. In fact with strong line concentration, convergence may not be possible from an initial guess constructed from unidirectional interpolation, while rapid convergence occurs from an initial guess formed with transfinite interpolation. With the slab and slit configurations, the interpolation must be unidirectional between the closest facing boundary segments as illustrated below:

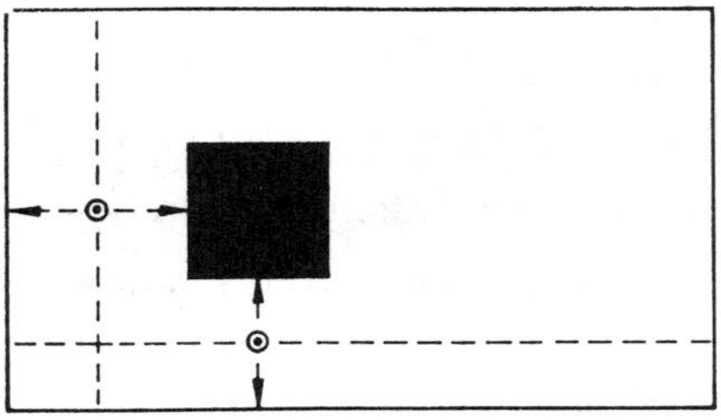

In a block structure, however, the slab/slit configuration can be avoided so that transfinite interpolation can be used.

Since the coordinate lines tend to concentrate near a convex boundary, very sharp convex corners may cause problems with the convergence of iterative solutions of the generation equations. These equations are nonlinear, and therefore convergence of an iterative procedure requires that the initial guess be within some neighborhood of the solution. With control functions designed to cause attraction to the boundary, it is possible for the coordinate lines to overlap a very sharp convex corner during the course of the iteration, even though a solution with no overlap exists:

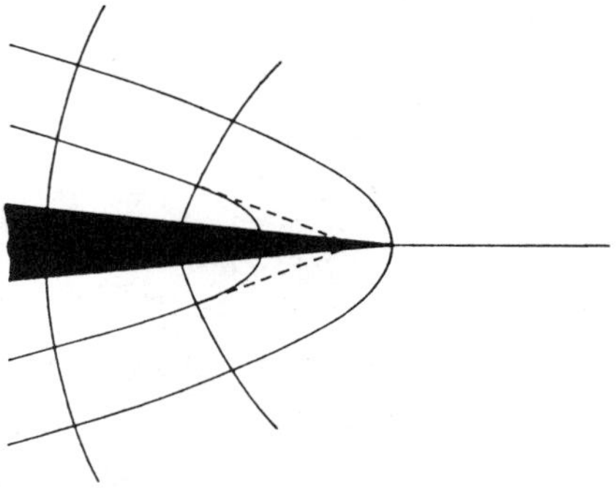

This problem may be handled by first converging the solution with the coordinate lines artificially locked off the corner. Thus, if newly calculated values of the cartesian coordinates at a point during the iteration would cause this point to move farther from its present location than the distance to the adjacent point on the curvilinear coordinate line running to the corner, then these new values are replaced by the average of the coordinates of the old point and the adjacent point. After convergence, this lock is removed and final convergence to the solution is obtained. Note that this problem does not arise when the curvilinear coordinate line emanating from the corner is the same as that on the boundary, as in the C-type configuration on p. 30 since then the lines do not wrap around the corner.

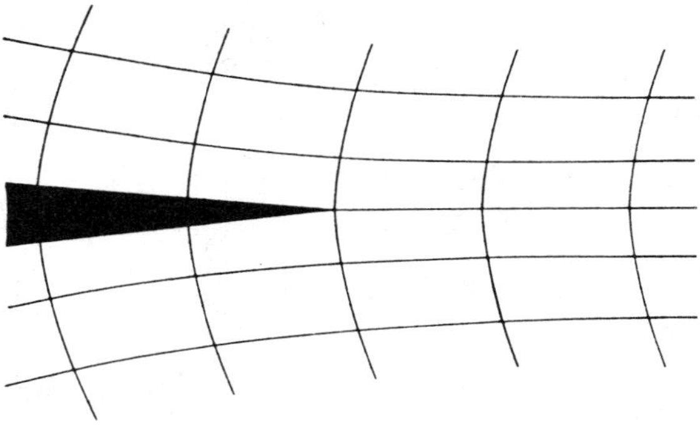

With very large cell aspect ratio, e.g., for $g_{11} \gg g_{22}$, the generation equation is dominated by the term containing the second derivative along the curvilinear coordinate line on which the shorter arc length lies. This causes the cartesian coordinates to tend strongly toward averages of adjacent points on this line during the course of the iteration. Therefore, when strong control functions are used to attract coordinate lines to the boundary in a C-type configuration,

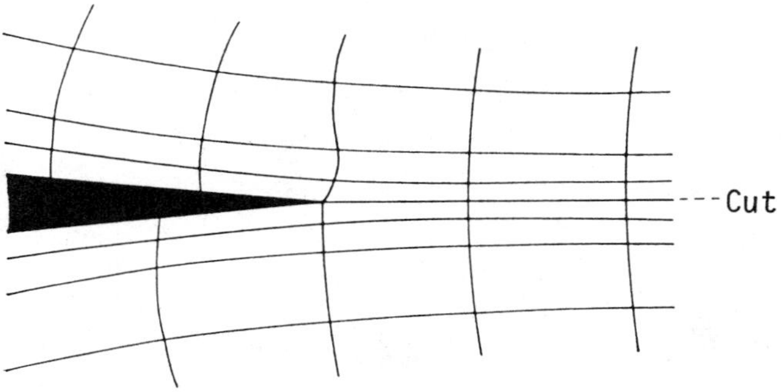

the points on the cut are very slow to move from the initial guess during the iteration. Convergence in such a case is very slow, and it is expedient to artificially fix the points on the cut as if it were a boundary. This will cause the coordinate lines crossing the cut to have discontinuous slopes at the cut, but since the spacing along these crossing lines is very small, the error thus incurred in difference solutions on the coordinate system is small.

B. Control functions

Several types of control functions have been discussed in Section 2 which serve to control the coordinate line spacing and orientation in the field. Most of these functions are set before the solution algorithm begins, either directly through input or by calculation from the boundary point distributions that have been input.

For the attraction to other coordinate lines/points, described in Section 2A, it is necessary to input the indices of the lines/points, i.e., the ξ_i and η_i of Eq. (30), to which attraction is to be made. In the case of attraction to lines, the line is identified by the single index which is constant thereon, while a point requires the specification of two indices (in 2D, with analogous generalization to 3D). The attraction amplitude and decay factor in Eq. (30) must also be input for each line/point. The control functions are then calculated at each point in the field (ξ,η) by performing the summations in Eq. (30), those summations being over all the attraction lines/points that have been input. As noted in Section 2A, these summations must also extend over some lines/points on other sheets across branch cuts in some cases.

This type of control function was used in the original TOMCAT code (cf. Ref. [1]), but is not really suitable as a primary means of control function definition because it only provides control--not control to achieve a specified spacing distribution, since the appropriate values of the various parameters involved can only be determined by experimentation. This form does, however, still serve as a useful addition to other types of control, in that it allows particular ad hoc concentrations or adjustments of line spacing and orientation to be made. This can be particularly useful near the special points discussed in Chapter II where the grid line configuration departs locally from the usual simple coordinate line intersections.

The attraction to lines/points in space, implemented through Eq. (31), requires input similar to that just described, except that here the location of the attraction lines must be defined in the physical region by inputing a set of points along the line sufficient for its definition in discrete form. For attraction to a point a unit vector must also be input with each point. Again, attraction amplitude and decay factors must be input.

More important is the evaluation of the control functions from the boundary point distribution that has been input, as described in Section 2E. With the point distribution specified on a boundary line, the control functions on this line can be evaluated from Eq. (45) - (47). Here the derviatives in Eq.(46) are best calculated from Eq. (36) and (37), using second-order, central difference expressions along the line:

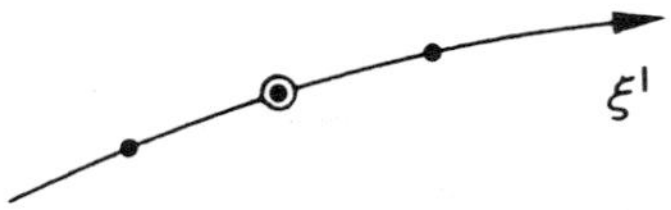

(Recall that $\sqrt{g_{11}} = |\underline{r}_{\xi^1}|$.) The curvature terms given by Eq. (47), if included, must either be input at each point on the line, or, as is more likely, must be interpolated from values on the ends of the line. In this latter case, the ξ^m and ξ^n derivatives are off the line and are evaluated from the point distribution on the other coordinate lines intersecting the line of interest at its ends, using first-order one-sided difference expressions along these intersecting lines:

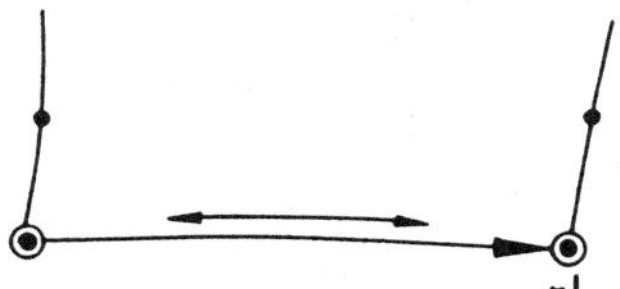

One-dimensional linear interpolation in ξ^1 then serves to define the curvature term quantities at each point on the line of interest. Recall that it is the entire curvature term, rather than the individual vectors involved, that should be interpolated.

This evaluation determines the P_1 control function on a boundary line on which ξ^1 varies. Such an evaluation can be made on each edge of a surface, corresponding to one face of a block in three dimensions (cf. Section 6 of Chapter II). If it is desired to generate a two-dimensional grid on this surface, control functions on the surface can be evaluated by interpolation from the function values on the edges, using linear interpolation between the two edges on which ξ^i is constant to evaluate P_j, and between the two edges on which ξ^j is constant to evaluate P_i (cf. the figure on p. 227). With the control functions thus defined on the surface, a two-dimensional grid on the surface can now be generated using a surface grid generation system described as in Section 3. If the surface is a portion of the physical boundary, then a parametric definition of the surface will need to be input, so that the system defined by Eq. (107) can be applied. If, however, the surface is simply an interface between blocks, then its position is arbitrary and either a plane two-dimensional generation system, such as

Eq. (20), can be used, or surface curvature values could be input at each point on the surface and the surface system Eq. (82) used. The former is the more likely choice.

With the grid points on all the block faces defined, either by surface generation systems or by direct input, two control functions on each face can be evaluated from the surface point distribution using Eq. (52). Here the m and n derivatives are along coordinate lines on the surface and thus can be represented by second-order central differences between points on the surface:

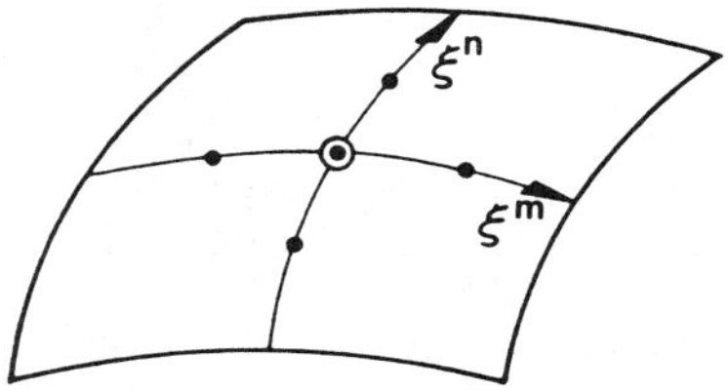

(Recall that $(\underset{\sim}{a}_m)_{\xi^m}$ can be expanded to $\underset{\sim}{r}_{\xi^m\xi^m}$ for evaluation.) The l-derivatives are off the surface and must either be specified by input at each point on the surface, or, as is more likely, must be interpolated from values evaluated along the coordinate lines intersecting the surface at its edges using first-order, one-sided difference expressions. The interpolation would here properly be two-dimensional transfinite interpolation discussed in Chapter VIII.

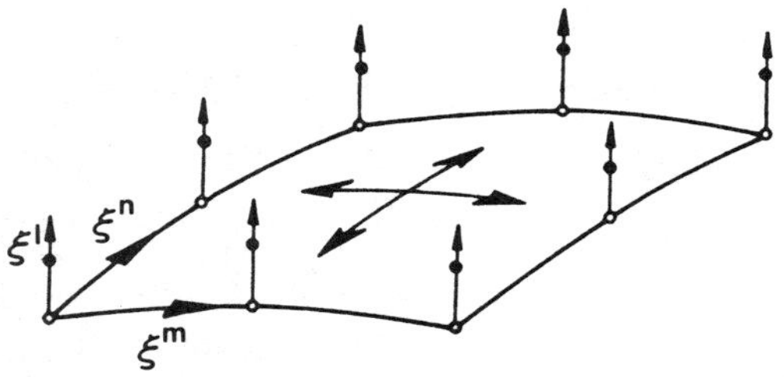

This then serves to determine the two control functions P_j and P_k, on a surface on which ξ^i is constant (cf. the figure on p. 226), so that each control function will be defined on four faces of the block. Transfinite interpolation among these four faces then determines this control function in the interior of the block (cf. p. 227).

Another possibility is to evaluate the radius of curvature, ρ, of the surface and to replace the curvature terms in Eq. (45) with $|\underline{a}_1|\rho$ (cf. Exercise 9). Here the radius of curvature should be interpolated unidirectionally between facing surfaces, and the same two-directional transfinite interpolation used for the first term of the control function should be used for the spacing $|\underline{a}_1|$.

Still another approach is to solve the three generation system equations for the three control functions at each point using an algebraic grid, but with the off-diagonal metric elements set to zero. This will produce a grid which will have a greater degree of smoothness and orthogonality than the algebraic grid and yet has the same general spacing distribution. Here the result of the Computer Exercise 6 in Appendix C must be considered since the algebraic grid influences the spacing distribution.

In generation systems that iteratively adjust the control functions during the course of the solution of the difference equations (Section 2F) to achieve a specified spacing and angle of intersection, e.g., orthogonality, at the boundary, this spacing and intersection angle are input for each boundary point and it is, of course, not necessary to calculate the control functions beforehand. Several references to discussion of such systems are given in Ref. [5]. The GRAPE code is based on this approach, cf. the users manual Ref. [24].

C. Surface generation systems

A boundary surface in the physical region will typically be input by giving the cartesian coordinates of points on a series of cross-sections, or other set of space curves:

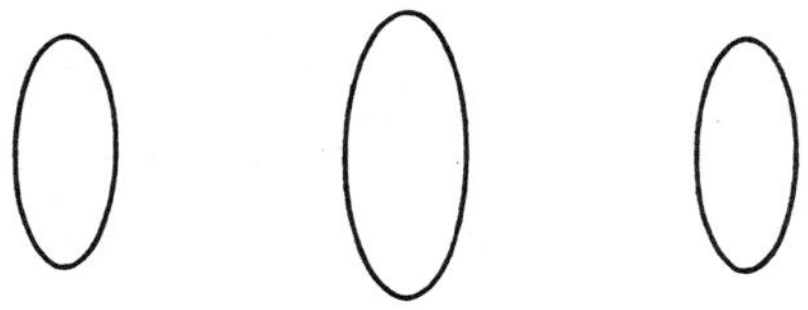

These input points may then be splined to provide a functional definition of these curves. These curves are then parameterized in terms of normalized arc length thereon, i.e., so that this normalized parameter varies over the same range on each curve.

This normalized arc length then provides one parametric coordinate on the surface. The other coordinate is defined by connecting points at the same value of the first coordinate on the successive curves, again using a spline fit:

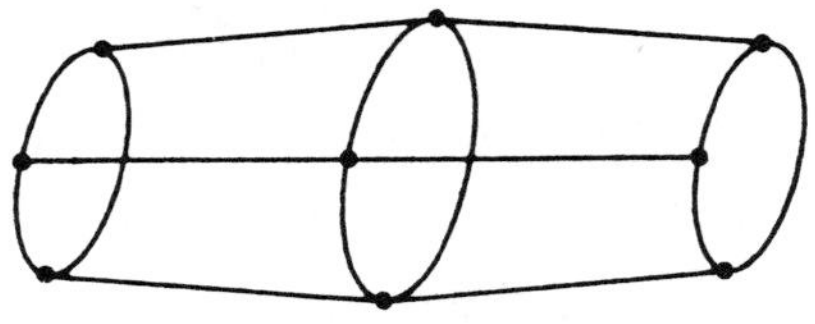

This second coordinate is then also expressed in terms of normalized arc length

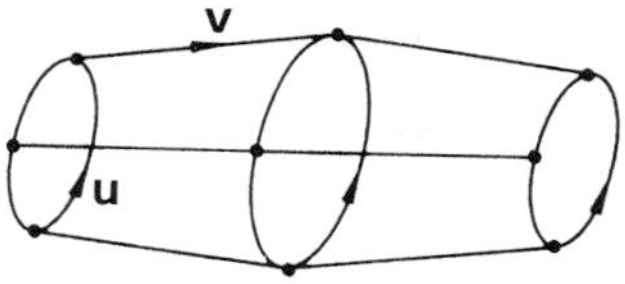

(On a sphere these two parametric surface coordinates could correspond to longitude and latitude, the latter arising from the cross-sections and the former from the connecting thereof.)

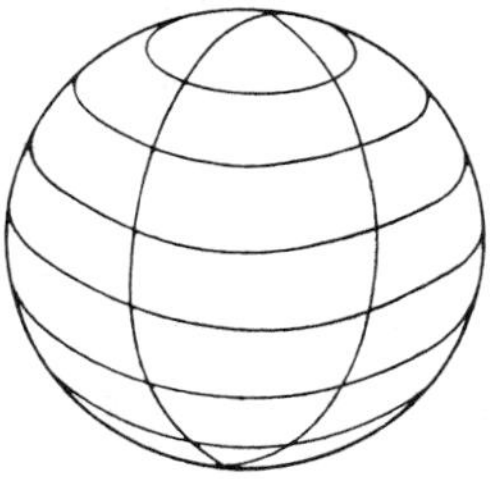

There are other techniques of surface definition and parameterization, cf. especially works on computer-aided design, but the above decription is representative. The end result of this stage in any case is $\underset{\sim}{r}(u,v)$, i.e., the cartesian coordinates on the surface in terms of two surface parametric coordinates.

The two parametric coordinates (u,v) used to define the surface can also be adopted as the curvilinear coordinates defining the surface grid. However it is more likely that these coordinates were selected for convenience of input definition of the surface than for the definition of an appropriate grid thereon. This is particularly true when two such intersection surfaces, e.g., a wing-body, are input, each with its own set of parametric coordinates. Therefore, the surface grid generation system defined by Eq. (107) or (104) is used to generate a new surface coordinate system (ξ^α,ξ^β) by generating values of the parametric coordinates (u,v) as functions of the curvilinear coordinates (ξ^α,ξ^β), analogous to the plane generation systems which generate values of the cartesian coordinates as functions of the curvilinear coordinates. In fact, as noted above, the surface generation system degenerates to the plane system when the surface curvature vanishes.

With $\underset{\sim}{r}(u,v)$ now available, as described above, the metric elements with overbars can be calculated from the definitions in Eq. (103), using second-order central differences for all derivatives as in the plane case. The quantities $\Delta_2 u$ and $\Delta_2 v$ are then calculated in the same manner from Eq. (106). Also the control fucntions are evaluated from the same relations given above for the plane case. All derivatives in the system (107) or (104) are represented by second-order central difference epxressions, and the resulting nonlinear difference equations are solved as in the plane case.

Exercises

1. Demonstrate the validity of Eq. (4) - (6).

2. For plane polar coordinates (r,θ) defined as

$$x = r\cos\theta,\ y = r\sin\theta$$

show that the curvilinear coordinates

$$\xi = \theta,\ \eta = \ln r$$

are solutions of the Laplace equations $\nabla^2\xi = 0$, $\nabla^2\eta = 0$.

3. Show that the one-dimensional control function in Eq. (13) that is equivalent to the use of a subsequent exponential stretching transformation by the function given by Eq. (VIII-26) is $P = -\alpha/I$. Hint: $x/L = \phi(\xi/I)$

4. Show that the one-dimensional control function in Eq. (13) that corresponds to a hyperbolic tangent stretching transformation by the function given by Eq. (VIII-32) is

$$P(\xi) = 2\,\frac{\delta}{I}\tanh[\delta(\frac{\xi}{I} - \frac{1}{2})] + 2\frac{(1 - A)u_\xi}{A + (1 - A)u}$$

where u is given by Eq. (VIII-33) and

$$u_\xi = \frac{\frac{\delta}{I}}{2\tanh\frac{\delta}{2}}\operatorname{sech}^2[\delta(\frac{\xi}{I} - \frac{1}{2})]$$

5. Show that the one-dimensional control function for the

generation system given by Eq. (23) corresponding to a distribution $x(\xi)$ is

$$P(\xi) = - x_{\xi\xi}/x_{\xi}^{3}$$

Note that this control function will be considerably larger than that for Eq. (12) because of the higher inverse power of x_{ξ}.

6. Show that a solution of Eq. (20), with $P = P(\xi)$ and $Q = Q(\eta)$, for a rectangular region with $0 \leq x \leq X$ and $0 \leq y \leq Y$, $0 \leq \xi \leq I$ and $0 \leq \eta \leq J$, is given by

$$\underset{\sim}{r} = \underset{\sim}{i}\, X \frac{F(\xi)}{F(I)} + \underset{\sim}{j}\, Y \frac{G(\eta)}{G(J)}$$

where

$$F(\xi) = \int_{0}^{\xi} \exp\left[-\int_{0}^{\xi''} P(\xi'')d\xi''\right]d\xi'$$

$$G(\eta) = \int_{0}^{\eta} \exp\left[-\int_{0}^{\eta''} Q(\eta'')d\eta''\right]d\eta'$$

7. Show that a solution of Eq. (20), with $P = P(\xi)$ and $Q = Q(\eta)$, for an annular region between two concentric circles of radius r_1 and r_2 is given, with $0 \leq \xi \leq I$ and $0 \leq \eta \leq J$, by

$$\underset{\sim}{r} = R(\eta)[\underset{\sim}{i} \cos\theta(\xi) + \underset{\sim}{j} \sin\theta(\xi)]$$

where

$$R(\eta) = r_1 \left(\frac{r_2}{r_1}\right)^{\left[\frac{G(\eta)}{G(J)}\right]}$$

$$\theta(\xi) = 2\pi \frac{F(\xi)}{F(I)}$$

with $F(\xi)$ and $G(\eta)$ given in the preceding exercise. Show also that for $P = p_{\xi\xi}/p_\xi$ and $Q = q_{\eta\eta}/q_\eta$, $R(\eta)$ and $\theta(\xi)$ become

$$R(\eta) = r_1 \left(\frac{r_2}{r_1}\right)^{\left[\frac{q(\eta) - q(0)}{q(J) - q(0)}\right]}$$

$$\theta(\xi) = 2\pi\left[\frac{p(\xi) - p(0)}{p(I) - p(0)}\right]$$

8. From the result of the preceding exercise show that the control function $Q(\eta)$ required to produce a specified radial distribution $r(\eta)$ is given by

$$Q(\eta) = -\frac{r_{\eta\eta}}{r_\eta} + \frac{r_\eta}{r}$$

9. Show that the first term in the control function $Q(\eta)$ given in the preceding exercise arises from the first term in Eq. (45), and that the second term arises from curvature term in (45).

10. Consider the generating system (23) for plane curvilinear coordinates. Let the control functions P and Q be defined as follows:

$$P = 0$$

$$Q = \frac{-g_{11}}{g} \frac{[2 + (\eta - \eta_i) \ln k] \ln k}{1 + (\eta - \eta_i) \ln k}$$

where $k > 0$ is a constant. Let it be desired to solve Eq. (23) for the generation of coordinates in the region of a circular annulus with $\eta = \eta_i (r = 1)$ as the inner circle and $\eta = \eta_o$ $(r = R)$ as the outer circle. Considering the clockwise traverse in the ξ-direction as positive, set

$$x = f(\eta)\cos a(\xi - \xi_o)$$

$$y = -f(\eta)\sin a(\xi - \xi_o)$$

where

$$0 \leq \xi - \xi_o \leq 2\pi/a$$

in Eq. (23), and show that

$$f(\eta) = \exp[A(\eta - \eta_i)k^{\eta}]$$

where

$$A = \frac{\ln R}{(\eta_o - \eta_i)k^{\eta_o}}$$

11. Show that the control function P in Exercise 4 has the following values at the boundaries:

$$P(0) = -\frac{\delta}{I}\left[\frac{1 - \frac{1}{A}\operatorname{sech}^2\left(\frac{\delta}{2}\right)}{\tanh\left(\frac{\delta}{2}\right)}\right]$$

$$P(I) = \frac{\delta}{I}\left[\frac{1 - A\operatorname{sech}^2\left(\frac{\delta}{2}\right)}{\tanh\left(\frac{\delta}{2}\right)}\right]$$

Note that an iterative procedure could be set up in the manner of Section 2F in which δ and A are determined from P(0) and P(I) and then these are used in the $P(\xi)$ of Exercise 4 to define the control function in the field, rather than interpolating from the boundary values.

12. Show that the Beltrami operator reduces to the Laplacian for a plane surface.

13. Verify Eq. (71).

14. Verify Eq. (77a).

15. Consider a sphere of unit radius in which it is desired to introduce a coordinate system (ζ,ξ) in such a way that (i) is orthogonal, and (ii) the resulting metric coefficients g_{33} and g_{11} are equal. (Such systems are known to be isothermic.)

    (a) Verify by inspection that for isothermic coordinates Eq. (80) is identically satisfied.

    (b) To obtain the isothermic coordinates on a sphere set

$$x = \psi(\zeta), \quad y = f(\zeta)\cos\xi, \quad z = f(\zeta)\sin\xi.$$

and show that

$$f(\zeta) = \frac{2e^{\zeta}}{1 + e^{2\zeta}}, \quad \psi(\zeta) = \frac{1 - e^{2\zeta}}{1 + e^{2\zeta}}$$

(c) Show that the relation between the standard longitude and latitude surface coordinates ϕ and θ where $0 \leq$

$\theta < 2\pi$ and $0 < \theta < \pi$, is

$$\xi = \phi, \ \zeta = \ln \tan \frac{\theta}{2}.$$

16. Using Eq. (15), (20) and (21) of Appendix A, show that the sum of the principal curvatures, $k_I^{(2)} + k_{II}^{(2)}$, of a prolate ellipsoid defined as

$$x = a \cos\zeta, \ y = b \sin\zeta \cos\xi, \ z = b \sin\zeta \sin\xi$$

is

$$k_I^{(2)} + k_{II}^{(2)} = \frac{-a\{a^2 \sin^2\zeta + b^2(1 + \cos^2\zeta)\}}{b(a^2 \sin^2\zeta + b^2 \cos^2\zeta)^{3/2}}$$

17. Verify the correspondence between Eq. (90) and (91).

18. Verify Eq. (92).

19. Let (ξ,η) be the surface coordinates in the surface on which ζ = constant. Then as shown in Appendix A, Eq. (21),

$$k_I^{(3)} + k_{II}^{(3)} = \frac{1}{G_3} (b_{11}g_{22} - 2b_{12}g_{12} + b_{22}g_{11})$$

Let a new coordinate system $(\bar{\xi},\bar{\eta})$ be introduced in the same surface such that $\bar{\xi} = \bar{\xi}(\xi,\eta)$, and $\bar{\eta} = \bar{\eta}(\xi,\eta)$ are admissible transformation functions.

(a) Use the chain rule of differentiation to show that the components of the normal $\underline{n}$ to the surface are coordinate invariants, i.e.,

$$X = \overline{X}, \quad Y = \overline{Y}, \quad Z = \overline{Z}$$

(b) Also show that on coordinate transformation

$$k_I + k_{II} = \overline{k}_I + \overline{k}_{II}$$

20. Let it be desired to obtain the 3D curvilinear coordinates in the region bounded by a prolate ellipsoid as an inner boundary ($\eta = \eta_i$) and a sphere as an outer boundary ($\eta = \eta_o$). The (x,y,z) for both the inner and outer bodies are given below in which τ and η_i are the parameters of the ellipsoid:

$$x = \tau \cosh\eta_i \cos\zeta, \qquad y = \tau \sinh\eta_i \sin\zeta \cos\xi,$$

$$z = \tau \sinh\eta_i \sin\zeta \sin\xi$$

$$x = e^{\eta_o} \cos\zeta, \quad y = e^{\eta_o} \sin\zeta \cos\xi, \quad z = e^{\eta_o} \sin\zeta \sin\xi.$$

(a) First write Eq. (91) as three equations in x, y and z for the generation of those surfaces on which ζ = constant. Also set P = Q = 0 and transform the three equations mentioned above from η to $\overline{\eta}$, where $\eta = \eta_i + \eta(\overline{\eta})$.

(b) Assume the solution is

$$x = f(\overline{\eta}) \cos\zeta, \; y = \phi(\overline{\eta}) \sin\zeta \cos\xi, \; z = \phi(\overline{\eta}) \sin\zeta \sin\xi,$$

and compute all the needed derivatives to find g_{11}, g_{12}, g_{22} while keeping ζ fixed. Also using Eq.

(15), (20) and (21) of Appendix A obtain the expressions for the components of $\underset{\sim}{n}^{(3)}$ and $R^{(3)}$.

(c) Use all the quantities obtained in (b) in the equations written in (a), and show that

$$f(\bar{\eta}) = A\, e^{B\eta(\bar{\eta})} + C$$

$$\phi(\bar{\eta}) = D\, e^{B\eta(\bar{\eta})}$$

where

$$A = \tau[(e^{\eta_o} - \tau \cosh\eta_i) \sinh\eta_i]/(e^{\eta_o} - \tau \sinh\eta_i)$$

$$B = (\eta_o - \ln \tau \sinh\eta_i)/(\eta_o - \eta_i)$$

$$C = \tau[e^{\eta_o}(\cosh\eta_i - \sinh\eta_i)]/(e^{\eta_o} - \tau \sinh\eta_i)$$

$$D = \tau \sinh\eta_i$$

21. Let a surface in the xyz-space by given as

$$z = f(x,y)$$

Show the following:

(a) The components of the unit normal vector to the surface are

$$\underset{\sim}{n}:\ X = \frac{-z_x}{\sqrt{1 + z_x^2 + z_y^2}}, \qquad Y = \frac{-z_y}{\sqrt{1 + z_x^2 + z_y^2}},$$

$$Z = \frac{1}{\sqrt{1 + z_x^2 + z_y^2}}$$

(b) The element of area dA on the surface is

$$dA = \sqrt{1 + z_x^2 + z_y^2}\, dxdy$$

(c) The element of length ds of a surface curve is given by

$$ds^2 = (1 + z_x^2)\, dx^2 + 2z_x z_y\, dxdy + (1 + z_y^2)\, dy^2$$

(d) The sum of the principal curvatures is given by

$$k_I + k_{II} = \frac{(1 + z_y^2)z_{xx} - 2z_x z_y z_{xy} + (1 + z_x^2)z_{yy}}{(1 + z_x^2 + z_y^2)^{3/2}}$$

22. (a) Show that the unit tangent vector $\underset{\sim}{t}$ to the curve of intersection of two surfaces $F(x,y,z) = 0$, $G(x,y,z) = 0$ is

$$\underset{\sim}{t} = \underset{\sim}{i}J_1 + \underset{\sim}{j}J_2 + \underset{\sim}{k}J_3)/(J_1^2 + J_2^2 + J_3^2)^{1/2}$$

where

$$J_k = F_{x_m} G_{x_n} - F_{x_n} G_{x_m}$$

and m,n,k are in the cyclic permutations of 1,2,3.
(b) Using the formula for the normal vector $\underset{\sim}{n}$ to F = constant, .i,e.,

$$\underset{\sim}{n} = \frac{\underset{\sim}{\nabla}F}{|\underset{\sim}{\nabla}F|}$$

find the Cartesian components of $\underset{\sim}{n}$.

23. Verify Eq. (104).

VII. PARABOLIC AND HYPERBOLIC GENERATION SYSTEMS

It is also possible to base a grid generation system on hyperbolic or parabolic partial differential equations, rather than elliptic equations. In each of these cases the grid is generated by numerically solving the partial differential equations, marching in the direction of one curvilinear coordinate between two boundary curves in two dimensions, or between two boundary surfaces in three dimensions. In neither case can the entire boundaries of a general region be specified - only the elliptic equations allow that.

The parabolic system can be applied to generate the grid between the two boundaries of a doubly-connected region with each of these boundaries specified. The hyperbolic case, however, allows only one boundary to be specified, and is therefore of interest only for use in calculation on physically unbounded regions where the precise location of a computational outer boundary is not important. Both parabolic and hyperbolic grid generation systems have the advantage of being generally faster than elliptic generation systems, but, as just noted, are applicable only to certain configurations. Hyperbolic generation systems can be used to generate orthogonal grids.

1. Hyperbolic Grid Generation

In two dimensions the condition of orthogonality is simply

$$g_{12} = 0 \tag{1}$$

If either the cell area, $\sqrt{g}$, or the cell diagonal length (squared), $g_{11} + g_{22}$, is a specified function of the curvilinear coordinates, i.e.,

$$\sqrt{g} = F(\xi,\eta) \tag{2a}$$

or

$$g_{11} + g_{22} = F(\xi,\eta) \tag{2b}$$

then the system consisting of Eq. (1) and either (2a) or (2b), as appropriate, is hyperbolic.

A hyperbolic generation system based on Eq. (1) and (2a) is constructed as follows (cf. Ref. [28-29]). Eq. (1) and (2a) become, with $\xi^1 = \xi$, $\xi^2 = \eta$, $x_1 = x$, $x_2 = y$,

$$x_\xi x_\eta + y_\xi y_\eta = 0 \tag{3a}$$

$$x_\xi y_\eta - x_\eta y_\xi = V(\xi,\eta) \tag{3b}$$

where the cell volume distribution, $V(\xi,\eta)$, is specified. This system is hyperbolic and therefore a non-iterative marching solution can be constructed proceeding in one coordinate direction, say η, away from a specified boundary.

The equations are first locally linearized about a known solution denoted x^o, y^o. Thus

$$A\underset{\sim}{r}_\xi + B\underset{\sim}{r}_\eta = \underset{\sim}{f} \tag{4}$$

where

$$\underset{\sim}{r} = \begin{bmatrix} x \\ y \end{bmatrix}, \quad \underset{\sim}{f} = \begin{bmatrix} 0 \\ V + V^o \end{bmatrix}$$

$$A = \begin{bmatrix} x_\eta^o & y_\eta^o \\ y_\eta^o & -x_\eta^o \end{bmatrix}, \quad B = \begin{bmatrix} x_\xi^o & y_\xi^o \\ -y_\xi^o & x_\xi^o \end{bmatrix}$$

Then with second-order central differences for the ξ-derivatives and first-order backward differences for the η-derivatives we have, with $\xi = i$ and $\eta = j$,

$$\underline{r}_{i,j+1} - \underline{r}_{ij} + \frac{1}{2} B^{-1} A(\underline{r}_{i+1,j+1} - \underline{r}_{i-1,j+1})$$

$$= B^{-1}\underline{f}_{i,j+1} + \varepsilon(\nabla_i \Delta_j)^2 \underline{r}_{ij} \qquad (5)$$

with $\nabla_i \underline{r}_{ij} = \underline{r}_{i+1,j} - \underline{r}_{ij}$ and $\Delta_j = \underline{r}_{ij} - \underline{r}_{i,j-1}$ and where A and B, and V^o in $\underline{f}$, are evaluated at j, and the last term is an added fourth-order dissipation term for stability. With x_ξ^o and y_ξ^o evaluated using central differences at j, x_η^o and y_η^o can be evaluated by simulatenous solution of Eq. (3a) and (3b). Eq. (5) then is a 2x2 block tridiagonal equation which is solved on each successive η-line, proceeding away from the specified boundary, to generate the grid.

The cell volume distribution in the field is controlled by the specified function, $V(\xi,\eta)$. One form of this specification is as follows. Let points be distributed on a circle having a perimeter equal to that of the specified boundary at the same arc length distribution as on that boundary. Then specify a radial distribution of concentric circles about this circle according to some distribution function, e.g., the hyperbolic tangent discussed in Chapter VIII. Then use the volume distribution from this unequally-spaced cylindrical coordinate system as $V(\xi,\eta)$, with ξ corresponding to the points around the circle, $\theta(\xi)$, and η corresponding to the radial distribution $r(\eta)$. An example of grids generated by this procedure follows:

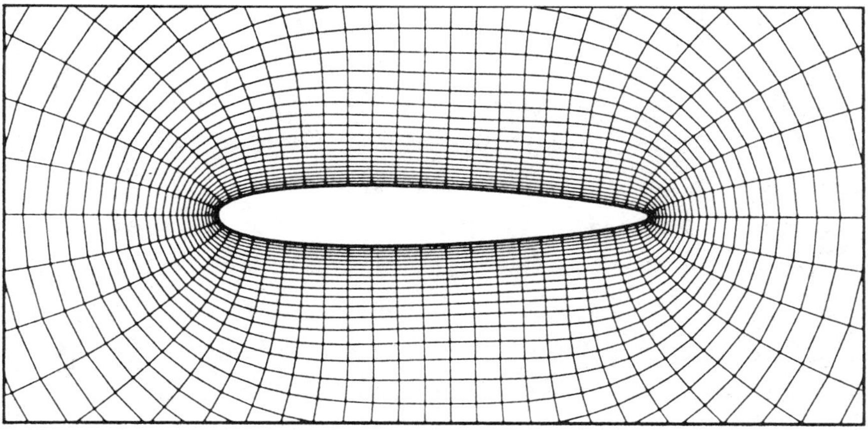

The specification of the cell volume prevents the coordinate system from overlapping even above a concave boundary. In this case the line spacing will expand rapidly away from the boundary in order to keep the cell volume from vanishing, as in the following figure.

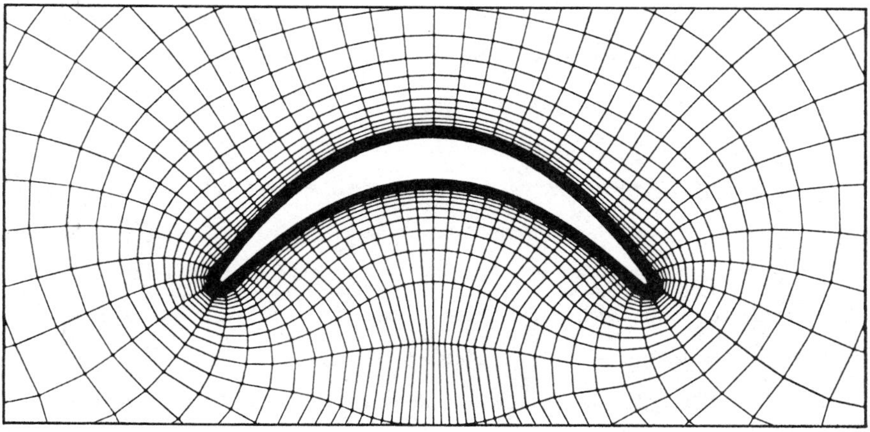

Although this prevents overlap, the rapid expansion that occurs can lead to problems with truncation error in some cases. This approach is extendable to 3-D with the coordinate lines emanating from the boundary being orthogonal to the other two coordinates, but the latter two lines not being orthogonal. There apparently is no system, hyperbolic or elliptic, that will give complete orthogonality in 3-D.

This hyperbolic grid generation system is faster than the elliptic generation systems by one or two orders of magnitude, the computational time required being equivalent to about that for one iteration in a solution of the elliptic system. The specification of the cell volume distribution avoids the grid line overlapping that otherwise can occur with concave boundaries in a method involving projection away from a boundary. The grid may, however, be somewhat distorted when concave boundaries are involved. The cell volume specification also allows control of the gird line spacing, of course, as in the upper part of the second figure on p. 275, but again concave boundaries may cause the intended spacing to occur in the wrong coordinate direction, as in the lower part of this figure, since it is only the volume, and not the spacing in the two separate coordinate directions, that is controlled. As has been noted, the grid is constructed to be orthgonal.

The hyperbolic generation system is not as general as the elliptic systems, however, since the entire boundary of the region cannot be specified. As noted above, boundary slope discontinuities are propagated into the field, so that the metric elements will be discontinuous along coordinate lines emanating from boundary slope discontinuities. Finally, since hyperbolic partial differential equations can have shock-like solutions in some circumstances, it is possible for very unsuitable grids to result with some specifications of boundary point and cell volume distributions. This is in contrast with the elliptic generation systems which tend to emphasize smoothness because of the nature of elliptic partial differential equations.

2. Parabolic Grid Generation

Parabolic grid generation sytems may be constructed by modifying elliptic generation systems so that the second derivatives in one coordinate direction do not appear. The solution then can be marched away from a boundary in much the same manner as described above for the hyperbolic systems. Here, however, some influence of the other boundary toward which the marching progresses is retained in the equations.

In Ref. [30] such a parabolic generation system is formed essentially by first representing all derivatives in an elliptic generation system with second-order central differences and then replacing all values on the forward line in one coordinate direction, say $\eta = j+1$, with values specified in some manner in terms of the values on the preceeding lines and specified values on the outer boundary. This reduces the difference equations to a set of 2x2 block tridiagonal equations to be solved on each coordinate line in succession, proceeding away from a specified boundary. Control of the coordinate line spacing can be achieved by certain control functions that are drawn from some analogy with the elliptic system. It is possible to use the functional specification of the forward values to cause the grid to be nearly-orthogonal.

The parabolic generation system is also faster than the elliptic generation systems to the same degree as is the hyperbolic system, since again only a succession of tridiagonal solutions is required. The functional specification of the forward values, with an influence of an outer boundary, introduces a smoothing effect from this second boundary not present in the hyperbolic system. Orthogonality is not achieved as directly as with the hyperbolic system, however.

The forms of the forward value specification, and of the control functions, have not yet been well-developed.

VIII. ALGEBRAIC GENERATION SYSTEMS

As noted earlier, the problem of generating a curvilinear coordinate system can be formulated as a problem of generating values of the cartesian coordinates in the interior of the rectangular transformed region from specified values on the boundaries. This, of course, can be done directly by interpolation from the boundaries, and such coordinate generation procedures are referred to as algebraic generation systems. Thus $\underset{\sim}{r}(\xi^1,\xi^2,\xi^3)$ is given as a specific function of the curvilinear coordinates. This function contains certain coefficients which are determined so that the function matches specified values of the cartesian coordinates, and perhaps derivatives also, on the boundary and perhaps elsewhere. Evaluation of this interpolation function at constant values of the curvilinear coordinates then defines the coordinate system. Algebraic grid generation is discussed in Ref. [31] and [8], as well as in the surveys, Ref. [1], [5] and [37], and in detail in Ref. [32-36].

1. Unidirectional Interpolation

Unidirectional interpolation means the interpolation is in one curvilinear coordinate direction only. In this section the cartesian coordinate vector $\underset{\sim}{r}$ will be shown as a function of the coordinate involved in the interpolation, as the unidirectional interpolation is fundamentally between points. These points can, however, lie on boundary (and perhaps interior) curves or surfaces, and in this sense the unidirectional interpolation can be considered to be between these curves or surfaces. Therefore the single-variable functional relationship $\underline{r}(\xi)$ used in this section can be considered to represent dependence on all coordinates, the interpolation points $\underline{r}_i$ being functions of the coordinates along the boundary curves or surfaces.

A. Lagrange interpolation

The simplest type of unidirectional interpolation is Lagrange interpolation, which is based on polynomials. In the linear form we have, with $0 \leq \xi \leq I$,

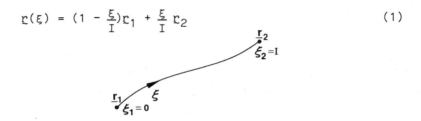

$$\underline{r}(\xi) = (1 - \frac{\xi}{I})\underline{r}_1 + \frac{\xi}{I}\underline{r}_2 \tag{1}$$

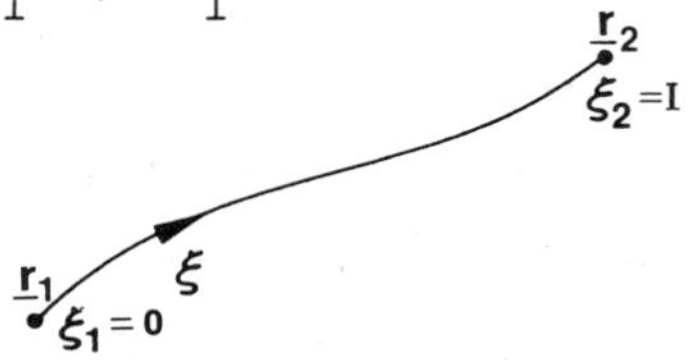

Here $\underline{r}_1=\underline{r}(0)$ and $\underline{r}_2=\underline{r}(I)$, so that $\underline{r}(\xi)$ is defined in terms of the two boundary values, $\underline{r}_1$ and $\underline{r}_2$. The grid points are located at the successive integer values of ξ from 0 to I. One family of grid lines will be straight lines connecting corresponding boundary points with this linear interpolation.

The general form is

$$\underline{r}(\xi) = \sum_{n=1}^{N} \phi_n(\frac{\xi}{I})\, \underline{r}_n \tag{2}$$

with $\underline{r}_n = \underline{r}(\xi_n)$, and the functions ϕ_n being polynomials defined on the entire interval $0 \leq \xi \leq I$ such that

$$\phi_n(\frac{\xi_m}{I}) = \delta_{nm} \tag{3}$$

In the linear case given above we have, with N=2,

$$\phi_1(\tfrac{\xi}{I}) = 1 - \tfrac{\xi}{I} \quad \text{and} \quad \phi_2(\tfrac{\xi}{I}) = \tfrac{\xi}{I}$$

From Eq. (2) and (3),

$$\underset{\sim}{r}(\xi_m) = \sum_{n=1}^{N} \phi_n(\tfrac{\xi_m}{I})\, \underset{\sim}{r}_n = \sum_{n=1}^{N} \delta_{nm}\, \underset{\sim}{r}_n = \underset{\sim}{r}_m$$

so that the interpolation function matches $\underset{\sim}{r}$ at the N points, $\xi=\xi_1, \xi_2, \ldots, \xi_N=I$:

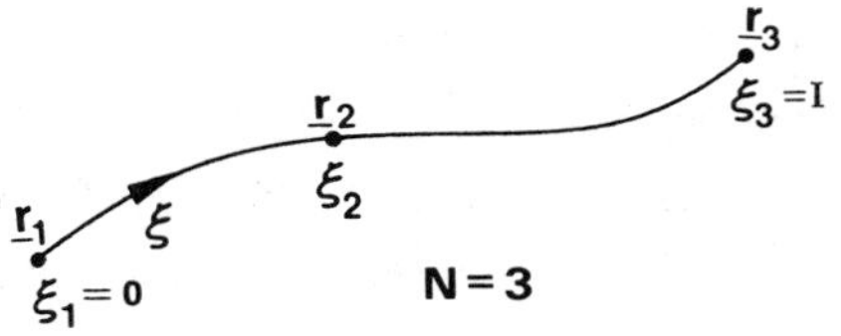

The specified interior points, $\underset{\sim}{r}_n$ for n=2,3,...N-1, are not necessarily grid points, since the grid points are defined by evaluating the interpolation formula at successive integer values of ξ, but are simply additional parameters that serve to control the distribution. It is possible to specify the locations of certain interior grids points, however, by taking the ξ_n corresponding to the specified $\underset{\sim}{r}_n$ to be the value of ξ at the grid point of interest.

The Lagrange interpolation polynomials, defined to satisfy by Eq. (3), are in general

$$\phi_n(\tfrac{\xi}{I}) = \prod_{l=1}^{N} \frac{\xi - \xi_l}{\xi_n - \xi_l} \quad (l \neq n) \qquad (4)$$

The quadratic forms thus are, with N=3, and $\xi_2 = I/2$

$$\phi_1(\frac{\xi}{I}) = 2\ (\frac{\xi}{I} - \frac{1}{2})\ (\frac{\xi}{I} - 1)$$

$$\phi_2(\frac{\xi}{I}) = 4\ \frac{\xi}{I}\ (1 - \frac{\xi}{I})$$

$$\phi_3(\frac{\xi}{I}) = 2\ \frac{\xi}{I}\ (\frac{\xi}{I} - \frac{1}{2})$$

for which $\underset{\sim}{r}(\xi)$ is defined in terms of the two boundary values, $\underset{\sim}{r}_1$ and $\underset{\sim}{r}_3$, and one interior value, $\underset{\sim}{r}_2$. It should be noted that the purpose of the inclusion of the interior points in grid generation is control of the grid point distribution, not to increase the accuracy of the interpolation as is normally the case. There is, in fact, no question of accuracy of the interpolation here, since the aim is just to generate a grid from the boundary values of the coordinates.

B. Hermite interpolation

Lagrange interpolation matches only function values. It is possible to match both function, $\underset{\sim}{r}$, and first-derivative, $\underset{\sim}{r}' = \underset{\sim}{r}_\xi$, values using Hermite interpolation defined by

$$\underset{\sim}{r}(\xi) = \sum_{n=1}^{N} \Phi_n(\frac{\xi}{I})\ \underset{\sim}{r}_n + \sum_{n=1}^{N} \Psi_n(\frac{\xi}{I})\ \underset{\sim}{r}_n' \qquad (5)$$

where the Hermite inerpolation polynomials are defined on $0 \leq \xi \leq I$ and satisfy the conditions

$$\Phi_n(\frac{\xi_m}{I}) = \delta_{nm}, \quad \Phi_n'(\frac{\xi_m}{I}) = 0$$

$$\Psi_n\left(\frac{\xi_m}{I}\right) = 0, \quad \Psi'_n\left(\frac{\xi_m}{I}\right) = \delta_{nm}$$

These polynomials can be obtained from the Lagrange interpolation polynomials by

$$\Phi_n\left(\frac{\xi}{I}\right) = \left[1 - 2\phi'_n\left(\frac{\xi_n}{I}\right)\left(\frac{\xi - \xi_n}{I}\right)\right]\phi_n^2\left(\frac{\xi}{I}\right) \tag{6a}$$

$$\Psi_n\left(\frac{\xi}{I}\right) = \left(\frac{\xi - \xi_n}{I}\right)\phi_n^2\left(\frac{\xi}{I}\right) \tag{6b}$$

where the prime here indicates differentiation of the polynomial with respect to the argument, $\frac{\xi}{I}$. With N=2 we have

$$\Phi_1\left(\frac{\xi}{I}\right) = \left(1 + 2\frac{\xi}{I}\right)\left(1 - \frac{\xi}{I}\right)^2$$

$$\Phi_2\left(\frac{\xi}{I}\right) = \left(3 - 2\frac{\xi}{I}\right)\left(\frac{\xi}{I}\right)^2$$

$$\Psi_1\left(\frac{\xi}{I}\right) = \left(1 - \frac{\xi}{I}\right)^2\frac{\xi}{I}$$

$$\Psi_2\left(\frac{\xi}{I}\right) = \left(\frac{\xi}{I} - 1\right)\left(\frac{\xi}{I}\right)^2$$

and the function matches the two boundary values, $\underline{r}_1$ and r_2, and the first derivatives, $\underline{r}'_1$ and $\underline{r}'_2$, at the two boundaries.

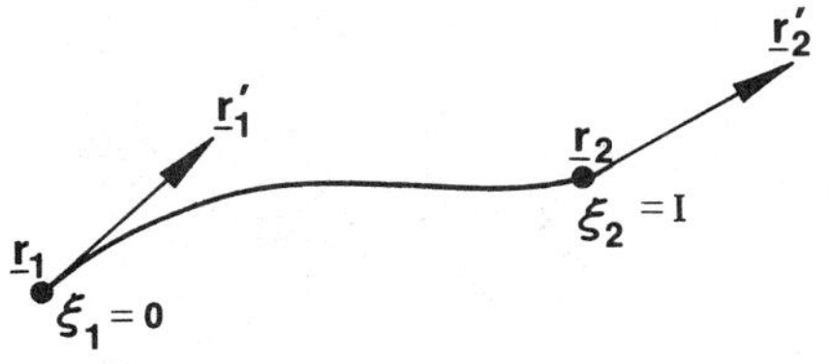

Extensions of polynomial interpolation to match higher-order derivatives is obviously possible, the degree of the polynomial increasing with each additional condition or point to be matched. The polynomials of high degree exhibit considerable oscillation, however, so such procedures are not of great importance to grid generation. The general form again includes matches at interior points, which can be used to control the coordinate line spacing, since the first derivative, $\underset{\sim}{r}' = \underset{\sim}{r}_\xi$, is a measure of the grid point spacing here, with $\Delta\xi$ being unity between points by construction. As with Lagrange interpolation, these specified interior points may or may not be grid points.

It is also possible, of course, to omit points from either of the summations in Eq. (5), so that $\underset{\sim}{r}$ and its first derivatives are not both matched at all points (deficient Hermite interpolation). Thus, with N=2 and the n=1 term omitted from the second summation, the two boundary values would be matched, but the first dervative at only the ξ=I boundary would be matched. Clearly, the Hermite interpolation form, Eq. (5), could be equivalently defined in terms of the Lagrangian interpolation form, Eq. (2), with 2N points, since both are polynomial representations. Obviously either approach can be used to control the grid point spacing in the field.

The capability of specifying $\underset{\sim}{r}_\xi$, as well as $\underset{\sim}{r}$, can be used to make the grid orthogonal at the boundary. From Eq. (III-33) the unit normal to a ξ^i-coordinate surface is given by

$$\underset{\sim}{n}^i = \frac{\underset{\sim}{a}^i}{|\underset{\sim}{a}^i|} = \frac{\underset{\sim}{a}_j \times \underset{\sim}{a}_k}{|\underset{\sim}{a}_j \times \underset{\sim}{a}_k|} \qquad (i,j,k) \text{ cyclic}$$

Using Eq. (III-10) this becomes

$$\underset{\sim}{n}^i = \frac{\underset{\sim}{a}_j \times \underset{\sim}{a}_k}{\sqrt{g_{jj}g_{kk} - g_{jk}^2}}$$

The condition for orthogonally at the boundary then is that $\underset{\sim}{r}_{\xi^i}$ be in the direction of the unit normal to the boundary:

$$\underset{\sim}{r}_{\xi^i} = |\underset{\sim}{r}_{\xi^i}|\underset{\sim}{n}^i = \sqrt{g_{ii}}\, \frac{\underset{\sim}{a}_j \times \underset{\sim}{a}_k}{\sqrt{g_{jj}g_{kk} - g_{jk}^2}} \tag{7}$$

where $s^i = |\underset{\sim}{r}_{\xi^i}| = \sqrt{g_{ii}}$ is the spacing off the boundary to be specified. Since all the quantities with j and k subscripts can be evaluated from the points on the boundary, it remains only to specify the spacing, s^i, off the boundary and to use Eq. (7) for $\underset{\sim}{r}_{\xi^i}$ on the boundary in the Hermite expressions.

C. Other forms of polynomial interpolation

As noted above, Hermite interpolation, which matches $\underset{\sim}{r}$ and $\underset{\sim}{r}_\xi$ at N points, can be equivalently constructed as an interpolant which matches $\underset{\sim}{r}$ at 2N points. Another form of expression of the polynomial interpolation uses the direct expression of the polynomial, so that

$$\underset{\sim}{r}(\xi) = \sum_{n=0}^{N-1} \underset{\sim}{a}_n \left(\frac{\xi}{I}\right)^n \tag{8}$$

Here we must have

$$\underset{\sim}{a}_o = \underset{\sim}{r}_o, \qquad \sum_{n=0}^{N-1} \underset{\sim}{a}_n = \underset{\sim}{r}_I$$

where $\underset{\sim}{r}_o$ and $\underset{\sim}{r}_I$ are the boundaries. This form is not as straightforward as the Lagrange form for use in grid generation, since in the latter form certain grid point locations can be specified directly, while in the former the coefficients must be evaluated in terms of these specified points.

Still another form is that of Bezier, using Bernstein polynomials:

$$\underset{\sim}{r}(\xi) = \sum_{n=o}^{N} \frac{N!}{(N-n)!n!} \left(\frac{\xi}{I}\right)^n \left(1 - \frac{\xi}{I}\right)^{N-n} \underset{\sim}{c}_n \tag{9}$$

with

$$\underset{\sim}{c}_o = \underset{\sim}{r}_o, \qquad \underset{\sim}{c}_N = \underset{\sim}{r}_I$$

Here we have

$$\underset{\sim}{c}_1 - \underset{\sim}{c}_o = \frac{I}{N} \underset{\sim}{r}_o', \qquad \underset{\sim}{c}_N - \underset{\sim}{c}_{N-1} = \frac{I}{N} \underset{\sim}{r}_I'$$

Thus the coefficients $\underset{\sim}{c}_1$ and $\underset{\sim}{c}_{N-1}$ specify the slopes at the boundaries. An advantage of the Bezier form is that the coefficients define the vertices of an open polygon to which the curve is an approximation. Thus the general shape of the curve can be inferred by considering the coefficients to represent points in the field, with the lines from $\underset{\sim}{c}_o$ to $\underset{\sim}{c}_1$, and from $\underset{\sim}{c}_{N-1}$ to $\underset{\sim}{c}_N$, defining the slopes at the two ends. The shape of the curve can then be designed by the placement of the vertices in the field as indicated below. Modifications of the curve can thus be made by adjusting the positions of these vertices.

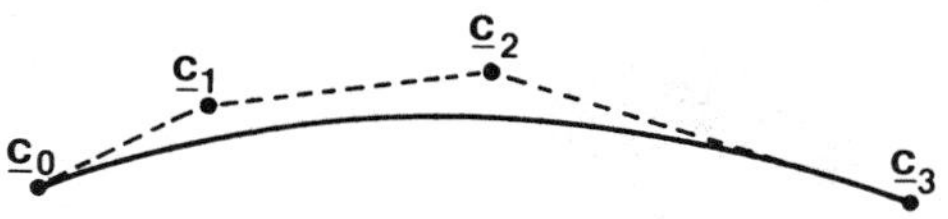

Still another form can be defined using piecewise polynomials for the interpolation functions. Some degree of continuity must be lost in this case, of course. Continuity of the grid lines can be achieved using the piecewise-linear polynomials shown below (truncated versions apply at or near the end points):

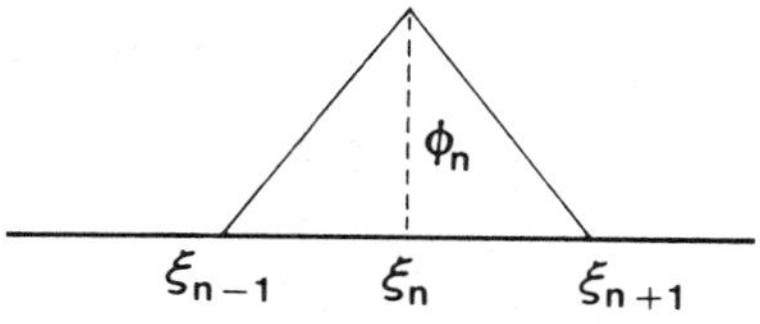

while slope continuity can be gotten with the following piecewise polynomials:

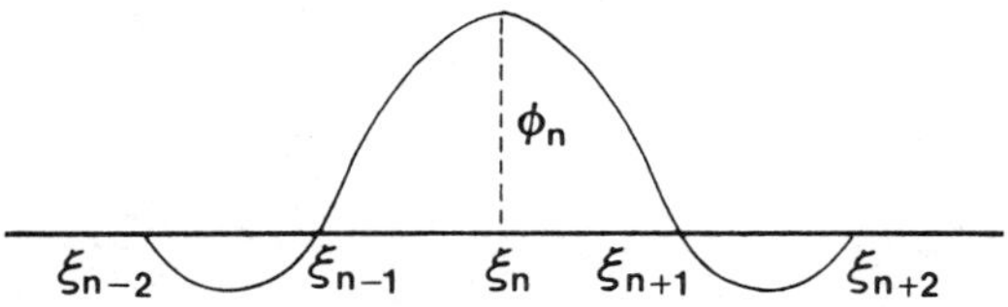

Such piecewise polynomials allow a greater degree of local adjustment to be made, since the polynomial ϕ_n which multiplies the interpolation point $\underline{r}_n$ vanishes except in the immediate vicinity of $\underline{r}_n$. By the conditions (3), any interpolation function ϕ_n must vanish at all the interpolation points except $\underline{r}_n$, but need not vanish between the points. Adding more interpolation points with global polynomials thus means increasing the degree of the polynomials, since the numbers of zeros must increase, and hence the polynomial becomes highly oscillitory.

D. Splines

The Lagrange and Hermite interpolation functions given above are completely continuous at all points. Complete continuity, however, may be attained at the price of oscillation. Both of these forms fit a single polynomial from one boundary to the other, matching specified values of the coordinates and perhaps the derivatives thereof (i.e., the point spacing). As more interior points are included, or as the first derivatives are included, the order of this global polynomial increases and thus oscillations become more likely. An alternative approach is to fit a low-order polynomial between each of the specified interior points, with continuity of as many derivatives as is possible enforced at the interior points. The interpolation function is then a piecewise-continuous polynomial.

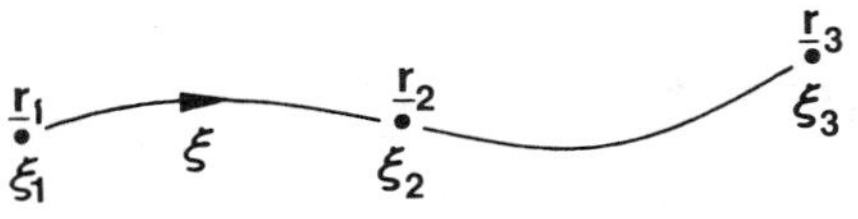

This type of interpolation function is called a spline and is formed as follows for the most common case of the cubic spline.

With a cubic polynomial fitted between points $\underset{\sim}{r}_i$ and $\underset{\sim}{r}_{i+1}$ we have a linear variation of the second derivative between these points and thus

$$\underset{\sim}{r}''(\xi) = \frac{\xi_{i+1} - \xi}{\xi_{i+1} - \xi_i}\underset{\sim}{r}_i'' + \frac{\xi - \xi_i}{\xi_{i+1} - \xi_i}r_{i+1}'' \qquad \xi_i \leq \xi \leq \xi_{i+1} \tag{10}$$

After two integrations and evaluation of the two constants of integration such that $\underset{\sim}{r}(\xi_i)=\underset{\sim}{r}_i$ and $\underset{\sim}{r}(\xi_{i+1})=\underset{\sim}{r}_{i+1}$, we have on $\xi_i \leq \xi \leq \xi_{i+1}$

$$\underset{\sim}{r}(\xi) = \frac{(\xi_{i+1} - \xi)^3}{6(\xi_{i+1} - \xi_i)} \underset{\sim}{r}_i'' + \frac{(\xi - \xi_i)^3}{6(\xi_{i+1} - \xi_i)} \underset{\sim}{r}_{i+1}''$$

$$+ \left(\frac{1}{\xi_{i+1} - \xi_i} \underset{\sim}{r}_i - \frac{\xi_{i+1} - \xi_i}{6} \underset{\sim}{r}_i''\right)(\xi_{i+1} - \xi)$$

$$+ \left(\frac{1}{\xi_{i+1} - \xi_i} \underset{\sim}{r}_{i+1} - \frac{\xi_{i+1} - \xi_i}{6} \underset{\sim}{r}_{i+1}''\right)(\xi - \xi_i) \tag{11}$$

Then, after differentiation and setting $\xi=\xi_i$, we have on $\xi_i \leq \xi \leq \xi_{i+1}$,

$$\underset{\sim}{r}_i' = \frac{\xi_{i+1} - \xi_i}{6}(2\underset{\sim}{r}_i'' + \underset{\sim}{r}_{i+1}'') + \frac{1}{\xi_{i+1} - \xi_i}(\underset{\sim}{r}_{i+1} - \underset{\sim}{r}_i) \tag{12}$$

Similar evaluation on the adjacent interval $\xi_{i-1} \leq \xi \leq \xi_i$ gives

$$\underset{\sim}{r}_i' = \frac{\xi_i - \xi_{i-1}}{6}(2\underset{\sim}{r}_i'' + \underset{\sim}{r}_{i-1}'') + \frac{1}{\xi_i - \xi_{i-1}}(\underset{\sim}{r}_i - \underset{\sim}{r}_{i-1}) \tag{13}$$

Equating these two expressions in order to produce continuity of r' at the interior points, we have

$$(\xi_i - \xi_{i-1})\,\underset{\sim}{r}_{i-1}'' + 2(\xi_{i+1} - \xi_{i-1})\,\underset{\sim}{r}_i'' + (\xi_{i+1} - \xi_i)\,\underset{\sim}{r}_{i+1}''$$

$$= 6\left(\frac{\underset{\sim}{r}_{i+1} - \underset{\sim}{r}_i}{\xi_{i+1} - \xi_i} - \frac{\underset{\sim}{r}_i - \underset{\sim}{r}_{i-1}}{\xi_i - \xi_{i-1}}\right) \tag{14}$$

which is a tridiagonal equation for $\underset{\sim}{r}''$ at the interior points. It is necessary to set some conditions on $\underset{\sim}{r}''$ on the boundaries in order to solve this system, and the "natural" spline uses $\underset{\sim}{r}_1'' = \underset{\sim}{r}_I'' = 0$. This choice minimizes the total curvature, and thus the natural spline is the smoothest interpolant. This solution defines the $\underset{\sim}{r}_i''$ in terms of the $\underset{\sim}{r}_i$, so that substitution of these values for $\underset{\sim}{r}_i''$ into Eq. (11) then gives the spline in the general form of Eq. (2), except that the interpolation functions, ϕ_n, are, of course, different from the Lagrange interpolation polynomials. It should be recalled again that the interior points may or may not be grid points, the latter being defined by the interpolation formula evaluated at successive integer values of ξ after the spline has been constructed over the entire field.

E. Tension splines

The spline tends to give a very smooth point distribution. Stronger localized curvature around the specified interior points can be obtained with the tension spline. Here Eq. (10) on $\xi_i \leq \xi \leq \xi_{i+1}$ is replaced by

$$\underset{\sim}{r}''(\xi) - \sigma^2 \underset{\sim}{r}(\xi) = \frac{\xi_{i+1} - \xi}{\xi_{i+1} - \xi_i}(\underset{\sim}{r}_i'' - \sigma^2 \underset{\sim}{r}_i) + \frac{\xi - \xi_i}{\xi_{i+1} - \xi_i}(\underset{\sim}{r}_{i+1}'' - \sigma^2 \underset{\sim}{r}_{i+1}) \tag{15}$$

where σ^2 is a constant to be specified. (The tension spline tends progressively toward a linear function for large values of σ, and toward a cubic spline for small values.) Integration and evaluation of constants then yields, on $\xi_i \leq \xi \leq \xi_{i+1}$,

$\underset{\sim}{r}(\xi) =$

$$\frac{1}{\sigma^2}\frac{\sinh[\sigma(\xi_{i+1} - \xi)]}{\sinh[\sigma(\xi_{i+1} - \xi_i)}\underset{\sim}{r}_i'' + \frac{1}{\sigma^2}\frac{\sinh[\sigma(\xi - \xi_i)]}{\sinh[\sigma(\xi_{i+1} - \xi_i)]}\underset{\sim}{r}_{i+1}''$$

$$+ (\underset{\sim}{r}_i - \frac{1}{\sigma^2}\underset{\sim}{r}_i'')\frac{\xi_{i+1} - \xi}{\xi_{i+1} - \xi_i} + (\underset{\sim}{r}_{i+1} - \frac{1}{\sigma^2}\underset{\sim}{r}_{i+1}'')\frac{\xi - \xi_i}{\xi_{i+1} - \xi_i} \qquad (16)$$

The requirement of continuity of first derivatives at the interior points then yields the tridiagonal equation

$$[\frac{1}{\Delta_{i-1}} - \frac{\sigma}{\sinh(\sigma\Delta_{i-1})}]\underset{\sim}{r}_{i-1}'' + [\frac{1}{\Delta_i} - \frac{\sigma}{\sinh(\sigma\Delta_i)}]\underset{\sim}{r}_{i+1}''$$

$$+ [\frac{\sigma}{\tanh(\sigma\Delta_{i-1})} - \frac{1}{\Delta_{i-1}} + \frac{\sigma}{\tanh(\sigma\Delta_i)} - \frac{1}{\Delta_i}]\underset{\sim}{r}_i''$$

$$= \sigma^2(\frac{\underset{\sim}{r}_{i+1} - \underset{\sim}{r}_i}{\Delta_i} - \frac{\underset{\sim}{r}_i - \underset{\sim}{r}_{i-1}}{\Delta_{i-1}}) \qquad (17)$$

where $\Delta_i = \xi_{i+1} - \xi_i$ and $\Delta_{i-1} = \xi_i - \xi_{i-1}$. Some application of tension splines are given in Ref. [33].

F. B-Splines

One further possibility is to use piecewise continuous functions which satisfy the cardinality conditions by vanishing identically outside some interval around ξ_n, as discussed in Section C above. This type of function allows the interpolation to be modified locally without affecting the interpolation function elsewhere. The B-splines are an example of this approach.

From Eq. (14) a cubic spline which matches the function at N points, with continuity of second-derivatives, requires N+2 items of data, i.e., the N values of $\underset{\sim}{r}_n$ (n = 1,2,...N) and the values of $\underset{\sim}{r}''$ at each boundary. Therefore a cubic spline which has $\underset{\sim}{r} = \underset{\sim}{r}' = \underset{\sim}{r}'' = 0$ at each boundary can be defined over five points if $\underset{\sim}{r}$ is specified at only a single interior point (since N+2=7 data items can be specified here). If such a spline over five points is joined to the line $\underset{\sim}{r} = 0$ outside these five points, we have a function which is non-zero only over four intervals and yet which has continuous second derivatives everywhere. Such a function is called a B-spline, denoted $N_{4n}(\frac{\xi}{I})$, where the end-points of the non-zero interval are ξ_{n-4} and ξ_n. Similarly, quadratic, linear and constant B-splines are non-zero over three, two, and one intervals, respectively, and are denoted N_{qn}, where q = 3, 2, and 1. The end-points of the interval of non-zero values for these splines are ξ_{n-q} and ξ_n. The specification of a single value in this interval is usually replaced by the specification of the integral over the interval so that

$$\int_{\xi_{n-q}}^{\xi_n} N_{qn}(\xi)d\xi = \frac{\xi_n - \xi_{n-q}}{q} \tag{18}$$

The practical importance of B-splines is that any spline of order q (the cubic spline is of order 4) can be expressed as a sum of multiples of B-splines. Thus the cubic spline can be written as

$$\underset{\sim}{r}(\xi) = \sum_{n=0}^{N} \underset{\sim}{c}_n \, N_{4n}(\xi) \tag{19}$$

Since the B-splines are non-zero only over four intervals, the modification of one coefficient here only affects the function over four intervals, thus allowing more localized control of the resulting grid.

The B-splines can be calculated from the recurrence relation

$$N_{qn}(\xi) = \frac{(\xi - \xi_{n-q})\, N_{q-1,n-1}(\xi) + (\xi_n - \xi)\, N_{q-1,n}(\xi)}{\xi_{n-1} - \xi_{n-q}} \tag{20}$$

Thus $N_{4n}(\xi)$ requires the successive calculation of $N_{1,n-1}$, $N_{2,n-1}$, $N_{2,n}$, $N_{3,n-1}$, $N_{3,n}$, and finally $N_{4,n}$. The constant B-spline, $N_{1,n-1}$, used to start this calculation, is given on the interval $\xi_{n-2}\leq\xi\leq\xi_{n-1}$ by $N_{1,n-1}=1$ and vanishes elsewhere.

For the point $\underset{\sim}{r}_n$ we have, in view of the vanishing of the B-splines outside four intervals,

$$\underset{\sim}{r}_n = \underset{\sim}{c}_{n-1}\, N_{4,n-1}(\xi_n) + \underset{\sim}{c}_n\, N_{4,n}(\xi_n) + \underset{\sim}{c}_{n+1}\, N_{4,n+1}(\xi_n)$$

$$n = 0,1,2...,N-1 \tag{21}$$

which is a tridiagonal relation (N+1 equations) for the coefficients $\underset{\sim}{c}_n$, with $\underset{\sim}{c}_o=\underset{\sim}{r}_o$ and $\underset{\sim}{c}_N=\underset{\sim}{r}_N$. Thus, even though the modification of a single coefficient only affects four intervals, the modification of an interpolation point requires a re-determination of all the coefficients and thus affects the function over the entire range.

The coefficients, $\underset{\sim}{c}_n$, in the B-spline representation may be interpreted as the vertices of an open polygon, to which the curve is an approximation, as for the Bezier form discussed above. The slopes at the ends are defined by the directions $\underset{\sim}{c}_1-\underset{\sim}{c}_o$ and $\underset{\sim}{c}_N-\underset{\sim}{c}_{N-1}$. The curve passes close to the mid-point of each side, with the exception of the first and last sides. The curve also passes through the points $(\underset{\sim}{r}_{k-1} + 4\underset{\sim}{r}_k + \underset{\sim}{r}_{k+1})/6$ for $k = 2,3,\ldots,N-2$. These points are one-third of the way along the straight line joining $\underset{\sim}{r}_k$ to the mid-point of the line joining $\underset{\sim}{r}_{k-1}$ and $\underset{\sim}{r}_{k+1}$. Since the B-splines are non-zero only on four intervals, the alteration of one vertex only affects the curve in its immediate vicinity. An application of B-splines in grid generation is given in Ref. [39].

G. Multi-surface interpolation

The multi-surface method, discussed in Ref. [32] - [36], is also a unidirectional interpolation procedure. This procedure is constructed from an interpolation of a specified vector field, followed by vector normalizations at each interpolation point in order to cause a desired telescopic collapse so that the boundaries are matched. The specified vector field is defined from piecewise-linear curves determined by the boundaries and successive intermediate control surfaces. Normals to such surfaces are special cases. Polynomial interpolants for the vector field yield all of the classical polynomial cases along with a rational method for avoiding disasters such as can occur with direct Hermite interpolation with excessively large or discontinuous derivatives. Here the immediate surfaces are not coordinate surfaces, but are used only to define the vector field. These vectors are taken to be tangents to the coordinate lines intersecting the surfaces, so that integration of this vector field produces the position vector field for the grid points.

A collection of subroutines which automatically perform the necessary parts of grid construction using this multi-surface procedure has been written and is described in Ref. [34]. Some of the automation features of this collection are applicable to other grid construction procedures as well. These subroutines can rotate and move curves, project one curve from another, normalize and parameterize curves, cluster points on a curve,and perform other such utilitarian functions to aid in the setup of an overall configuration.

In the multi-surface interpolation we have

$$\underset{\sim}{r}(\xi) = \underset{\sim}{r}_o + \sum_{n=1}^{N-1} \frac{G_n(\xi)}{G_n(I)} (\underset{\sim}{P}_{n+1} - \underset{\sim}{P}_n) \tag{22a}$$

where

$$G_n(\xi) = \int_o^{\xi} \psi_n(\frac{\xi'}{I}) d\xi' \tag{22b}$$

and where the $\underset{\sim}{P}_n$ are specified points, with $\underset{\sim}{P}_1=\underset{\sim}{r}(0)$ and $\underset{\sim}{P}_N=\underset{\sim}{r}(I)$, on the boundary surfaces. (Recall the discussion at the beginning of this section, i.e., that the points $\underset{\sim}{P}_n$ can be considered to lie on curves or surfaces and thus the interpolation, while being fundamentally between points, can be considered to be between the surfaces on which those points lie.) Here the telescopic collapse for the series for $\xi=I$ matches the boundary at $\underset{\sim}{r}_I$. The intermediate points here, $\underset{\sim}{P}_2,\underset{\sim}{P}_3,\ldots,\underset{\sim}{P}_{N-1}$, are not grid points, but serve only to define the slopes $\underset{\sim}{r}_\xi$, as given by Eq. (24) below. Eq. (22) is a polynomial if the functions ψ_n are polynomials, but such is not required.

Since, by differentiation of Eq. (22),

$$\underset{\sim}{r}_\xi = \sum_{n=1}^{N-1} \frac{\psi_n(\frac{\xi}{I})}{G_n(I)} (\underset{\sim}{P}_{n+1} - \underset{\sim}{P}_n) \tag{23}$$

we have, for $0=\xi_1<\xi_2<....<\xi_{n-1}=I$,

$$\underset{\sim}{r}_\xi(\xi_n) = \frac{\underset{\sim}{P}_{n+1} - \underset{\sim}{P}_n}{G_n(I)} \tag{24}$$

if the functions ψ_n satisfy the cardinality conditions

$$\psi_n(\frac{\xi_m}{I}) = \delta_{nm} \qquad n = 1,2,..,N-1, \qquad m = 1,2,..,N-1 \tag{25}$$

The polynomials that satisfy these conditions are simply the Lagrange polynomials given by Eq. (4), here stated as

$$\psi_n(\frac{\xi}{I}) = \prod_{l=1}^{N-1} \frac{\xi - \xi_l}{\xi_n - \xi_l} \qquad (l \neq n) \tag{26}$$

Using Eq. (24), Eq. (23) can be written as

$$\underset{\sim}{r}_\xi(\xi) = \sum_{n=1}^{N-1} \psi_n(\frac{\xi}{I})\, \underset{\sim}{r}_\xi(\xi_n) \tag{27}$$

This form thus is based on an interpolation of the first-derivatives $\underset{\sim}{r}_\xi$, instead of $\underset{\sim}{r}$, the interpolation expression for $\underset{\sim}{r}$ coming from an integration of the interpolation function for $\underset{\sim}{r}_\xi$. Note, however, that this amounts to the specification of the slope $\underset{\sim}{r}_\xi$ at particular values of the curvilinear coordinate ξ_n, and not at a specified position in space as is done in the Hermite form. It is clear from Eq. (24) that the intermediate points, $\underset{\sim}{P}_n$ for $n=2,3,....,N-2$, serve to define the slopes $\underset{\sim}{r}_\xi(\xi_n)$:

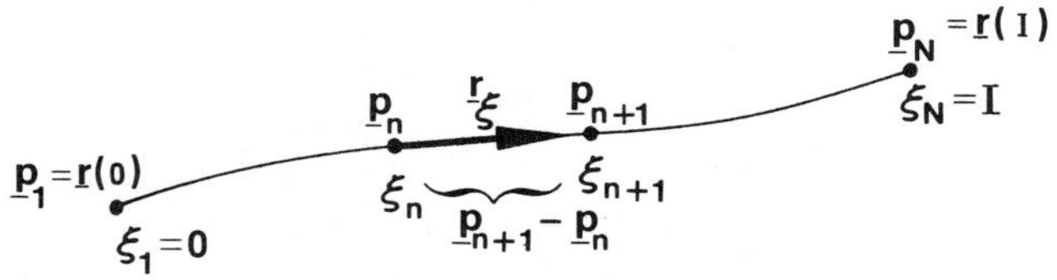

Because of the integration involved, the degree of the interpolation polynomial will be one greater than that of the functions ψ_n.

Also with Eq. (24), the interpolation for $\underset{\sim}{r}$, Eq. (22), can be written

$$\underset{\sim}{r}(\xi) = \underset{\sim}{r}(0) + \sum_{n=1}^{N-1} G_n(\xi)\, \underset{\sim}{r}_\xi(\xi_n) \tag{28}$$

and thus is equivalent to a form of deficient Hermite interpolation. In implementation, however, it is the points $\underset{\sim}{P}_n$ that are specified, as in Eq. (22). Again it should be recalled that the point $\underset{\sim}{P}_n$ is not the grid point at ξ_n, except for the boundaries $\xi_1=0$ and $\xi_N=I$.

As has been noted, $\underset{\sim}{P}_1$ and $\underset{\sim}{P}_N$ are determined by the boundaries:

$$\underset{\sim}{P}_1 = \underset{\sim}{r}(0), \qquad \underset{\sim}{P}_N = \underset{\sim}{r}(I) \tag{29a}$$

For $N \geq 4$, $\underset{\sim}{P}_2$ and $\underset{\sim}{P}_{N-1}$ are determined by the intended values of $\underset{\sim}{r}_\xi$ at the boundaries through Eq. (24):

$$\underset{\sim}{P}_2 = \underset{\sim}{r}(0) + \underset{\sim}{r}_\xi(0)G_1(I)$$

$$\underset{\sim}{P}_{N-1} = \underset{\sim}{r}(I) - \underset{\sim}{r}_\xi(I)G_{N-1}(I) \tag{29b}$$

For N=3 only one of the above equations can be used, i.e., $\underset{\sim}{r}_\xi$ can be specified at either boundary but not at both. The use of the intermediate surfaces, instead of direct specification of the derivatives $\underset{\sim}{r}_\xi$ as in classical Hermite interpolation, provides a geometric interpretation that serves to help avoid the overlapping of grid lines that can occur if too large a value is given for $\underset{\sim}{r}_\xi$.

Following Ref. [35], consider now for the ψ_n the piecewise-linear functions diagramed below:

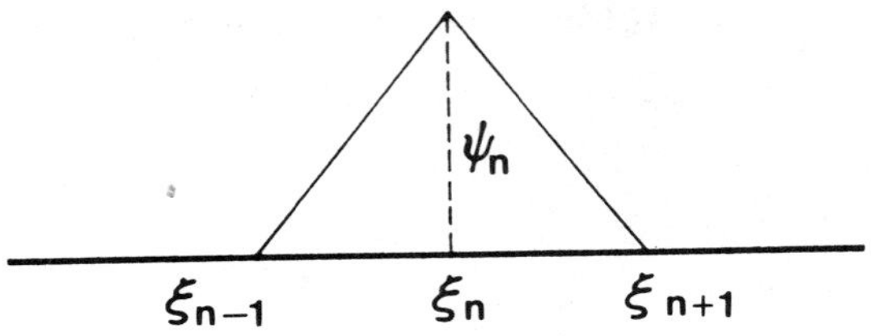

with the normalization

$$\psi_1(\xi_1) = \frac{2}{\Delta\xi_1}, \qquad \psi_{N-1}(\xi_{N-1}) = \frac{2}{\Delta\xi_{N-2}}$$

$$\psi_n(\xi_n) = \frac{2}{\Delta\xi_{n-1} + \Delta\xi_n} \qquad n = 2,3,\ldots,N-2$$

where $\Delta\xi_n = \xi_{n+1} - \xi_n$ so that each ψ_n integrates to unity. These functions are given by, for $n = 2,3,\ldots,N-2$,

$$\psi_1(\xi) = \begin{cases} \dfrac{2}{\Delta\xi_1^2}(\xi_2 - \xi) & \xi_1 \le \xi \le \xi_2 \\ 0 & \xi_2 \le \xi \le \xi_{N-1} \end{cases}$$

$$\psi_{N-1}(\xi) = \begin{cases} 0 & \xi_1 \le \xi \le \xi_{N-2} \\ \dfrac{2}{\Delta\xi_{N-2}^2}(\xi - \xi_{N-2}) & \xi_{N-2} \le \xi \le \xi_{N-1} \end{cases}$$

$$\psi_n(\xi) = \begin{cases} 0 & \xi_1 \le \xi \le \xi_{n-1} \\ \dfrac{2}{\Delta\xi_{n-1} + \Delta\xi_n}\left(\dfrac{\xi - \xi_n}{\Delta\xi_{n-1}} + 1\right) & \xi_{n-1} \le \xi \le \xi_n \\ \dfrac{2}{\Delta\xi_{n-1} + \Delta\xi_n}\left(\dfrac{\xi_n - \xi}{\Delta\xi_n} + 1\right) & \xi_n \le \xi \le \xi_{n+1} \\ 0 & \xi_{n+1} \le \xi \le \xi_{N-1} \end{cases} \tag{30}$$

With these functions we have

$$G_1(\xi) = \begin{cases} 1 - \dfrac{(\xi_2 - \xi)^2}{\Delta\xi_1^2} & \xi_1 \le \xi \le \xi_2 \\ 1 & \xi_2 \le \xi \le \xi_{N-1} \end{cases}$$

$$G_{N-1}(\xi) = \begin{cases} 0 & \xi_1 \le \xi \le \xi_{N-2} \\ \dfrac{(\xi - \xi_{N-2})^2}{\Delta\xi_{N-2}^2} & \xi_{N-2} \le \xi \le \xi_{N-1} \end{cases}$$

$$G_n(\xi) = \begin{cases} 0 & \xi_1 \le \xi \le \xi_{n-1} \\ \dfrac{(\xi - \xi_{n-1})^2}{(\Delta\xi_{n-1} + \Delta\xi_n)\Delta\xi_{n-1}} & \xi_{n-1} \le \xi \le \xi_n \\ 1 - \dfrac{(\xi_{n+1} - \xi)^2}{(\Delta\xi_{n-1} + \Delta\xi_n)\Delta\xi_n} & \xi_n \le \xi \le \xi_{n+1} \\ 1 & \xi_{n+1} \le \xi \le \xi_{N-1} \end{cases} \tag{31}$$

Note that $G_n(I) = 1$ here for all n and that $G_1(\xi_1) = 0$,

$$G_{N-1}(\xi_{N-1}) = 1 \text{ and } G_n(\xi_n) = \frac{\Delta\xi_{n-1}}{\Delta\xi_{n-1} + \Delta\xi_n}$$

These interpolation functions have the form

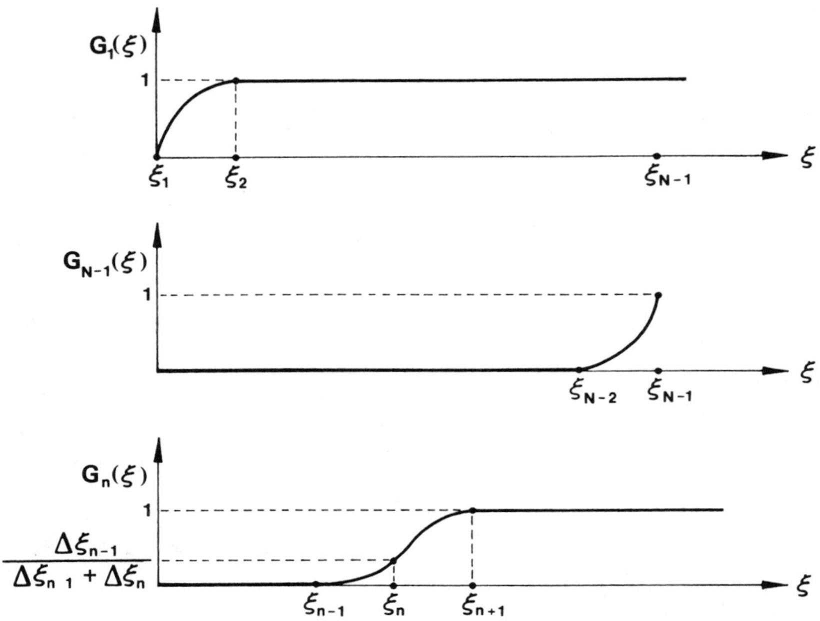

Now on the interval $\xi_n \le \xi \le \xi_{n+1}$ we have

$G_m(\xi)=1$ for $m=1,2,\ldots\ldots,n-1$ and $G_m(\xi)=0$ for $m=n+2,n+3,\ldots\ldots,N-1$. Therefore, on this interval

$$\underline{r}(\xi) = \underline{r}(0) + \sum_{i=1}^{n-1} (\underline{P}_{i+1} - \underline{P}_i) + G_n(\xi)\,(\underline{P}_{n+1} - \underline{P}_n)$$

$$+ \; G_{n+1}(\xi)\,(\underline{P}_{n+2} - \underline{P}_{n+1})$$

which, because of the telescopic collapse of the summation, reduces to, for $\xi_n \le \xi \le \xi_{n+1}$,

$$\underline{r}(\xi) = \underline{P}_n + G_n(\xi)\,(\underline{P}_{n+1} - \underline{P}_n) + G_{n+1}(\xi)\,(\underline{P}_{n+2} - \underline{P}_{n+1}) \qquad (32)$$

Then

$$\underset{\sim}{r}(\xi_n) = \frac{(\Delta\xi_n)\underset{\sim}{P}_n + (\Delta\xi_{n-1})\underset{\sim}{P}_{n+1}}{\Delta\xi_{n-1} + \Delta\xi_n} \tag{33a}$$

and

$$\underset{\sim}{r}(\xi_{n+1}) = \frac{(\Delta\xi_{n+1})\underset{\sim}{P}_{n+1} + (\Delta\xi_n)\underset{\sim}{P}_{n+2}}{\Delta\xi_n + \Delta\xi_{n+1}} \tag{33b}$$

Also, from Eq. (32), for $\xi_n \le \xi \le \xi_{n+1}$

$$\underset{\sim}{r}_\xi(\xi) = \frac{2(\xi_{n+1} - \xi)}{(\Delta\xi_{n-1} + \Delta\xi_n)\Delta\xi_n}(\underset{\sim}{P}_{n+1} - \underset{\sim}{P}_n) + \frac{2(\xi - \xi_n)}{(\Delta\xi_n + \Delta\xi_{n+1})\Delta\xi_n}(\underset{\sim}{P}_{n+2} - \underset{\sim}{P}_{n+1}) \tag{34}$$

so that

$$\underset{\sim}{r}_\xi(\xi_n) = \frac{2}{\Delta\xi_{n-1} + \Delta\xi_n}(\underset{\sim}{P}_{n+1} - \underset{\sim}{P}_n) \tag{35a}$$

and

$$\underset{\sim}{r}_\xi(\xi_{n+1}) = \frac{2}{\Delta\xi_n + \Delta\xi_{n+1}}(\underset{\sim}{P}_{n+2} - \underset{\sim}{P}_{n+1}) \tag{35b}$$

We thus have on this interval

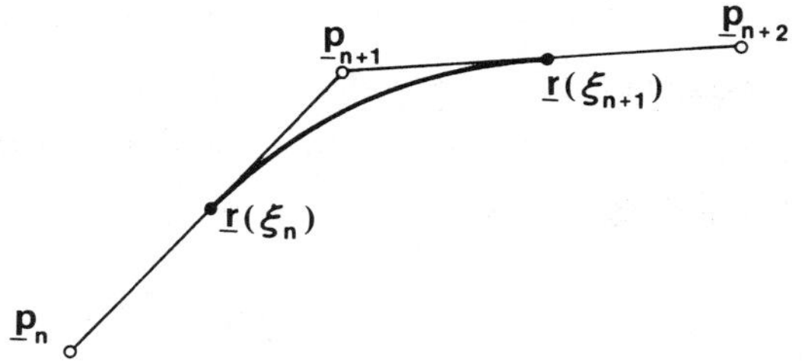

Thus on the interval $\xi_n \leq \xi \leq \xi_{n+1}$, $\underline{r}(\xi)$ is affected only by $\underline{P}_n$, $\underline{P}_{n+1}$, and $\underline{P}_{n+2}$. Conversely, $\underline{P}_n$ affects only the grid point locations on the interval $\xi_{n-2} \leq \xi \leq \xi_{n+1}$. Therefore local adjustments in the grid point locations can be made without affecting all of the points.

With the grid points located at unity increments of ξ, so that I is the total number of points, we have, from Eq. (33), the grid points given by

$$\underline{r}_n = \frac{1}{2}(\underline{P}_n + \underline{P}_{n+1})$$

The local control provided by these piecewise linear interpolants can be used to restrict undesirable mesh forms or to embed desirable ones within a global system with continuous first derivatives (cf. Ref. [34] - [35]). The second derivatives are, however, discontinuous. As examples, the propagation of boundary slope discontinuities can be arbitrarily restricted and general rectilinear Cartesian systems can be embedded to simplify problems over a large part of their domain.

In a further development (cf. Ref. [36]) the procedure is extended to use piecewise quadratic local interpolants, thus achieving continuity of second derivatives, with discontinuous third derivatives. The conceptual extension to higher order piecewise polynomial local interpolants, with consequent higher degree of continuity, is also discussed. Note that because of the integration of ψ_n, the level of derivative continuity is always one greater than that of the piecewise polynomials.

H. Uniformity

It may be desirable for purposes of control of the grid point distribution to have a uniform distribution of the relative projection of $\underset{\sim}{r}(\xi) - \underset{\sim}{r}(0)$ along the straight line connecting the boundary points, i.e., $\underset{\sim}{r}(I) - \underset{\sim}{r}(0)$. This property has been called "uniformity" by Eiseman, cf. Ref. [32] - [36], and can be realized as follows: The unit vector along this straight line is

$$\frac{\underset{\sim}{r}(I) - \underset{\sim}{r}(0)}{|\underset{\sim}{r}(I) - \underset{\sim}{r}(0)|}$$

so that the relative projection of $\underset{\sim}{r}(\xi) - \underset{\sim}{r}(0)$ along this line is given by

$$S(\xi) = \frac{[\underset{\sim}{r}(\xi) - \underset{\sim}{r}(0)] \cdot \frac{\underset{\sim}{r}(I) - \underset{\sim}{r}(0)}{|\underset{\sim}{r}(I) - \underset{\sim}{r}(0)|}}{|\underset{\sim}{r}(I) - \underset{\sim}{r}(0)|} = [\underset{\sim}{r}(\xi) - \underset{\sim}{r}(0)] \cdot \underset{\sim}{I} \tag{36}$$

where

$$\underset{\sim}{I} = \frac{\underset{\sim}{r}(I) - \underset{\sim}{r}(0)}{|\underset{\sim}{r}(I) - \underset{\sim}{r}(0)|^2} \tag{37}$$

Uniformity is then achieved by choosing the interpolation parameters such that $S(\xi)$ is linear. This does not completely determine all the interpolation parameters, however, so that some remain to be specified as desired. Uniformity is trivially assumed for linear interpolation of course.

For Lagrange interpolation we have, from Eq. (2),

$$S(\xi) = \sum_{n=2}^{N} \phi_n\left(\frac{\xi}{I}\right) [\underset{\sim}{r}_n - \underset{\sim}{r}(0)] \cdot \underset{\sim}{I} \tag{38}$$

so that uniformity is achieved by selecting the $\underline{r}_n$, for n=2,3,....,N-1, to cause all of the terms in S that are quadratic or higher to vanish. For Hermite interpolation, (5),

$$S(\xi) = \sum_{n=2}^{N} \Phi_n(\frac{\xi}{I}) \, [\underline{r}_n - \underline{r}(0)] \cdot \underline{I} + \sum_{n=1}^{N} \Psi_n(\frac{\xi}{I}) \, \underline{r}_n' \cdot \underline{I} \tag{39}$$

For the multi-surface interpolation defined by Eq. (22) we have

$$S(\xi) = \sum_{n=1}^{N-1} \frac{G_n(\xi)}{G_n(I)} \, (\underline{P}_{n+1} - \underline{P}_n) \cdot \underline{I} \tag{40}$$

For $S(\xi)$ to be linear we must have

$$S(\xi) \sim \xi$$

or

$$\sum_{n=1}^{N-1} \frac{\psi_n(\frac{\xi}{I})}{G_n(I)} \, (\underline{P}_{n+1} - \underline{P}_n) \cdot \underline{I} \sim \text{constant} \tag{41}$$

But, using Eq. (25), we then have

$$(\underline{P}_{n+1} - \underline{P}_n) \cdot \underline{I} \sim \frac{G_n(I)}{\psi_n(\frac{\xi_n}{I})} \tag{42}$$

as the uniformity condition on the $\underline{P}_n$'s (cf. Ref. [35]). Both the P_n (for n=2,3,....,N-1) and the ξ_n (for n=2,3,....,N-2) are free to be chosen in order to satisfy the uniformity conditions (42). Thus a one-parameter family of cubic forms (N=4) results, a two-parametric family of quartic forms, etc. Substitution of Eq. (42) back into (41) yields a restriction on the choice of the functions ψ_n since these must satisfy the relation

$$\sum_{n=1}^{N-1} \frac{\psi_n(\frac{\xi}{I})}{\psi_n(\frac{\xi_n}{I})} = 1 \qquad (43)$$

Uniformity is particularly useful when the distribution function, such as those discussed in the next section, is used to redistribute the points on the grid lines set up by the interpolation (cf. Ref. [34] - [36]). Thus the interpolation is first applied with I=1 and with the uniformity conditions enforced. The final grid points then are placed according to the distribution function on the grid lines set up by the interpolation variable in place of the arc length, s, in the distribution function $s(\xi)$.

I. Functions other than polynomials

The interpolation functions in the general forms given by Eq. (2), (5), and (22) do not have to be polynomials, and, in fact, if the variation in spacing over the field is large, other functions are better suited for grid generation. With N = 2, Eq. (2) can be written in the form

$$\underset{\sim}{r}(\xi) = \phi(\tfrac{\xi}{I})\, \underset{\sim}{r}_2 + [1 - \phi(\tfrac{\xi}{I})]\, \underset{\sim}{r}_1 \qquad (44)$$

where ϕ can be any function such that $\phi(0)=0$ and $\phi(1)=1$. Here we have taken $\phi_1=1-\phi$ and $\phi_2=\phi$. The linear polynomial case is obtained here with $\phi(\frac{\xi}{I}) = \frac{\xi}{I}$. The function ϕ in this form may contain parameters which can be determined so as to match the slope at the boundary, or to match interior points and slopes.

The interpolation function, ϕ, in this form is often referred to as a "stretching" function, and the most widely used function has been the exponential:

$$\phi\left(\frac{\xi}{I}\right) = \frac{\exp\left(\frac{\alpha\xi}{I}\right) - 1}{\exp(\alpha) - 1} \tag{45}$$

where α is a parameter that can be determined to match the slope at a boundary. Thus, since, from Eq. (44)

$$\underline{r}_\xi = \frac{1}{I}(\underline{r}_2 - \underline{r}_1)\,\phi' \tag{46}$$

we can determine α from the equation

$$(\underline{r}_\xi)_1 = \frac{\underline{r}_2 - \underline{r}_1}{I}\,\frac{\alpha}{\exp(\alpha) - 1} \tag{47}$$

with $(\underline{r}_\xi)_1$ specified.

As noted in Chapter V, the truncation error is strongly affected by the point distribution, and studies of distribution functions have been made in that regard. The exponential, while reasonable, is not the best choice when the variation of spacing is large, and polynomials are not suitable in this case. The better choices are the hyperbolic tangent and the hyperbolic sine. The hyperbolic sine gives a more uniform distribution in the immediate vicinity of the minimum spacing, and thus has less error in this region, but the hyperbolic tangent has the better overall distribution (cf. Section 3 of Chapter 5). These functions are implemented as follows (following Ref. [18]), with the spacing specified at either or both ends, or a point in the interior, of a point distribution on a curve.

Let arc length, s, vary from 0 to 1 as ξ varies from 0 to I: s(0)=0, s(I)=1. Then let the spacing be specified at ξ=0 and ξ=I:

$$s_\xi(0) = \Delta s_1, \qquad s_\xi(I) = \Delta s_2 \tag{48}$$

The hyperbolic tangent distribution is then constructed as follows.

First,

$$A = \frac{\sqrt{\Delta s_2}}{\sqrt{\Delta s_1}} \tag{49}$$

$$B = \frac{1}{I\sqrt{\Delta s_1 \Delta s_2}} \tag{50}$$

Then the following nonlinear equation is solved for δ:

$$\frac{\sinh\delta}{\delta} = B \tag{51}$$

The arc length distribution then is given by

$$s(\xi) = \frac{u(\xi)}{A + (1-A)u(\xi)} \tag{52}$$

where

$$u(\xi) = \frac{1}{2}\left\{1 + \frac{\tanh[\delta(\frac{\xi}{I} - \frac{1}{2})]}{\tanh(\frac{\delta}{2})}\right\} \tag{53}$$

If this is applied to a straight line on which $\underset{\sim}{r}$ varies from $\underset{\sim}{r}_0$ to $\underset{\sim}{r}_I$ we have for the point locations:

$$\underset{\sim}{r}(\xi) = \underset{\sim}{r}_0 + (\underset{\sim}{r}_I - \underset{\sim}{r}_0)\, s(\xi) \tag{54}$$

The points are then located by taking integer values of ξ:

$$\xi = 0, 1, 2 \ldots, I$$

Clearly the arc length distribution, $s(\xi)$, here is the function ϕ of Eq. (44).

With the spacing Δs specified at only $\xi=0$, the construction proceeds as follows. First B is calculated from

$$B = \frac{1}{I\Delta s} \tag{55}$$

and Eq. (51) is solved for δ. The arc length distribution then is given by

$$s(\xi) = 1 + \frac{\tanh[\frac{\delta}{2}(\frac{\xi}{I} - 1)]}{\tanh(\frac{\delta}{2})} \tag{56}$$

With the spacing specified only at $\xi=I$ the procedure is the same, except that Eq. (56) is replaced by

$$s(\xi) = \frac{\tanh\ (\frac{\delta}{2}\frac{\xi}{I})}{\tanh\ (\frac{\delta}{2})} \tag{57}$$

If the spacing Δs is specified at only an interior point $s = \sigma$, B is again calculated from Eq. (55), and then δ is determined as the solution of

$$1 + (\frac{1}{B\sigma\delta})^2 = (\frac{\cosh\delta - 1 + \frac{1}{\sigma}}{\sinh\ \delta})^2 \tag{58}$$

The value of ξ at which $s = \sigma$ is obtained by solving the nonlinear equation

$$\chi = \frac{I}{\delta} \tanh^{-1}\left(\frac{\sinh \delta}{\frac{1}{\sigma} + \cosh \delta - 1}\right) \tag{59}$$

The arc length distribution then is given by

$$s(\xi) = \sigma\left\{1 + \frac{\sinh\left[\delta\left(\frac{\xi - \chi}{I}\right)\right]}{\sinh\left(\delta\frac{\chi}{I}\right)}\right\} \tag{60}$$

This last distribution is based on the hyperbolic sine. From this a distribution based on the hyperbolic sine with the spacing specified at one end can be derived. Here B is evaluated from Eq. (55), and then δ is determined as the solution of

$$\frac{\sinh\delta}{\delta} = B \tag{61}$$

The arc length distribution then is given by

$$s(\xi) = \frac{\sinh\left(\delta\frac{\xi}{I}\right)}{\sinh \delta} \tag{62}$$

if the spacing is specified at $\xi=0$. With the specification at $\xi=I$, the distribution is

$$s(\xi) = 1 - \frac{\sinh\left[\delta\left(1 - \frac{\xi}{I}\right)\right]}{\sinh \delta} \tag{63}$$

It is also possible to construct a distribution based on the hyperbolic sine with specified spacing on each end. Here A and B are again calculated from Eq. (49) and (50), but δ is determined from

$$\frac{\tanh(\frac{\delta}{2})}{\frac{\delta}{2}} = B \tag{64}$$

The distribution is then given by Eq. (52), but with

$$u(\xi) = \frac{1}{2}\{1 + \frac{\sinh[\delta(\frac{\xi}{I} - \frac{1}{2})]}{\sinh(\frac{\delta}{2})}\} \tag{65}$$

Finally, a procedure for incorporating the effect of curvature into the distribution function is given in Ref. [38], where the arc length distribution is given in the inverse form by

$$\xi(s) = I\,\frac{F(s)}{F(s_t)} \tag{66}$$

where

$$F(s) = \int_o^s \sqrt{1 + A(\bar{s})\ K(\bar{s})]^2 + [A'(\bar{s})]^2}\ d\bar{s} \tag{67}$$

s_t is the total arc length, $K(\bar{s})$ is the curvature, and $A(\bar{s})$ is any distribution function (in the inverse form) that would be used without consideration of curvature.

2. Multi-Directional Interpolation

A. Transfinite interpolation

In two directions we may write a linear Lagrange interpolation function individually in each curvilinear direction:

$$\underset{\sim}{r}(\xi,\eta) = \sum_{n=1}^{2} \phi_n(\tfrac{\xi}{I})\, \underset{\sim}{r}(\xi_n,\eta) \qquad (68a)$$

and

$$\underset{\sim}{r}(\xi,\eta) = \sum_{m=1}^{2} \psi_m(\tfrac{\eta}{J})\, \underset{\sim}{r}(\xi,\eta_m) \qquad (68b)$$

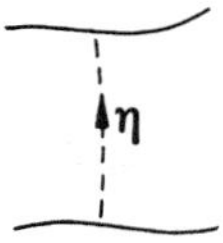

This interpolation is now called "transfinite" since it matches the function on the entire boundary defined by $\xi=0$ and $\xi=I$ in the first equation, or by $\eta=0$ and $\eta=J$ in the second, i.e., at a nondenumerable number of points,cf. Ref. [40] and [41]).

The tensor product form

$$\underset{\sim}{r}(\xi,\eta) = \sum_{n=1}^{2} \sum_{m=1}^{2} \phi_n(\tfrac{\xi}{I})\, \psi_m(\tfrac{\eta}{J})\, \underset{\sim}{r}_{nm} \qquad (69)$$

where $\underset{\sim}{r}_{nm} = \underset{\sim}{r}(\xi_n,\eta_m)$ matches the function at the four corners:

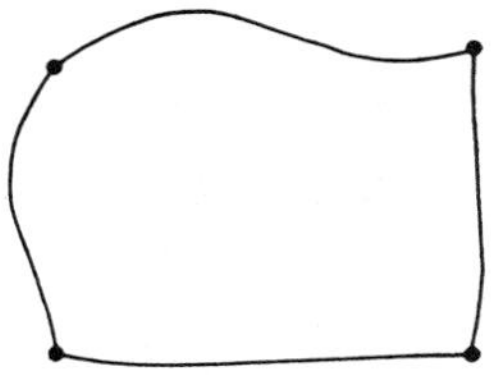

It does not, however, match the function on all the boundary.

The sum of Eq. (68a) and (68b),

$$\underset{\sim}{S}(\xi,\eta) = \sum_{n=1}^{2} \phi_n(\tfrac{\xi}{I})\, \underset{\sim}{r}(\xi_n,\eta) + \sum_{m=1}^{2} \psi_m(\tfrac{\eta}{J})\, \underset{\sim}{r}(\xi,\eta_m) \qquad (70)$$

when evaluated on the $\xi=0$ boundary gives

$$\underset{\sim}{S}(0,\eta) = \underset{\sim}{r}(0,\eta) + \sum_{m=1}^{2} \psi_m(\tfrac{\eta}{J})\, \underset{\sim}{r}(0,\eta_m) \tag{71}$$

This does not match the function on the $\xi=0$ boundary because of the second term on the right, which is an interpolation between the ends of this boundary:

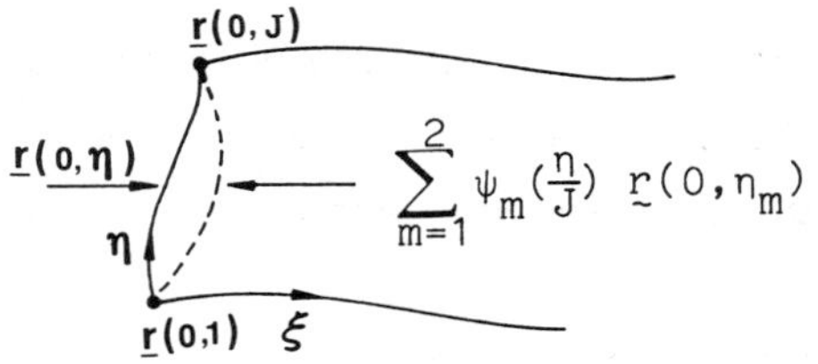

Similar effects occur on all the other boundaries, and the discrepancy on the $\xi=I$ boundary is

$$\sum_{m=1}^{2} \psi_m(\tfrac{\eta}{J})\, \underset{\sim}{r}(I,\eta_m)$$

The discrepancies on both of these boundaries can be removed by subtracting from $\underset{\sim}{S}(\xi,\eta)$ a function formed by interpolating the discrepancies between the two boundaries:

$$\underset{\sim}{R}(\xi,\eta) = \sum_{n=1}^{2} \phi_n(\tfrac{\xi}{I}) \left[\sum_{m=1}^{2} \psi_m(\tfrac{\eta}{J})\, \underset{\sim}{r}(\xi_n,\eta_m)\right] \tag{72}$$

But this is simply the tensor product form given by Eq. (69), which matches the function at the four corners.

The function $\underset{\sim}{S}-\underset{\sim}{R}$ then matches the function on all four sides of the boundary, so that we have the transfinite interpolation form,

$$\underset{\sim}{r}(\xi,\eta) = \sum_{n=1}^{2} \phi_n(\tfrac{\xi}{I})\, \underset{\sim}{r}(\xi_n,\eta) + \sum_{m=1}^{2} \psi_m(\tfrac{\eta}{J})\, \underset{\sim}{r}(\xi,\eta_m)$$

$$- \sum_{n=1}^{2} \sum_{m=1}^{2} \phi_n(\tfrac{\xi}{I})\, \psi_m(\tfrac{\eta}{J})\, \underset{\sim}{r}(\xi_n,\eta_m) \qquad (73)$$

which matches the function on the entire boundary. By contrast, the tensor product form

$$\underset{\sim}{r}(\xi,\eta) = \sum_{n=1}^{2} \sum_{m=1}^{2} \phi_n(\tfrac{\xi}{I})\, \psi_m(\tfrac{\eta}{J})\, \underset{\sim}{r}(\xi_n,\eta_m) \qquad (74)$$

matches the function only at the four corners on the boundary. This generalizes to the interpolation from a set of N + M intersecting curves for which the univariate interpolation is given by

$$\underset{\sim}{r}(\xi,\eta) = \sum_{n=1}^{N} \phi_n(\tfrac{\xi}{I})\, \underset{\sim}{r}(\xi_n,\eta) \qquad (75a)$$

and

$$\underset{\sim}{r}(\xi,\eta) = \sum_{m=1}^{M} \psi_m(\tfrac{\eta}{J})\, \underset{\sim}{r}(\xi,\eta_m) \qquad (75b)$$

where now the "blending" functions, ϕ_n and ψ_m, are any functions which satisfy the cardinality conditions

$$\phi_n(\tfrac{\xi_l}{I}) = \delta_{nl} \qquad n = 1,2,\ldots,N, \qquad l = 1,2,\ldots,N$$

$$\psi_m(\tfrac{\eta_l}{J}) = \delta_{ml} \qquad m = 1,2,\ldots,M, \qquad l = 1,2,\ldots,M \qquad (76)$$

The general form of the transfinite interpolation then is

$$\underset{\sim}{r}(\xi,\eta) = \sum_{n=1}^{N} \phi_n(\tfrac{\xi}{I})\, \underset{\sim}{r}(\xi_n,\eta) + \sum_{m=1}^{M} \psi_m(\tfrac{\eta}{J})\, \underset{\sim}{r}(\xi,\eta_m)$$

$$- \sum_{n=1}^{N} \sum_{m=1}^{M} \phi_n(\tfrac{\xi}{I})\, \psi_m(\tfrac{\eta}{J})\, \underset{\sim}{r}(\xi_n,\eta_m) \qquad (77)$$

while the tensor-product form is

$$\underset{\sim}{r}(\xi,\eta) = \sum_{n=1}^{N} \sum_{m=1}^{M} \phi_n(\tfrac{\xi}{I})\, \psi_m(\tfrac{\eta}{J})\, \underset{\sim}{r}(\xi_n,\eta_m) \qquad (78)$$

Eq. (77) can be written in the form

$$\underset{\sim}{r}(\xi,\eta) = \sum_{m=1}^{M} \psi_m(\tfrac{\eta}{J})\underset{\sim}{r}(\xi,\eta_m) +$$

$$\sum_{n=1}^{N} \phi_n(\tfrac{\xi}{I})\, [\underset{\sim}{r}(\xi_n,\eta) - \sum_{m=1}^{M} \psi_m(\tfrac{\eta}{J})\, \underset{\sim}{r}(\xi_n,\eta_m)]$$

But here the first term is the result at each point in the field of the unidirectional interpolation in the η-direction, and the bracket is the difference between the specified values on the $\xi=\xi_n$ lines and the result of the unidirectional interpolation on those lines. The two-directional transfinite interpolation can thus be implemented in two unidirectional interpolation steps by first performing the unidirectional interpolation in one direction, say η, over the entire field, calling the result $\underset{\sim}{F}_1(\xi,\eta)$:

$$\underset{\sim}{F}_1(\xi,\eta) = \sum_{m=1}^{M} \psi_m(\tfrac{\eta}{J})\, \underset{\sim}{r}(\xi,\eta_m) \qquad (79a)$$

then interopolating the discrepancy on the $\xi=\xi_n$ lines over the entire field in the other direction, ξ here, calling the result $\underset{\sim}{F}_2(\xi,\eta)$:

$$\underset{\sim}{F}_2(\xi,\eta) = \sum_{n=1}^{N} \phi_n(\tfrac{\xi}{I})[\underset{\sim}{r}(\xi_n,\eta) - \underset{\sim}{F}_1(\xi_n,\eta)] \qquad (79b)$$

and then adding $\underset{\sim}{F}_1$ and $\underset{\sim}{F}_2$:

$$\underset{\sim}{r}(\xi,\eta) = \underset{\sim}{F}_1(\xi,\eta) + \underset{\sim}{F}_2(\xi,\eta) \qquad (79c)$$

The transfinite interpolation form given by Eq. (77) is the algebraically best approximation, while the tensor product from of Eq. (78) is the algebraically worst (cf. Ref. [40]). The difference between these two forms should be fully understood. The transfinite interpolation form, Eq. (77), interpolates to the entirety of a set of intersecting arbitrary curves, while the tensor product form, Eq. (78), interpolates <u>only</u> to the intersections of these curves. The interpolation function defined by Eq. (77) with N=M=2, using the Lagrange interpolation polynomials as the blending functions, is termed the transfinite bilinear interpolant. With N=M=3, this form is the transfinite biquadratic interpolant. Other immediate candidates for the blending functions are the Hermite interpolation polynomials and the splines, since these all can be expressed in the form of Eq. (75). The spline-blended form gives the smoothest grid with continuous second derivatives.

B. Projectors

Now let $P_\xi(\underset{\sim}{r})$ be a one-dimensional interpolation function in the ξ-direction which matches $\underset{\sim}{r}$ on the N lines, $\xi=\xi_n$ (n=1,2,...N),:

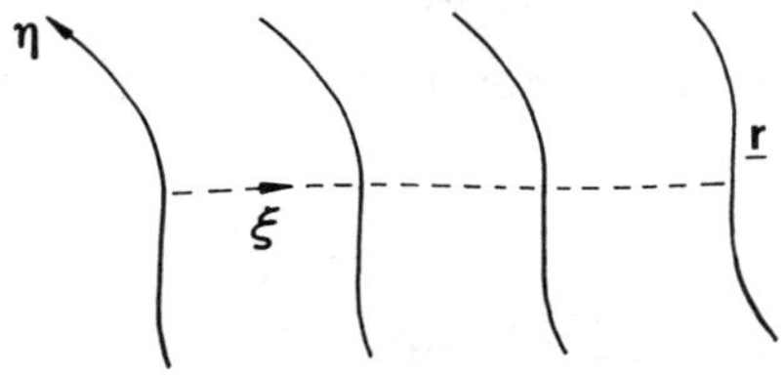

(Note that the subscript ξ here does not denote differentiation.) Similarly, let $P_\eta(\underset{\sim}{r})$ match $\underset{\sim}{r}$ on the M lines, $\eta=\eta_m$ (m = 1,2,...M). These interpolations are performed by projectors, P_ξ and P_η, which are assumed to be idempotent linear operators. Projectors are discussed in more detail in Ref. [40]. Some discussion is also given in Ref. [37]. The product projector, $P_\xi[P_\eta(\underset{\sim}{r})]$, then matches the function $P_\eta(\underset{\sim}{r})$, instead of $\underset{\sim}{r}$, on the N lines, $\xi=\xi_n$:

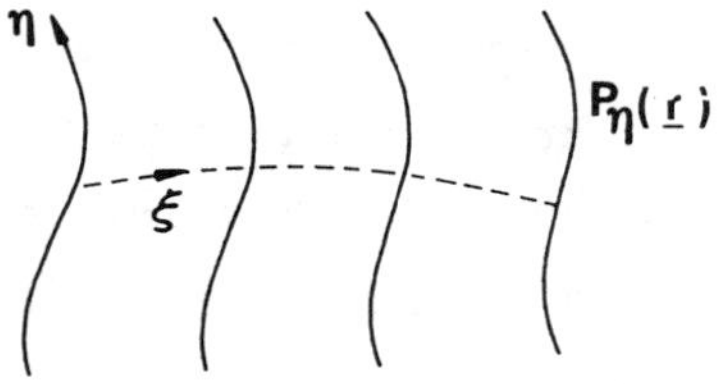

Then, since $P_\eta(\underset{\sim}{r})$ matches $\underset{\sim}{r}$ on the M lines, $\eta=\eta_m$, it follows that the product projector will match $\underset{\sim}{r}$ at the NxM points (ξ_n,η_m):

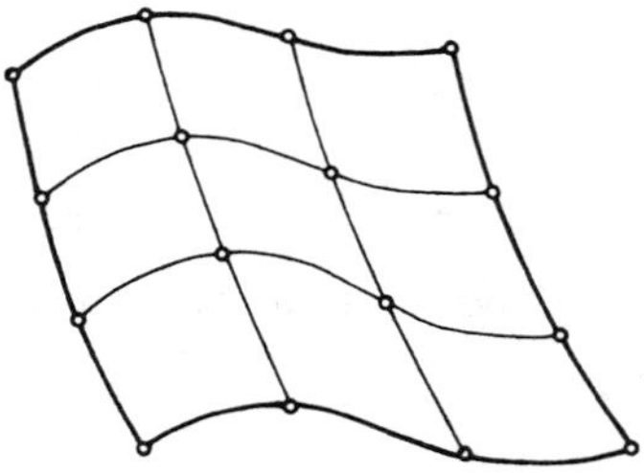

Clearly the same conclusion is reached for the product projector $P_\eta[P_\xi(\underset{\sim}{r})]$, so that the projectors P_ξ and P_η commute.

The sum projector, $P_\xi(\underset{\sim}{r})+P_\eta(\underset{\sim}{r})$ matches $\underset{\sim}{r}+P_\eta(\underset{\sim}{r})$ on the N lines $\xi=\xi_n$, and matches $\underset{\sim}{r}+P_\xi(\underset{\sim}{r})$ on the M lines $\eta=\eta_m$. It should be clear then that the projector, $P_\xi(\underset{\sim}{r})+P_\eta(\underset{\sim}{r})-P_\xi[P_\eta(\underset{\sim}{r})]$ will match $\underset{\sim}{r}$ on the N lines $\xi=\xi_n$, since $P_\xi[P_\eta(\underset{\sim}{r})]$ matches $P_\eta(\underset{\sim}{r})$ on these lines. Similarly, the projector $P_\xi(\underset{\sim}{r})+P_\eta(\underset{\sim}{r})-P_\eta[P_\xi(\underset{\sim}{r})]$ matches $\underset{\sim}{r}$ on the M lines $\eta=\eta_m$. Therefore, since $P_\xi P_\eta = P_\eta P_\xi$, the Boolean sum projector, $P_\xi \oplus P_\eta = P_\xi + P_\eta - P_\xi P_\eta$, will match $\underset{\sim}{r}$ on the entirety of the N+M lines $\xi=\xi_n$ and $\eta=\eta_m$ which includes, of course, the entire boundary of the region.

In summary, the individual projectors, P_ξ and P_η, interpolate undirectionally between two opposing boundaries:

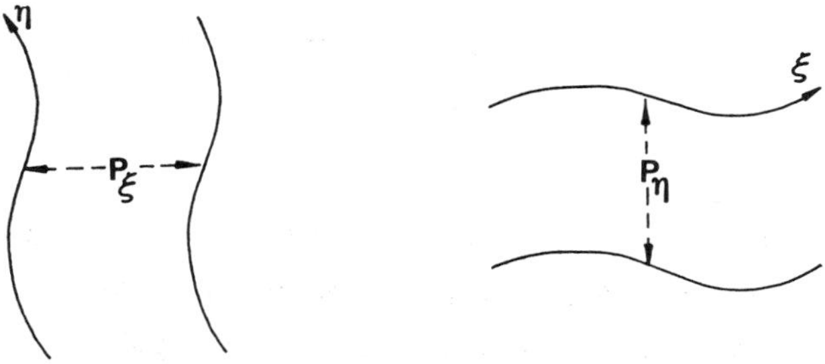

The product projector, $P_\xi P_\eta$, interpolates in two directions from the four corners:

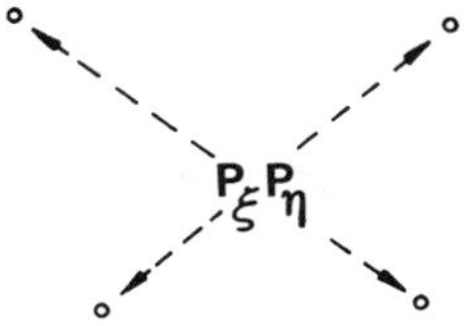

The Boolean sum projector, $P_\xi \oplus P_\eta$, interpolates from the entire boundary:

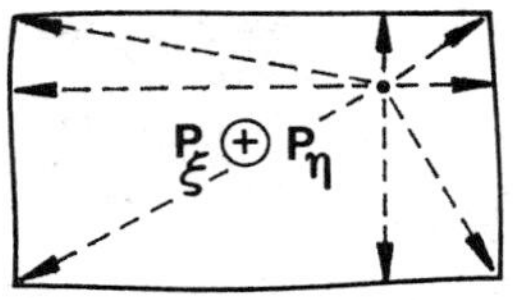

In three dimensions, the individual projectors, P_ξ, P_η, and P_ζ, interpolate undirectionally between two opposing faces of the six-sided region:

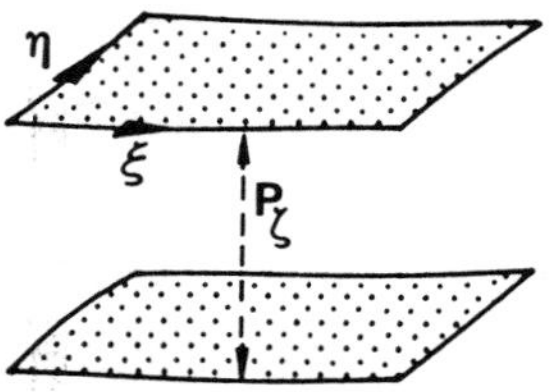

(matching $\underset{\sim}{r}$ on each of the two faces in each case). The double product projector, $P_\xi P_\eta$, interpolates in two directions from the four edges along which ξ and η are constant:

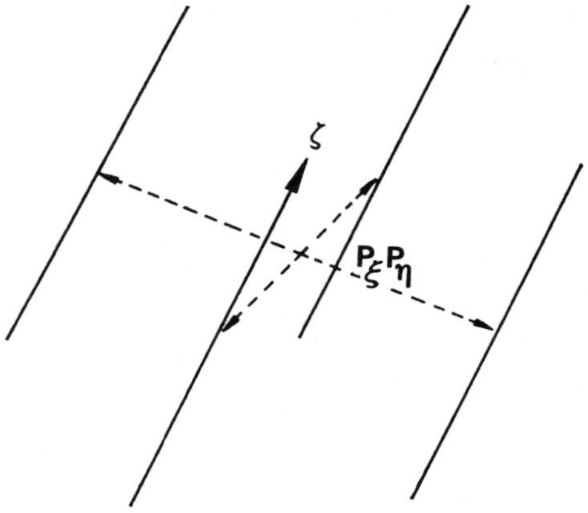

(matching $\underset{\sim}{r}$ on each of these edges). The Boolean sum projector

$$P_\xi \oplus P_\eta = P_\xi + P_\eta - P_\xi P_\eta \tag{80}$$

interpolates in two directions from the four faces on which either ξ or η is constant:

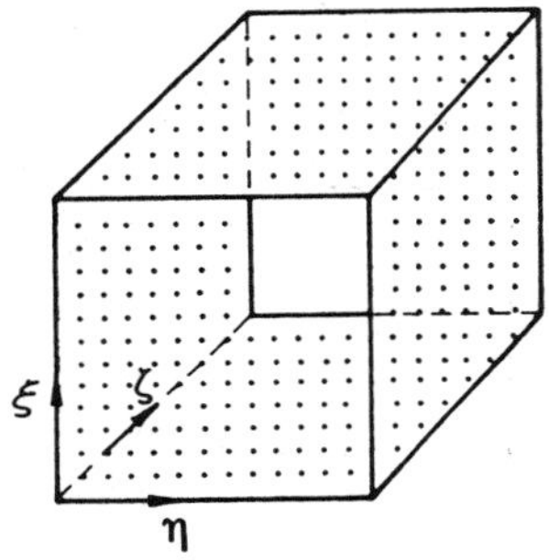

(matching $\underset{\sim}{r}$ on all of these faces).

The Boolean sum projector

$$P_\xi P_\eta \oplus P_\zeta = P_\xi P_\eta + P_\zeta - P_\xi P_\eta P_\zeta \tag{81}$$

interpolates in three directions, matching $\underset{\sim}{r}$ on the four edges on which ξ and η are constant and also on the two faces on which ζ is constant:

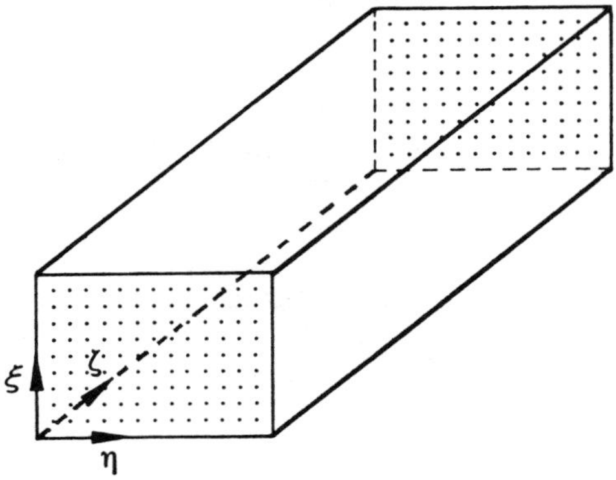

The Boolean sum projector

$$P_\xi P_\eta \oplus P_\eta P_\zeta \oplus P_\zeta P_\xi = P_\xi P_\eta + P_\eta P_\zeta + P_\zeta P_\xi - 2P_\xi P_\eta P_\zeta \qquad (82)$$

interpolates in three directions, with $\underline{r}$ matched on all twelve edges:

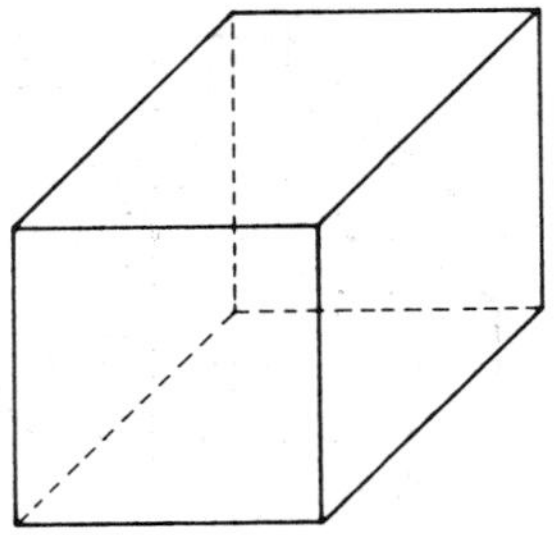

The triple product projector, $P_\xi P_\eta P_\zeta$, interpolates $\underline{r}$ from the eight corners:

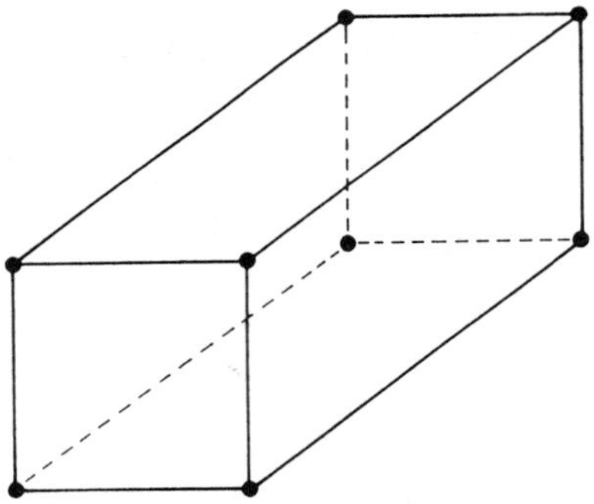

Finally the Boolean sum projector

$$P_\xi \oplus P_\eta \oplus P_\zeta = P_\xi + P_\eta + P_\zeta - P_\xi P_\eta - P_\eta P_\zeta - P_\zeta P_\xi + P_\xi P_\eta P_\zeta \qquad (83)$$

matches $\underset{\sim}{r}$ on the entire boundary.

Much cancellation occurs in the algebraic manipulation of the projectors involved in developing the above relations, since $P_\xi P_\xi = P_\xi$, etc. Thus, for example,

$$P_\xi \oplus P_\xi P_\eta = P_\xi + P_\xi P_\eta - P_\xi P_\xi P_\eta = P_\xi + P_\xi P_\eta - P_\xi P_\eta = P_\xi$$

This is to be expected since interpolation by P_ξ matches the function on all of the two sides on which ξ is constant, while $P_\xi P_\eta$ matches the function on the four edges on which ξ and η are constant. But these edges are contained on the two sides cited, so that nothing is changed by adding $P_\xi P_\eta$ to P_ξ in the Boolean sense. The projector formed as the Boolean sum of all three of the individual projectors is algebraically maximum, while the triple product projector is algebraically minimal.

The importance of the projectors is that the structure given above allows multi-directional interpolation to be constructed systematically from unidirectional forms. With one-dimensional interpolation of the form of Eq. (75) we have

$$P_\xi(\underset{\sim}{r}) = \sum_{n=1}^{N} \phi_n\left(\frac{\xi}{I}\right) \underset{\sim}{r}(\xi_n, \eta) \tag{84a}$$

$$P_\eta(\underset{\sim}{r}) = \sum_{m=1}^{M} \psi_m\left(\frac{\eta}{J}\right) \underset{\sim}{r}(\xi, \eta_m) \tag{84b}$$

so that

$$P_\xi P_\eta(\underset{\sim}{r}) = P_\xi[P_\eta(\underset{\sim}{r})]$$

$$= \sum_{n=1}^{N} \phi_n(\tfrac{\xi}{I}) \left[\sum_{m=1}^{M} \psi_m(\tfrac{\eta}{J}) \underset{\sim}{r}(\xi_n, \eta_m) \right]$$

$$= \sum_{n=1}^{N} \sum_{m=1}^{M} \phi_n(\tfrac{\xi}{I}) \psi_m(\tfrac{\eta}{J}) \underset{\sim}{r}(\xi_n, \eta_m) \tag{85}$$

which is just the tensor product form given previously in Eq. (78), so that the two-directional transfinite interpolation corresponding to the projector $P_\xi \oplus P_\eta$ is just that given by Eq. (77). As noted above, spline interpolation also falls directly into this form, so that the multi-directional transfinite interpolation based on splines requires only the determination of the splines separately in the individual directions.

Although Hermite interpolation can be defined in terms of additional points, and thus be put in this same form also, the use of projectors allows a more direct statement as follows. For the projectors we have, following Eq. (5),

$$P_\xi(\underset{\sim}{r}) = \sum_{n=1}^{N} \phi_n(\tfrac{\xi}{I}) \underset{\sim}{r}(\xi_n, \eta) + \sum_{n=1}^{N} \Phi_n(\tfrac{\xi}{I}) \underset{\sim}{r}_\xi(\xi_n, \eta) \tag{86a}$$

and

$$P_\eta(\underset{\sim}{r}) = \sum_{m=1}^{M} \psi_m(\tfrac{\eta}{J}) \underset{\sim}{r}(\xi, \eta_m) + \sum_{m=1}^{M} \Psi_m(\tfrac{\eta}{J}) \underset{\sim}{r}_\eta(\xi, \eta_m) \tag{86b}$$

Now

$$
\begin{aligned}
P_\xi P_\eta(\underset{\sim}{r}) &= P_\xi[P_\eta(\underset{\sim}{r})] \\
&= \sum_{n=1}^{N} \phi_n(\tfrac{\xi}{I}) \, [\sum_{m=1}^{M} \psi_m(\tfrac{\eta}{J}) \, \underset{\sim}{r}(\xi_n,\eta_m) \\
&\quad + \sum_{m=1}^{M} \Psi_m(\tfrac{\eta}{J}) \, \underset{\sim}{r}_\eta(\xi_n,\eta_m)] \\
&\quad + \sum_{n=1}^{N} \Phi_n(\tfrac{\xi}{I}) \, [\sum_{m=1}^{M} \psi_m(\tfrac{\eta}{J}) \, \underset{\sim}{r}_\xi(\xi_n,\eta_m) \\
&\quad + \sum_{m=1}^{M} \Psi_m(\tfrac{\eta}{J}) \, \underset{\sim}{r}_{\xi\eta}(\xi_n,\eta_m)] \\
&= \sum_{n=1}^{N} \sum_{m=1}^{M} [\phi_n(\tfrac{\xi}{I}) \, \psi_m(\tfrac{\eta}{J}) \, \underset{\sim}{r}(\xi_n,\eta_m) \\
&\quad + \phi_n(\tfrac{\xi}{I}) \, \Psi_m(\tfrac{\eta}{J}) \, \underset{\sim}{r}_\eta(\xi_n,\eta_m) + \Phi_n(\tfrac{\xi}{I}) \, \psi_m(\tfrac{\eta}{J}) \, \underset{\sim}{r}_\xi(\xi_n,\eta_m) \\
&\quad + \Phi_n(\tfrac{\xi}{I}) \, \Psi_m(\tfrac{\eta}{J}) \, \underset{\sim}{r}_{\xi\eta}(\xi_n,\eta_m)] \qquad (87)
\end{aligned}
$$

Then the two-directional transfinite interpolation can be constructed by substitution of Eq. (86) and (87) into the projector $P_\xi \oplus P_\eta$. Here the tensor product form, $P_\xi P_\eta$, interpolates from the values of the function, its two first derivatives, and the cross-derivative at the four corners of the boundary. The transfinite interpolation form, $P_\xi \oplus P_\eta$, however, interpolates from the value of the function and its normal derivative on the entire boundary.

The triple product corresponding to Eq. (84) is simply

$$P_\xi P_\eta P_\zeta(\underset{\sim}{r}) = \sum_{n=1}^{N} \sum_{m=1}^{M} \sum_{l=1}^{L} \phi_n(\tfrac{\xi}{I})\ \psi_m(\tfrac{\eta}{J})\ \theta_l(\tfrac{\zeta}{K})\ \underset{\sim}{r}(\xi_n,\eta_m,\zeta_l) \tag{88}$$

Recall that with the unidirectional form given by Eq. (44), we have in these relations L=M=N=2 and

$$\phi_1 = 1 - \phi, \qquad \phi_2 = \phi$$

$$\psi_1 = 1 - \psi, \qquad \psi_2 = \psi$$

$$\theta_1 = 1 - \theta, \qquad \theta_2 = \theta$$

The above evaluations of the product projectors serve to illustrate the evaluation of such products for general projectors, i.e., that the effect of the product projectors is simply an interpolation in one-direction of an interpolant in another direction. This allows the multi-directional transfinite interpolation to be constructed from the Boolean sums of the projectors given above using any appropriate unidirectional interpolation forms as the basis projectors. It should also be noted that the unidirectional interpolation does not have to be of the same form in all the directions. Thus Lagrange interpolation could be used in one of the directions while Hermite is used in the other direction of a two-directional construction. As noted above, the blending functions do not have to be polynomials. In fact, all of the unidirectional interpolation that was discussed earlier in this chapter can be applied in the context of multi-directional interpolation based on the projectors. This freedom to combine different types of univariate interpolation gives considerable flexibility to transfinite interpolation based on the projector structure, and allows attention to be focused on developing appropriate unidirectional interpolations, the multi-directional format then following automatically.

The projectors allow the transfinite interpolation to be easily set up as a sequence of unidirectional interpolations, in the manner discussed above. Thus in the two-directional case, Eq. (80) can be written as

$$P_\xi \oplus P_\eta = P_\xi + P_\eta\, (I - P_\xi) \tag{89}$$

where I indicates the identity operation. But here the first term is clearly the unidirectional interpolation in the ξ-direction, while the parenthesis $(I-P_\xi)$ is the discrepancy on the η-lines on which $\underline{r}$ is specified that results from this ξ-interpolation. The second term then is the unidirectional interpolation of this descrepancy interpolated in the η-direction.

The two-directional interpolation thus can be implemented by: (1) interpolating $\underline{r}$ in the ξ-direction, (2) calculating the discrepancy between $\underline{r}$ and this result on the η-lines that are to be used in the η-interpolation, (3) interpolating this discrepancy in the η-direction, and (4) adding the result of this η-interpolation to that of the ξ-interpolation. Symbolically these steps can be stated as the following:

$$\underline{F}_1 = P_\xi\, \underline{r}$$

$$\underline{F}_2 = P_\eta\, (\underline{r} - \underline{F}_1)$$

$$\underline{r}(\xi,\eta) = \underline{F}_1 + \underline{F}_2 \tag{90}$$

Obviously, the order of the unidirectional interpolation is immaterial.

Similarly, Eq. (83) can be written as

$$P_\xi \oplus P_\eta \oplus P_\zeta = P_\xi + (P_\eta + P_\zeta - P_\eta P_\zeta)(I - P_\xi) \tag{91}$$

The three-directional interpolation thus can be implemented by (1) interpolating $\underset{\sim}{r}$ in the ξ-direction, (2) calculating the discrepancy between $\underset{\sim}{r}$ and this result on the η-surfaces and ζ-surfaces that are to be used in the interpolation in those directions, (3) interpolating this discrepancy by two-directional interpolation, and (4) adding this result to that of the ξ-interpolation. These operations can be stated as

$$\underset{\sim}{F}_1 = P_\xi \underset{\sim}{r}$$

$$\underset{\sim}{F}_2 = P_\eta (\underset{\sim}{r} - \underset{\sim}{F}_1)$$

$$\underset{\sim}{F}_3 = P_\zeta (\underset{\sim}{r} - \underset{\sim}{F}_1 - F_2)$$

$$\underset{\sim}{r}(\xi,\eta) = \underset{\sim}{F}_1 + \underset{\sim}{F}_2 + \underset{\sim}{F}_3 \tag{92}$$

Exercises

1. Show that with N=2 and ψ_n=constant, the multi-surface interpolation is equivalent to the linear Lagrange interpolation. Note that the Lagrange polynomials here satisfy Eq. (25) with $\frac{\xi_2}{I} = \frac{1}{2}$. For other choices of ξ_2, other quadratic polynomials result from Eq. (25), so that there exists a one-parameter family of cubic forms of the multi-surface interpolation. Similarly, a two-parameter family of quadratic forms exists, etc.

2. Show that the quadratic form of the multi-surface interpolation is given by

$$\underset{\sim}{r}(\xi) = (1 - \frac{\xi}{I})^2 \underset{\sim}{r}(0) + 2 \frac{\xi}{I} (1 - \frac{\xi}{I}) \underset{\sim}{P}_2 + (\frac{\xi}{I})^2 \underset{\sim}{r}(I)$$

3. Show that the quadratic forms of the multi-surface and Bezier interpolations are equivalent.

4. Show that with N=4 and ψ_n the quadradic Lagrange interpolation polynomials given on p. 282, the interpolation functions for the multi-surface interpolation are given by

$$G_1(\xi) = \frac{\xi}{I} - \frac{3}{2} (\frac{\xi}{I})^2 + \frac{2}{3} (\frac{\xi}{I})^3$$

$$G_2(\xi) = 2 (\frac{\xi}{I})^2 - \frac{4}{3} (\frac{\xi}{I})^3$$

$$G_3(\xi) = \frac{2}{3} (\frac{\xi}{I})^3 - \frac{1}{2} (\frac{\xi}{I})^2$$

the multi-surface interpolation is equivalent to the cubic Hermite interpolation.

5. Show that $S(\xi)$ is given by Eq. (38) for Lagrange interpolation. (Hint: If all the $\underline{r}_n$ in Eq. (2) are the same, the interpolation must reproduce this value, hence the Lagrange interpolation polynomials satisfy

$$\sum_{n=1}^{N} \phi_n \left(\frac{\xi}{I}\right) = 1$$

6. Show that for quadratic Lagrange interpolation, uniformity requires that $\underline{r}_2$ be selected such that

$$[\underline{r}_2 - \underline{r}(0)] \cdot \underline{\iota} = \frac{1}{2}$$

Note that this does not completely determine $\underline{r}_2$.

7. Show that uniformity is achieved with cubic Hermite interpolation with $\frac{\xi_2}{I} = \frac{1}{2}$ with orthogonality at the boundaries if the spacings at the boundaries are given by

$$s = \frac{1}{\underline{n} \cdot \underline{\iota}}$$

where $\underline{n}$ is the unit normal to the boundary. This completely specifies the interpolation in this case. However, as noted in Exercise 1, ξ_2 is a free parameter.

8. Show that for multi-surface interpolation with N=4 and $\frac{\xi_2}{I} = \frac{1}{2}$ uniformity is achieved with

$$[\underline{P}_2 - \underline{r}(0)] \cdot \underline{\iota} = [\underline{r}(I) - \underline{P}_3] \cdot \underline{\iota} = \frac{1}{6}$$

9. Show also that with orthogonality at the boundary the result of Exercise 8 completely determines all of the interpolation parameters, i.e., that

$$\underset{\sim}{P}_2 = \underset{\sim}{r}(0) + \frac{1}{6\underset{\sim}{n}(0) \cdot \underset{\sim}{I}} \underset{\sim}{n}(0)$$

$$\underset{\sim}{P}_3 = \underset{\sim}{r}(I) - \frac{1}{6\underset{\sim}{n}(I) \cdot \underset{\sim}{I}} \underset{\sim}{n}(I)$$

where $\underset{\sim}{n}$ is the unit normal to the boundary. Hint: Use Eq. (7) and (29). For general ξ_2 the 1/6 is replaced by $1/2 - 1/[6(\xi_2/I)]$ and $1/2 - 1/[6(1-\xi_2/I)]$ in the above expressions involving $\underset{\sim}{P}_2$ and $\underset{\sim}{P}_3$, respectively. Some effects of the choice of ξ_2 are shown in Ref. [32].

10. Show that local uniformity on the interval $\xi_n \leq \xi \leq \xi_{n+1}$ for the multi-surface interpolation based on piecewise-linear functions requires that

$$\frac{C_n}{C_{n+1}} = \frac{\xi_{n+1} - \xi_{n-1}}{\xi_{n+2} - \xi_n}$$

where $C_n = (\underset{\sim}{P}_{n+1} - \underset{\sim}{P}_n) \cdot \underset{\sim}{I}$, with $\underset{\sim}{I} = \dfrac{\underset{\sim}{P}_{n+2} - \underset{\sim}{P}_n}{|\underset{\sim}{P}_{n+2} - \underset{\sim}{P}_n|^2}$

11. Consider a rectangular physical region with equally-spaced points on the bottom and top, but with unequal spacing on the left and right sides (but with the same point distribution on both of these sides). Show that horizontal interpolation will reflect the unequal spacing of the horizontal grid lines in the field, but that vertical interpolation will not. Show also that the unequal spacing is reflected with transfinite interpolation.

12. Show that transfinite interpolation based on linear blending functions will reflect the unequal boundary point spacing in the field for the rectangular physical region of Exercise 11, but will not for a C-grid. From the consideration of transfinite interpolation as a sequence of unidirectional interpolations, explain why this is so.

13. Show that with cubic Lagrange interpolation the locations of the two intermediate surfaces, $\mathbf{r}_2$ and $\mathbf{r}_3$, are related to the slopes at both ends. Note the contrast between this and the multi-surface interpolation where each of the intermediate surfaces depends on only the slope at one end.

14. Give the cubic form of Lagrange interpolation.

15. Show that in two dimensions transfinite interpolation is equivalent to a generation system based on the fourth - order partial differential equation

$$\mathbf{r}_{\xi\xi\eta\eta} = 0$$

(This is also equivalent to the quadralaterial isoparemetric elements often used to construct finite element meshes.)

IX. ORTHOGONAL SYSTEMS

Orthogonal coordinate systems produce fewer additional terms in transformed partial differential equations, and thus reduce the amount of computation required. Also, as has been noted in Chapter V, severe departure from orthogonality will introduce truncation error in difference expressions. A general discussion of orthogonal systems on planes and curved surfaces is given in Ref. [42], and various generation procedures are surveyed in Ref. [42] and Ref. [1].

In numerical solutions, the concept of numerical orthogonality, i.e., that the off-diagonal metric coefficients vanish when evaluated numerically, is usually more important than strict analytical orthogonality, especially when the equations to be solved on the system are in the conservative law form.

There are basically two types of orthogonal generation systems, those based on the construction of an orthogonal system from a non-orthogonal system, and those involving field solutions of partial differential equations. The first approach involves the construction of orthogonal trajectories on a given non-orthogonal system. Here one set of coordinate lines of the non-orthogonal system is retained, while the other set is replaced by lines emanating from a boundary and constructed by integration across the field so as to cross each line of the retained set orthogonally. Control of the line spacing is exercised through the generation of the non-orthogonal system and through the point distribution on the boundary from which the trajectories start. The point distributions on only three of the four boundaries can be specified. Several methods for the construction of orthogonal trajectories are discussed in Ref. [42] and Ref. [1]. If point distributions are to be specified on all boundaries, the field approach must be taken, and it is to this approach that this chapter is primarily directed.

1. General Formulation

The characteristic criterion for orthogonal coordinates is the vanishing of the off-diagonal elements of the metric tensor, i.e., $g_{ij} = g^{ij} = 0$ for $i \neq j$. Thus the Jacobian of the transformation is simply

$$\sqrt{g} = \sqrt{g_{11}g_{22}g_{33}} \tag{1}$$

For brevity, writing

$$h_i = \sqrt{g_{ii}} \qquad i = 1,2,3$$

it is easy to show from Eq. (III-74) that

$$\nabla^2\xi^i = \frac{1}{\sqrt{g}}\frac{\partial}{\partial\xi^i}\left(\frac{h_j h_k}{h_i}\right) \qquad \begin{array}{l}(i,j,k) \text{ cyclic}\\ i=1,2,3\end{array} \tag{2}$$

The general differential equations satisfied in the transformed region are, from Eq. (VI-10),

$$\sum_{i=1}^{3}\sum_{j=1}^{3} g^{ij}\,\underline{r}_{\xi^i\xi^j} + \sum_{k=1}^{3}(\nabla^2\xi^k)\,\underline{r}_{\xi^k} = 0 \tag{3}$$

Substituting Eq. (2) in (3) for the Laplacians, these grid generation equations take the following simpler form for an orthogonal system:

$$\sum_{i=1}^{3}\left(\frac{h_j h_k}{h_i}\,\underline{r}_{\xi^i}\right)_{\xi^i} = 0 \qquad (i,j,k) \text{ cyclic} \tag{4}$$

where $\underset{\sim}{r}$ is the cartesian coordinate vector.

On the other hand, starting from Eq. (2), by writing

$$H_i = \frac{h_i}{h_j h_k} \qquad (i,j,k) \text{ cyclic} \quad i = 1,2,3$$

and using the chain rule of differentiation, we get the generation equations in the physical region as

$$\sum_{i=1}^{3} \frac{\partial}{\partial x_i} \left(H_i \frac{\partial \xi^m}{\partial x_i}\right) = 0 \qquad m = 1,2,3 \tag{5}$$

Another fundamental set of equations for orthogonal coordinates are known as Lame's equations, stated as

$$\frac{\partial}{\partial \xi^j}\left(\frac{1}{h_j}\frac{\partial h_k}{\partial \xi^j}\right) + \frac{\partial}{\partial \xi^k}\left(\frac{1}{h_k}\frac{\partial h_j}{\partial \xi^k}\right) + \frac{1}{h_i^2}\frac{\partial h_j}{\partial \xi^i}\frac{\partial h_k}{\partial \xi^i} = 0 \tag{6a}$$

$$\frac{\partial^2 h_i}{\partial \xi^j \partial \xi^k} = \frac{1}{h_j}\frac{\partial h_i}{\partial \xi^j}\frac{\partial h_j}{\partial \xi^k} + \frac{1}{h_k}\frac{\partial h_i}{\partial \xi^h}\frac{\partial h_k}{\partial \xi^j} \tag{6b}$$

where (i,j,k) are cyclic. Equations (6) express essentially the condition that the curvilinear coordinates are to be introduced in an Euclidean space. (cf. Ref. [27]). In three dimensions, Eq. (6) represents six equations, although there are only three distinct metric coefficients, h_1, h_2, h_3.

In summary the equations (2), (4), (5) and (6), together with the vanishing of the off-diagonal metric elements, are the fundamental equations which any orthogonal coordinate system must satisfy.

2. Two-Dimensional Orthogonal Coordinates

The fundamental equations for two-dimensional orthogonal coordinates are collected below as a particular case of the equations (2) - (6):

I. Transformed plane: $g_{12} = 0$ and (7a)

$$\frac{\partial}{\partial \xi^1}\left(\frac{h_2}{h_1}\underset{\sim}{r}_{\xi^1}\right) + \frac{\partial}{\partial \xi^2}\left(\frac{h_1}{h_2}\underset{\sim}{r}_{\xi^2}\right) = 0 \tag{7b}$$

$$\frac{\partial}{\partial \xi^1}\left(\frac{1}{h_1}\frac{\partial h_2}{\partial \xi^1}\right) + \frac{\partial}{\partial \xi^2}\left(\frac{1}{h_2}\frac{\partial h_1}{\partial \xi^2}\right) = 0 \tag{7c}$$

II. Physical plane: $g^{12} = 0$ and (8a)

$$\frac{\partial}{\partial x_1}\left(\frac{h_1}{h_2}\xi^1_{x_1}\right) + \frac{\partial}{\partial x_2}\left(\frac{h_1}{h_2}\xi^1_{x_2}\right) = 0 \tag{8b}$$

$$\frac{\partial}{\partial x_1}\left(\frac{h_2}{h_1}\xi^2_{x_1}\right) + \frac{\partial}{\partial x_2}\left(\frac{h_2}{h_1}\xi^2_{x_2}\right) = 0 \tag{8c}$$

Also, the Laplacians (2) take the simple forms

$$\nabla^2 \xi^1 = \frac{1}{h_1 h_2}\frac{\partial}{\partial \xi^1}\left(\frac{h_2}{h_1}\right) \tag{9a}$$

$$\nabla^2 \xi^2 = \frac{1}{h_1 h_2}\frac{\partial}{\partial \xi^2}\left(\frac{h_1}{h_2}\right) \tag{9b}$$

Considering Eq. (7a) and (8a), either of which provide the orthogonality condition, it is a straightforward matter to conclude that there exists a positive function F such that

$$\frac{\partial x_1}{\partial \xi^2} = -F\,\frac{\partial x_2}{\partial \xi} \quad , \qquad \frac{\partial x_2}{\partial \xi^2} = F\,\frac{\partial x_1}{\partial \xi^1} \tag{10}$$

and the Eq. (7a) is identically satisfied. In the same manner, from Eq. (8a),

$$\frac{\partial \xi^1}{\partial x_2} = -F\,\frac{\partial \xi^2}{\partial x_1}\,, \qquad \frac{\partial \xi^1}{\partial x_1} = F\,\frac{\partial \xi^2}{\partial x_2} \tag{11}$$

It is obvious that the positive function F is related to the grid aspect ratio:

$$F = h_2/h_1 = \sqrt{g_{22}/g_{11}} \tag{12}$$

The choice of the sign in Eq. (10) and (11) follows from the right-handedness of the system ξ^1, ξ^2.

Introducing (12) into Eq. (7b), while using Eq. (9), we get

$$g_{22}\mathbf{r}_{\xi^1\xi^1} + g_{11}\mathbf{r}_{\xi^2\xi^2} + g_{11}g_{22}(\mathbf{r}_{\xi^1}\nabla^2\xi^1 + \mathbf{r}_{\xi^2}\nabla^2\xi^2) = 0 \tag{13}$$

which forms the basic generation system for plane orthogonal coordinates. Though the generating equations (7b) and (13) are completely equivalent, nevertheless, the apparent difference in their structures must be taken into consideration to decide about the type of boundary conditions for their solution.

With Eq. (7b) as the generating system then the two options are: (i) Specify $F=h_2/h_1$ as a known function of ξ^1,ξ^2. This case covers the cases $F=\alpha$ and $F=\phi_1(\xi^1)\cdot\phi_2(\xi^2)$, where α=constant. For any constant α, Eq. (9) reduce to the Laplace equations $\nabla^2\xi^1=0$, $\nabla^2\xi^2=0$, and Eq. (7b) becomes

$$\alpha^2 \mathbf{r}_{\xi^1\xi^1} + \mathbf{r}_{\xi^2\xi^2} = 0 \tag{14}$$

For $\alpha=1$, the coordinates ξ^1,ξ^2 are isothermic, i.e., $h_2=h_1$, and so are conformal. Cases in which $\alpha\neq1$ have also been considered, and specific references are given in Ref. [1]. It is also of interest to state that starting from a conformal system (ξ^1,ξ^2), yet another system $(\bar{\xi}^1,\bar{\xi}^2)$ can be established by transforming the Laplace equations $\nabla^2\xi^1=0$, $\nabla^2\xi^2=0$, such that $\bar{F}\neq1$ and $\bar{F}$ is a product of a function of $\bar{\xi}^1$ and a function of $\bar{\xi}^2$. (cf. Ref. [1]). (ii) The other option is to calculate F iteratively. In this case the field values of F are updated by iteratively changing its values at the boundaries under the orthogonality condition $g_{12}=0$.

With Eq. (13) as the generating system, the two Laplacians $\nabla^2\xi^1$ and $\nabla^2\xi^2$ have to be specified. Following the nonorthogonal case, let

$$\nabla^2\xi^1 = \frac{1}{g_{11}g_{22}}(g_{11}P_1 + g_{22}P_2) \tag{15a}$$

$$\nabla^2\xi^2 = \frac{1}{g_{11}g_{22}}(g_{11}Q_1 + g_{22}Q_2) \tag{15b}$$

where $P_1,\ldots,Q_2$ are arbitrary specified functions of ξ^1,ξ^2. Using Eqs. (9) and (12) one can rewrite these equations as

$$\frac{\partial F}{\partial \xi^1} = P_1/F + FP_2 \tag{16a}$$

$$\frac{\partial F}{\partial \xi^2} = -FQ_1 - F^3Q_2 \tag{16b}$$

Thus, if $P_1,\ldots,Q_2$ are specified, the above equations provide a way to determine F. (Using the condition

$$\frac{\partial^2 F}{\partial\xi^1 \partial\xi^2} = \frac{\partial^2 F}{\partial\xi^2 \partial\xi^1}$$

one can establish a fourth order algebraic equation in F.) It is therefore concluded that the use of Eq. (13) with $P_1,\ldots,Q_2$ specified is equivalent to using Eq. (7b) in which F has explicitly been specified.

The above noted considerations are important in deciding about the type of boundary data needed for the solution of either Eq. (7b) or Eq. (13). The solution of Eq. (7b) with specified F, or the solution of Eq. (13) with specified $P_1,\ldots,Q_2$, does not allow an arbitrary point distribution on the domain boundaries. The reason for this is as follows: For example, on a boundary segment $\xi^2=\xi^2_o$=constant if $x_1(\xi^1,\xi^2_o)$ is prescribed, then from Eq. (10) the normal derivative $\partial x_2/\partial\xi^2$ becomes available. If in addition to $x_1(\xi^1,\xi^2_o)$, one also specifies $x_2(\xi^1,\xi^2_o)$, which amounts to specifying the complete boundary point distribution, then the problem becomes overdetermined. Thus for the cases under consideration, specification of the complete boundary point distribution is not possible. That is, Eq. (7b) with F specified, or Eq. (13) with specified $P_1,,\ldots,Q_2$, cannot be solved when the complete boundary point distribution is prescribed. The appropriate boundary conditions for such problems are discussed in the context of conformal coordinates in Section A.

The specification of the complete boundary point distribution is possible in the case when Eq. (7b) is solved without specifying F. An iterative approach can be used to update the values of F based on the changed values at the boudnaries. (cf. Section B).

A. Conformal systems

Considering first conformal systems, i.e., with $h_2=h_1$ and F=1, the basic equations from (9a,b) are

$$\nabla^2\xi^1 = 0, \qquad \nabla^2\xi^2 = 0 \tag{17a}$$

$$\xi^1_{x_1} = \xi^2_{x_2}, \qquad \xi^1_{x_2} = -\xi^2_{x_1} \tag{17b}$$

Let the domain in which the conformal coordinates are to be generated be bounded by a piecewise-smooth curve on which s is the arc length and n the outward normal. The Cauchy-Riemann equations (17b) on the boundary take the form

$$\xi^1_s = \xi^2_n, \qquad \xi^1_n = -\xi^2_s \tag{18}$$

Referring to the figure below, let the curves Γ_1 and Γ_2 be those portions on which ξ^1=constant, and the curves Γ_3 and Γ_4 be those on which ξ^2= constant. From Eq. (18) we readily find that on Γ_1 and Γ_2 the condition $\xi^2_n=0$, and on Γ_3 and Γ_4 the condition $\xi^1_n=0$, are to be imposed, where the subscript n indicates the normal derivative.

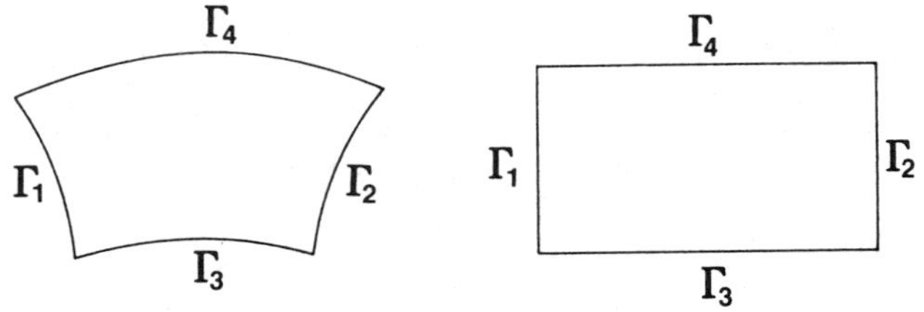

Therefore, for the generation of conformal coordinates, the properly posed boundary value problems are

$$\nabla^2 \xi^1 = 0$$

$$\text{on } \Gamma_1 \text{ and } \Gamma_2\text{: } \xi^1 = \xi^1_{(1)},\ \xi^1 = \xi^1_{(2)}, \text{ respectively}$$

$$\text{on } \Gamma_3 \text{ and } \Gamma_4\text{: } \xi^1_n = 0 \tag{19}$$

$$\nabla^2 \xi^2 = 0$$

$$\text{on } \Gamma_1 \text{ and } \Gamma_2\text{: } \xi^2_n = 0$$

$$\text{on } \Gamma_3 \text{ and } \Gamma_4\text{: } \xi^2 = \xi^2_{(1)},\ \xi^2 = \xi^2_{(2)}, \text{ respectively} \tag{20}$$

In the transformed plane the governing equations for conformal coordinates are obtained from (13):

$$\underline{r}_{\xi^1\xi^1} + \underline{r}_{\xi^2\xi^2} = 0 \tag{21a}$$

$$\frac{\partial x_1}{\partial \xi^1} = \frac{\partial x_2}{\partial \xi^2}, \qquad \frac{\partial x_1}{\partial \xi^2} = -\frac{\partial x_2}{\partial \xi^1} \tag{21b}$$

Taking ξ^1 and ξ^2 as monotonically increasing parameters having the ranges, $\xi^1_{(1)} \le \xi^1 \le \xi^1_{(2)}$, $\xi^2_{(1)} \le \xi^2 \le \xi^2_{(2)}$, the given equations of the curves Γ_1, Γ_2, Γ_3, Γ_4, respectively, can be expressed in parametric form as

$$\begin{aligned}
\Gamma_1&: \quad x_1 = x_1(\xi^1_{(1)}, \xi^2), \qquad x_2 = x_2(\xi^1_{(1)}, \xi^2)\\
\Gamma_2&: \quad x_1 = x_1(\xi^1_{(2)}, \xi^2), \qquad x_2 = x_2(\xi^1_{(2)}, \xi^2)\\
\Gamma_3&: \quad x_1 = x_1(\xi^1, \xi^2_{(1)}), \qquad x_2 = x_2(\xi^1, \xi^2_{(1)})\\
\Gamma_4&: \quad x_1 = x_1(\xi^1, \xi^2_{(2)}), \qquad x_2 = x_2(\xi^1, \xi^2_{(2)})
\end{aligned} \tag{22}$$

The specification of the boundary data in the form of (21) should at best be regarded as a statement of the problem, rather than as a procedure, since the exact boundary point-distribution in this form is not possible a'priori. To develop the procedure itself we regard the specification in (22) as an initial guess. However, this type of specification produces an overdetermined situation. For example, if on Γ_1 both $x_1(\xi^1_{(1)},\xi^2)$ and $x_2(\xi^1_{(1)},\xi^2)$ are specified, then from the first equation in (21b), $\partial x_i/\partial\xi^1$ can be calculated on this boundary. Thus both

$$(x_1)_{\xi^1=\xi^1_{(1)}} \quad \text{and} \quad (\partial x_1/\partial\xi^1)_{\xi^1=\xi^1_{(1)}}$$

become specified, which makes the problem overdetermined. Following this logic, we can isolate the proper arbitrarily specifed boundary values for Eq. (21) as follows: specifying $x_1(\xi^1_{(1)},\xi^2)$ on Γ_1, $x_1(\xi^1_{(2)},\xi^2)$ on Γ_2, $x_2(\xi^1,\xi^2_{(1)})$ on Γ_3, and $x_2(\xi^1,\xi^2_{(2)})$ on Γ_4. Thus, for the x_1-equation the normal derivative conditions on Γ_3 and Γ_4 are provided by the second equation in (21b) through the specified x_2-values. Similarly, for the x_2-equation the normal derivative conditions on Γ_1 and Γ_2 are provided by the second equation in (21b) through the specified x_1-values.

In any numerical procedure, the values of x_1 are determined by integration through the formula

$$(x_1)_j - (x_1)_{j-1} = -\int_{j-1}^{j} \frac{\partial x_2}{\partial\xi^1}\, d\xi^2 \tag{23}$$

and these values in turn give the new values of x_2 through the exact functional relations between x_1 and x_2 for these curves. Similarly, the values of x_2 are calculated by the formula

$$(x_2)_i - (x_2)_{i-1} = - \int_{i-1}^{i} \frac{\partial x_1}{\partial \xi^2} d\xi^1 \tag{24}$$

and then the new values of x_1 are determined by the functional relations between x_1 and x_2 for these curves. Further discussion of conformal systems is given in Chapter X.

B. Other systems

For general orthogonal systems, the basic equations for x_1 and x_2 remain Eq. (13). As noted earlier, the other constraint besides orthogonality ($g_{12}=0$) is now to specify the function F defined in Eq.(12), which is the ratio of the scale factors, i.e., the grid aspect ratio. One approach is to specify the function F explicitly, in which case, as with the conformal coordinates, it is not possible to specify an arbitrary point distribution on the boundaries. The set of equations in (7a) must be used to find the proper x_1 and x_2 values by integration on the appropriate boundaries. Another alternative is to specify an arbitrary point distribution on the boundaries, and leave the function F to be determined iteratively in the course of the solution for the grid. This is done in a manner similar to that used in the GRAPE code, discussed in Chapter VI, with new boundary values of the function F being calculated from the present iterate for the coordinates. The function F in the field is then determined from these boundary values by either transfinite interpolation or as the solution of Laplace's equations, the former being found preferable in the cases considered. (With more distorted boundaries the Laplace solution might be more reliable than the interpolation.) Different forms of interpolation, or an equation other than the Laplace, for the determination of the control function in the field would allow some control of coordinate line spacing in the field. However, since only a single control function is involved, it is not possible to exercise control of the coordinate line spacing in the field in both directions.

Another approach in which the boundary point distribution can only be fixed in a specified manner is to take the basic generation equation to be Eq. (7c) which for conformal coordinates ($h_2=h_1$) takes the form

$$P_{\xi^1\xi^1} + P_{\xi^2\xi^2} = 0 \tag{25}$$

where $P = 2\ln(h_1)$. An exact solution of Eq. (25) can be obtained if appropriate values of P are known at the boundaries. The important problem then becomes the choice of those points at the inner and outer boundaries which can be put in orthogonal correspondence with one another. This can be accomplished if the ξ^1-coordinate, both at the inner and outer boundaries is selected to satisfy the Laplace equation $\nabla^2\xi^1=0$. This condition can be satisfied by taking ξ^1 as the angle traced out by the common radii of those concentric circles which are the conformal maps of the contours in the physical plane. The solution of Eq. (25) under these conditions then can be used to generate non-conformal coordinates by a coordinate transformation of the other coordinate ξ^2.

An orthogonal grid can be generated by solving the Laplace equations (21a) provided that the boundary point distribution is compatible. Since a conformal mapping generates an orthogonal grid, a compatible boundary point distribution can be obtained by conformally mapping the boundary contour as follows (cf. Ref. [43]): Consider an open physical boundary contour

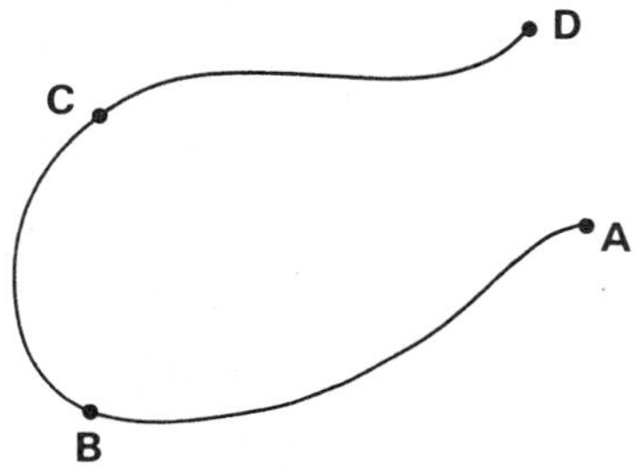

where $\overline{BA}$ and $\overline{CD}$ are to be lines of constant ξ^2, while $\overline{BC}$ and a connecting line $\overline{AD}$ to be generated are to be lines of constant ξ^1.

Each point of the set that defines this contour is successively mapped onto the real axis in the complex plane by a hinge point transformation (such a transformation has the effect of mapping one point onto the real axis while points already on the real axis remain there):

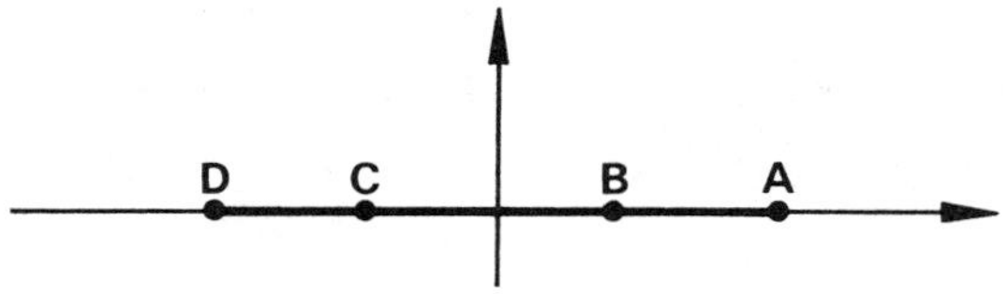

The straight line $\overline{DCBA}$ on the real axis is then mapped conformally onto an open rectangle in the complex plane:

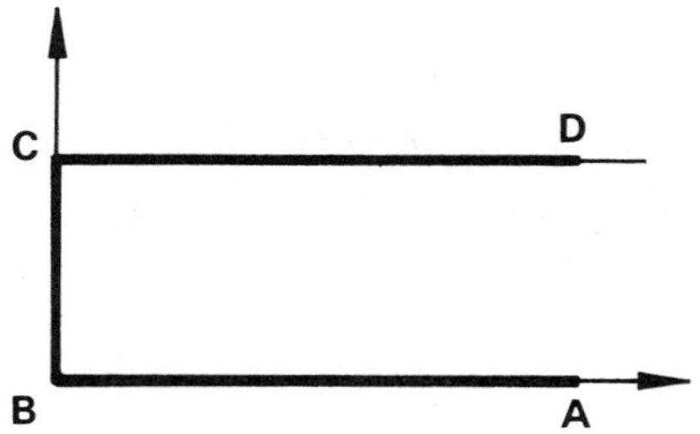

Points are then placed as desired along the sides $\overline{BA}$ and $\overline{BC}$ of this rectangle, these points on $\overline{BA}$ and $\overline{BC}$ being assigned successive integer values of ξ^1 and ξ^2, respectively. (This placement of points on these two sides is arbitrary and may be done by any distribution function desired.) The key to the construction of a compatible boundary point distribution is then that the points on the other sides of the rectangle, i.e., $\overline{CD}$ and $\overline{AD}$, are placed with the same distributions chosen for $\overline{BA}$ and $\overline{BC}$. The points in the physical plane that correspond to these boundary points on the rectangle in the complex plane are then determined by exponential spline interpolation among the values at the original set of points defining the contour, except for the open side of the rectangle where the points in the conformal transformations. Finally the orthogonal grid is generated by solving the Laplace equations (21a) with this fixed boundary point distribution.

C. Systems based on first-order equations

Equations (10) are formally related to the Cauchy-Riemann equations (with F=1), but otherwise form a set of first order nonlinear partial differential equations. In order to preserve the orientation of coordinates, the sign of F is taken to be positive throughout the domain. For certain choices of the function F the system is hyperbolic, and the complete initial-value problem is then

$$x_\eta = -Fy_\xi, \qquad y_\eta = Fx_\xi \qquad F > 0 \tag{26}$$

$$x(\xi,\eta_o) = x(\xi), \qquad y(\xi,\eta_o) = y(\xi)$$

Here $\eta = \eta_o$ is the given body contour, and, unlike the elliptic problem, the data on another boundary cannot be specified.

This system may be shown to exhibit the following important properties:

(i). First, g_{22} in principle can be expressed as a function of g_{11}.

(ii) Because of (i), $F > 0$ is a function of ξ^1, ξ^2, and g_{11}, i.e.,

$$F = F(\xi^1, \xi^2, g_{11})$$

For brevity, writing

$$z^2 = g_{11}$$

we have

$$F = F(\xi^1, \xi^2, z)$$

(iii). For a well-posed initial value problem the system of equations in (26) must be hyperbolic.

A test for the well-posedness is that small perturbations produce small effects. Using this test, for Eqs. (26) to be hyperbolic, the function f(z), defined as

$$f(z) = zF$$

must be a strictly decreasing function of z.

3. Three-Dimensional Orthogonal Coordinates

The problem of three-dimensional orthogonal coordinate generation, though of much importance in many practical problems, has received little attention in comparison to its two-dimensional counterpart. The reason is not so much in the complicated form of the governing equations but rather in the prescription of the boundary conditions and in their numerical implementation.

Orthogonality in three dimensions is difficult to achieve, and only exists when the coordinate lines on the bounding surfaces follow lines of curvature, i.e., lines in the direction of maximum or minimum curvature of the surface. Therefore, three dimensional orthogonal coordinates will not be available in most cases with nontrivial geometry. It is possible, however, to have the system locally orthogonal at boundaries, and/or to have orthogonality of surface coordinates.

The governing equations for generation of orthogonal coordinates are obtained in a straightforward manner and have been listed above as Eq. (4) - (6). The set of equations which are to be solved for x_1, x_2, x_3 and h_1, h_2, h_3 has Eq. (4) and (6). The set (6) has six equations for the three unknowns. On the other hand, without imposing the orthogonality condition, $g_{ij} = 0$ $(i \neq j)$, there are six equations for the determination of six unknowns. Thus the orthogonality does not reduce the number of the equations which govern the distribution of the metric coefficients, and it would be wrong to try to select a set of three equations out of the available six.

4. Nearly-Orthogonal Systems

Since a part of the truncation error is decreased as the grid becomes more orthogonal, it is of interest to generate grids which are "nearly-orthogonal". Such grids do not approximate orthogonality sufficiently well, however, for the terms arising from nonorthogonality in transformation relations to be dropped. The generation of nearly-orthogonal grids naturally follows some of the procedures discussed above in this chapter, but with the conditions for orthogonality only partially satisfied. Several procedures are discussed in Ref. [1] and Ref. [42].

A simple procedure for generating a nearly-orthogonal system from a nonorthogonal system is to first generate curves of a nonorthogonal system by connecting points obtained by any specified distribution function along straight lines connecting boundary points on two arbitrary closed boundaries. Coordinate lines connecting points on each succeeding pair of curves from the original coordinate system then are constructed as follows: At selected points on the inner curve, normals are constructed, and the points of intersection with the next curve outward are determined. Normal directions form the intersection point are determined and translated to the original point in the inner curve. Then a second point on the outer curve is determined as before. Finally, the new coordinate lines are constructed as straight lines joining the selected points on the inner curve with points located halfway between the corresponding pair of points on the outer curve located as described above. The resulting lines will not actually be orthogonal to either the inner or outer curve, and the slopes of these lines will, in fact, be discontinuous at each curve. The observed departures from orthgonality, however, have been small and the departure may be made arbitrarily small by the addition of more curves. Since the procedure is applied successively between pairs of coordinate lines, concave bodies can be treated as well.

Exercises

1. The unit tangent vector on a curve C defined in the parametric form $\mathbf{r} = \mathbf{r}(s)$, with s as the arc length along C, is given by $\mathbf{t} = d\mathbf{r}/ds$. Let C be a plane curve in the xy-plane having $\mathbf{n}$ as the unit normal vector. Using the condition $\mathbf{n}\cdot\mathbf{t} = 0$ and the convention that $(\mathbf{t},\mathbf{n},\mathbf{k})$, in the order shown, form a right-handed triad of vectors, find the components of $\mathbf{n}$. Here $\mathbf{k}$ is the constant unit vector along the z-axis.

2. Let $\xi(x,y)$ and $\eta(x,y)$ be the conformal coordinates in the xy-plane so that the Cauchy-Riemann equations

$$\xi_x = \eta_y, \; \xi_y = -\eta_x$$

are satisfied. Consider the curve C defined in excersise 1 and the normal derivative operator

$$\frac{\partial}{\partial n} = \mathbf{n} \cdot \nabla$$

and show that the Cauchy-Riemann equations in the natural coordinates (s,n) are

$$\xi_s = \eta_n, \; \xi_n = -\eta_s$$

3. Let $F(\xi^i)$ be a scalar function of position and $\phi(\xi^i)$=constant be a surface.
(a) Show that the unit normal vector $\mathbf{n}$ to the surface ϕ=constant in curvilinear coordinates is given by

$$\mathbf{n} = \frac{1}{|\text{grad } \phi|} \sum_i \frac{\partial \phi}{\partial \xi^i} \mathbf{a}^i$$

(b) Prove that the normal derivative of F on the surface ϕ=constant is

$$\left(\frac{\partial F}{\partial n}\right)_{\phi=\text{constant}} = \frac{1}{|\text{grad }\phi|} \sum_i \sum_j g^{ij} \frac{\partial \phi}{\partial \xi^i} \frac{\partial F}{\partial \xi^j}$$

(c) In particular, for two-dimensional curvilinear coordinates show that

$$\left(\frac{\partial F}{\partial n}\right)_{\xi^1=\text{constant}} = \frac{1}{\sqrt{g g_{22}}}\left(g_{22}\frac{\partial F}{\partial \xi^1} - g_{12}\frac{\partial F}{\partial \xi^2}\right)$$

$$\left(\frac{\partial F}{\partial n}\right)_{\xi^2=\text{constant}} = \frac{1}{\sqrt{g g_{11}}}\left(g_{11}\frac{\partial F}{\partial \xi^2} - g_{12}\frac{\partial F}{\partial \xi^1}\right)$$

(d) Particularize the results in (c) for orthogonal curvilinear coordinates. Write the partial derivative operator $\frac{\partial}{\partial n}$ for orthogonal coordinates.

4. Consider Eq. (26) of this chapter, which form a system of first-order partial differential equations for two-dimensional orthogonal coordinates. It was stated subsequently that these equations form a hyperbolic system if the initial value problem is well-posed. To prove this assertion consider the perturbed state $x+\delta x$, $y+\delta y$, $F(\xi,\eta,z+\delta z)$, where $z=\sqrt{g_{11}}$. Retaining only the first order terms, develop a system of algebraic equations in $(\delta x)_\xi, (\delta y)_\xi$, $(\delta x)_\eta$, $(\delta y)_\eta$, and show that the resulting matrix has eigenvalues given by

$$\lambda^2 = -F(F + zF_z)$$

Show from the preceding result that the eigenvalues are real only when zF is a strictly decreasing function of z.

X. CONFORMAL MAPPING

Innovations in conformal mapping continue to extend this classical technique to more complicated configurations, and surveys of the various techniques available are given in Ref. [7] and Ref. [1]. Some specific recommendations of techniques and tools are given in Ref. [7]. Conformal systems have advantage the of introducing the fewest additional terms in transformed partial differential equations. Considerable understanding of the theory of functions of a complex variable may be necessary for effective applications, though.

Although the complex variable techniques by which conformal transformations are usually generated are inherently two-dimensional, certain more general cases can be treated by rotating or stacking two-dimensional systems:

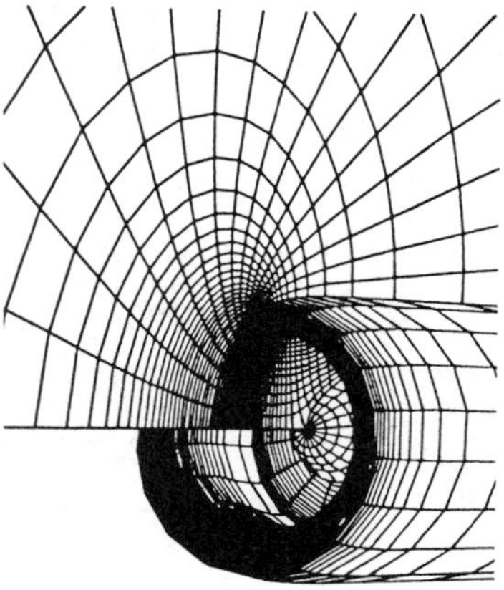

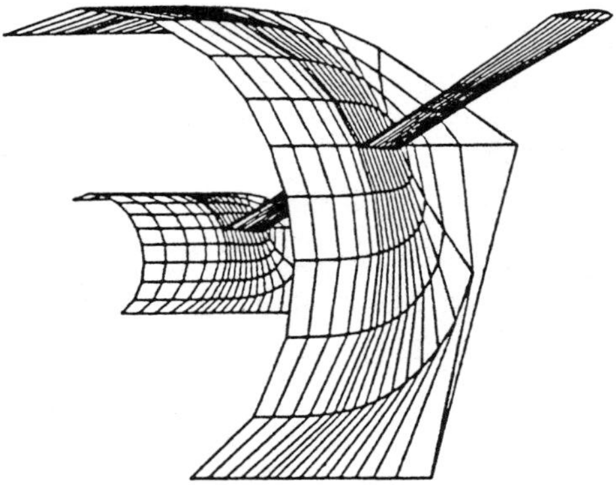

Systems can also be generated on curved surfaces, as has been done by cartographers, for stacking. Examples of the use of conformal mapping in the construction of three-dimensional configurations are noted in Ref. [5].

A curvilinear coordinate system generated by a conformal mapping is very rigid in the sense that little control can be exerted over the distribution of the grid points. Conformal mappings also do not exist in three dimensions (except for trivial cases). Furthermore, the coordinate system tends to be more difficult to construct than when using algebraic or elliptic systems. In spite of these facts, conformal mappings continue to play a significant role in grid generation. A number of recent developments and applications of conformal transformations are noted in Ref. [1], [5], and [7].

The desirability of a coordinate system generated by a conformal transformation lies in the form of the transformed equations. For example, consider the diffusion equation

$$A_t = \mu \nabla^2 A \tag{1}$$

Now ξ and η satisfy the Cauchy-Riemann equations

$$\begin{aligned} \xi_x &= \eta_y \\ \xi_y &= -\eta_x \end{aligned} \tag{2}$$

or equivalently

$$\begin{aligned} x_\xi &= y_\eta \\ x_\eta &= -y_\xi \end{aligned} \tag{3}$$

It follows that in this case $g_{12}=g_{21}=0$ and $g_{11}=g_{22}=\sqrt{g}$. Equation (1) can be written in curvilinear coordinates, using Eq. (III-46), as

$$A_t = (\mu/g)\ \nabla^2 A \tag{4}$$

where the Laplacian is defined in terms of the curvilinear coordinates. Therefore it is observed that the diffusion equation remains essentially unchanged. The only effect of the transformation is a change in the diffusion coefficient. Neumann boundary conditions are also unchanged in conformal coordinates. The boundary condition

$$A_n = B$$

where $\underset{\sim}{n}$ is normal to a ξ=constant coordinate line, is expressed in curvilinear coordinates as

$$A_\xi = \sqrt{g}\, B$$

1. Construction by Finite-Differences

The literature abounds with methods for constructing conformal mappings. As can be seen in the review article, Ref. [1], these methods may include the construction of Schwarz-Christoffel transformations, the solution of integral equations, or expansions in terms of power series or Fourier series. Since this chapter is not intended to be a comprehensive treatment of conformal mapping, only the simple, yet frequently used, finite difference method based on elliptic systems is discussed here.

Consider the problem of conformally mapping the interior of the contour Γ onto the interior of a rectangle. The Riemann Mapping Theorem states that such a mapping exists, and it also implies that the mapping is uniquely determined by specifying three real parameters. Suppose we wish to indicate four specific points on Γ which are to map to the vertices of the rectangle. If the rectangle is fixed, then the problem is over-determined and no conformal mapping exists. Therefore, the mapping must determine one of the dimensions of the rectangular region which we will now denote as the set

$$0 \le \xi \le 1, \qquad 0 \le \eta \le M$$

Rather than allow a rectangle with variable width, one can equivalently introduce the parameter M in (3) so that

$$\begin{aligned} Mx_\xi &= y_\eta \\ x_\eta &= -My_\xi \end{aligned} \tag{5}$$

where

$$0 \le \xi \le 1, \qquad 0 \le \eta \le 1$$

The mapping is no longer conformal, but the conformal mapping can be easily obtained by simply multiplying the η coordinate by M. On the unit square the functions x and y now satisfy

$$\begin{aligned} M^2x_{\xi\xi} + x_{\eta\eta} &= 0 \\ M^2y_{\xi\xi} + y_{\eta\eta} &= 0 \end{aligned} \tag{6}$$

Two boundary conditions are needed in order to determine a unique solution for this elliptic system. One condition is derived from the equation of the boundary curve Γ which might be

$$F(x,y) = 0 \tag{7}$$

The other condition comes from applying the orthgonality equation, $g_{12}=0$. This condition also follows on eliminating the parameter M in Eq. (5). The implementation of the boundary conditions is done in the following order. First a boundary value for x or y is computed from the orthogonality constraint. If the boundary point lies along $\xi = 0$, then we may use

$$x_\xi = -y_\xi y_\eta / x_\eta \text{ or } y_\xi = -x_\xi x_\eta / y_\eta \tag{8}$$

A forward difference is used to approximate the ξ-derivatives and central differences for the η-derivatives. The same equations are used along $\xi = 1$ with backward differences for the ξ-derivatives. Once an x or y value has been computed from Eq. (8), the other coordinate value is given by writing (7) in the form

$$y = G(x) \text{ or } x = H(y) \tag{9}$$

Although either equation in (8) could be used, it is advisable to choose either the first or second equation, depending on whether x_η or y_η has the largest absolute value. This avoids not only the possibility of division by zero but also problems with the solvability of the implicit equation (7). The same techniques are used along an η=constant coordinate line. In this case the orthogonality constraint can be written as

$$x_\eta = -y_\xi y_\eta / x_\xi \text{ or } y_\eta = -x_\xi x_\eta / y_\xi \tag{10}$$

Now the parameter M must also be determined. It follows from Eq. (5) that

$$M^2 = g_{22}/g_{11} \tag{11}$$

An iterative algorithm is used to construct the mapping, and any algorithm which can be used for the elliptic systems in Chapter VI can also be used here. At each iteration a new set of boundary values for x and y are computed using Eq. (8)-(10). There are two options in computing a value for M. Either a different value at each point can be computed from Eq. (11), or a constant value can be computed from a relation such as

$$M = [x(1,\eta) - x(0,\eta)]^{-1} \int_0^1 y_\eta(\xi,\eta)d\xi \qquad (12)$$

where $0 \leq \eta \leq 1$. Eq. (12) is derived from the first equation in Eq. (5) by integrating along an η=constant coordinate line. This same technique can also be used to derive an alternate formula for finding the boundary values at x and y. Along $\eta = 0$, for example, we have

$$x(\xi,0) = x(0,0) + 1/M \int_0^\xi y_\eta(\xi,0)d\xi$$

The constant M, called the conformal module of the region by complex analysts, has a simple geometric interpretation. From Eq. (11) it is noted that M is simply the aspect ratio of the grid cells. There exist highly accurate numerical methods for computing both M and the boundary values for x and y. If these values are computed first, then the system (6) can be solved by a direct elliptic solver.

The only control over the distribution of grid points with a conformal mapping is by changing the points which map to the vertices of the rectangular region. However, most of the advantageous features are retained when the conformal mapping is combined with one-dimensional stretching transformations. Thus we will consider a new set of computational variables, χ and ζ, with ξ and η serving as intermediate variables defined by the one-dimensional equations

$$\xi = f(\chi), \quad \eta = h(\zeta) \qquad (13)$$

If x and y are solutions of Eq. (6), then in terms of the new computational variables,

$$\frac{M^2}{[f'(\chi)]^2} \underset{\sim}{r}_{\chi\chi} + \frac{1}{[h'(\zeta)]^2} \underset{\sim}{r}_{\zeta\zeta} - \frac{M^2 f''(\chi)}{[f'(\chi)]^3} \underset{\sim}{r}_{\chi} - \frac{h''(\zeta)}{[h'(\zeta)]^3} \underset{\sim}{r}_{\zeta} = 0 \qquad (14)$$

$$M^2 = \frac{g_{22}}{g_{11}} \left[\frac{f'(\chi)}{h'(\zeta)}\right]^2 \qquad (15)$$

In Eq. (15), the covariant metric tensor components are defined relative to the transformation from the physical x,y variables to the computational χ,ζ variables.

The application of this transformation to the diffusion equation (1) results in the following transformed equation:

$$\frac{g_{22}}{[h'(\zeta)]^2} A_t = \mu\left\{\frac{M^2}{[f'(\chi)]^2} A_{\chi\chi} + \frac{1}{[h'(\zeta)]^2} A_{\zeta\zeta} - \frac{M^2 f''(\chi)}{[f'(\chi)]^3} A_\chi - \frac{h''(\zeta)}{[h'(\zeta)]^3} A_\zeta\right\} \qquad (16)$$

Note that the coefficients of the χ and ζ derivatives in Eq. (16) are functions of χ and ζ, respectively. Therefore, only one-dimensional arrays are needed to store these coefficients. It can be further noted that the steady-state equation (A_t=0) is a separable elliptic equation which can be solved using a direct elliptic solver.

In the above development the stretching functions given by Eq. (13) are used to control the grid point distribution. Clearly the derivatives of these functions must be nonvanishing, and these derivatives may as well be taken to be positive so that the orientation of the physical boundary is preserved. The function $f(\chi)$ is a contraction mapping if $f'(\chi) < 1$ and an expansion mapping if $f'(\chi) > 1$. Therefore, relative to a conformal mapping, the χ=constant coordinate lines will be closer together where $f'(\chi) < 1$ and farther apart when $f'(\chi) > 1$. The same control over the ζ=constant coordinate lines is exerted by the function $h(\zeta)$. Several one-dimensional functions are discussed in Chapter VIII.

On any particular boundary segment, say $\zeta=0$, it is in theory possible to match any desired distribution of grid points provided the correct stretching function $f(\chi)$ can be determined. However, there is no known way of generating this stretching function so that Eq. (14), together with the boundary conditions given by Eq. (7) and $g_{12}=0$, will have a solution with the prescribed boundary values along $\zeta=0$. The solution to this problem lies in the implicit determination of the stretching function in the solution of the elliptic system. Suppose that $h(\zeta)=\zeta$, so that Eq. (15) can be written as

$$\left[\frac{M}{f'(\chi)}\right]^2 = \frac{g_{22}}{g_{11}} \qquad (17)$$

This equation allows Eq. (14) to be written as

$$\frac{g_{22}}{g_{11}} \underset{\sim}{r}_{\chi\chi} + \underset{\sim}{r}_{\zeta\zeta} + \frac{1}{2}\left(\frac{g_{22}}{g_{11}}\right)_\chi \underset{\sim}{r}_\chi = 0 \qquad (18)$$

This quasilinear system can be solved with the orthogonality condition on all boundary components except $\zeta = 0$ where we will now impose the Dirichlet condition

$$\underset{\sim}{r}(\chi,0) = \underset{\sim}{R}(\chi)$$

where $\underset{\sim}{R}$ defines the desired distribution of grid points.

The ability to specify grid points along a boundary component extends the usefulness of conformal mappings. For example, one can assign coordinates around an airfoil and along the branch cut in a C-type coordinate system so that the coordinate lines pass smoothly through the cut. In many segmented systems the grid points can be chosen so that coordinate lines pass smoothly from one sub-region into the next. One disadvantage of this method is the reported slow convergence in the iterative solution of (18) for certain problems. An alternate method of achieving the same result would be to generate a conformal mapping from Eq. (6) and then use interpolation to redistribute the grid lines. Note that the interpolation scheme may affect the orthogonality of the coordinate system to some degree.

2. Schwarz-Christoffel Transformation

Conformal mappings of circular disks or half-planes onto polygonal regions are defined by the Schwarz-Christoffel formula. Suppose the points $\zeta_1, \zeta_2, \ldots. \zeta_n$ lie on the real axis of the ζ-plane. Then the mapping defined by

$$z = A + B \int_0^{\zeta} \prod_{i=1}^{n} (\zeta_i - \zeta)^{-\frac{\alpha_i}{\pi}} \, d\zeta \tag{19}$$

transforms the upper half plane onto a polygonal region with interior angles of $\pi - \alpha_i = \theta_i$. However this is not exactly what is needed in most grid generation problems. Presumably one would be given a polygonal region with vertices $z_1, z_2, \ldots z_n$. Thus the parameters $A, B, \zeta_1, \zeta_2, \ldots \zeta_n$ must be determined so that the real axis maps onto the given polygon:

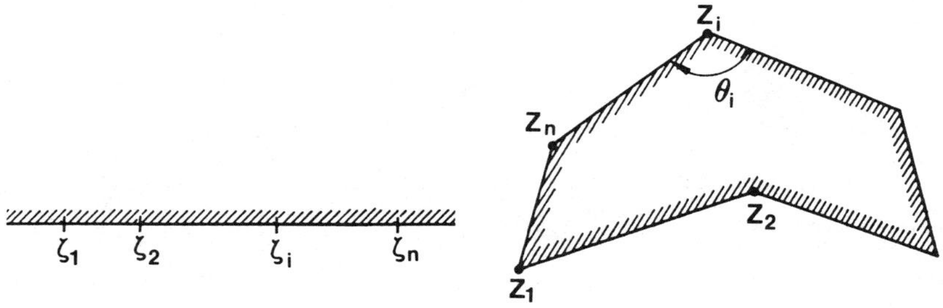

There are several numerical techniques for the approximation of the parameters in the Schwarz-Christoffel transformations. Since a conformal mapping of a simply-connected region has three degrees of freedom, three of these parameters must be given in order for the mapping to be uniquely determined. In certain infinite regions, the value of B can be calculated from the asymptotic behavior of the mapping function. We can also set $z_n = \zeta_n = 0$, which implies that $A = 0$ from Eq. (19). The remaining parameters to be determined are $\zeta_1, \zeta_2, \ldots, \zeta_{n-1}$. Alternately, as is commonly done in bounded regions, we can choose the values ζ_1, ζ_2, ζ_3 which are to map the points z_1, z_2, z_3. In this case the parameters to be determined are $A, B, \zeta_4, \zeta_5, \ldots \zeta_n$. The basic algorithm for determining the unknown parameters consists of computing the distances $|z_{i+1} - z_i|$, using Eq. (19) and a quadrature formula to approximate the integral, and then iterating on the parameters until these distances are correct. Once these parameters in the transformation have been computed to the desired accuracy, the image of any point ζ in the upper half-plane is formed by numerically evaluating the integral in Eq. (19).

Schwarz-Christoffel transformations are not limited to regions with polygonal boundaries. They can be used in composition with other conformal mapping methods to map regions with curved boundaries onto various computational regions. For example, an integral equation method can be used to map a physical region with curved boundary components onto the unit disk, which can be easily transformed onto the upper half-plane. Now the upper half-plane can be mapped onto the computation region, which may consist of several rectangular blocks, by Eq. (19). There are also direct generalizations of the Schwarz-Christoffel transformation for regions with curved boundaries. These are obtained by considering the limiting case of Eq. (19) as $n \to \infty$.

Recent extensions of the Schwarz-Christoffel transformation to curved contours have made this procedure a powerful tool for treating complicated internal and other configurations. These improvements also lead to smoother metric coefficients for boundaries with slope discontinuities than in older methods for the Schwarz-Christoffel transformation. This procedure for the Schwarz-Christoffel transformation may also be more efficient than other conformal procedures involving an intermediate mapping of a near-circle for mapping contours and circles in some cases. Several sources on the recent developments and applications of the Schwarz-Christoffel transformation are cited in Ref. [1] and [5].

3. Construction from Integral Equations

Integral equations have played a major role in the solution of partial differential equations. Mathematicians have often resorted to integral equations when attempting to prove the existence and uniqueness of solutions. Numerical analysts turned to the so-called panel methods for solving partial differential equations in two and three-dimensional regions. These mehtods replaced the partial differential equations by a set of integral equations and thereby reduced the dimension of the problem, since panel methods only involve boundary integrals. The application of integral equations depends on the availability of fundamental solutions of the partial differential equation. Therefore they are especially useful in the solution of Laplace's equation. Numerous solutions of Laplace's equation can be generated by determining the real and imaginary parts of analytic functions. As most conformal mappings can be reduced to the solution of boundary-value problems for Laplace's equation, it should come as no surprise that integral equations can be a valuable tool in the construction of conformal mappings. Only the basic integral equation method of Symm (cf. Ref. [1]) will be presented here. This method has proven to be robust, yet is easily derived and involves only the solution of a system of linear equations.

Suppose the simply-connected region D, bounded by the contour Γ, is to be $|\zeta|<1$. Let $z = z_o$ be the point in D which maps to the origin $\zeta = 0$. If the Dirichlet problem

$$\nabla^2 q = 0 \text{ in } D$$

$$q = -\log|z - z_o| \text{ on } \Gamma \tag{20}$$

can be solved and the harmonic conjugate h of q can be found, then it can be directly verified that the analytic function

$$\zeta = (z - z_o) \exp[q(z) + ih(z)] \tag{21}$$

maps Γ onto $|\zeta| = 1$. Due to the form of the series expansion for the exponential, it can also be shown that this function has a nonvanishing derivative, and hence the conformal mapping of D onto the unit disk is given by Eq. (21). We now turn to the problem of solving the boundary value problem in Eq. (20). Suppose there exists a solution of the form

$$q(z) = \int_\Gamma \sigma(\zeta) \log|z - \zeta| ds \tag{22}$$

for z on Γ. Regardless of the value of the function $\sigma(\zeta)$, the function q(z) is harmonic on D. In order that q(z) satisfy the boundary condition, it is clear that we need to choose $\sigma(\zeta)$ such that, for z on Γ,

$$-\log|z - z_o| = \int_\Gamma \sigma(\zeta) \log|z - \zeta| ds \tag{23}$$

This is then the integral equation for determining the unknown function $\sigma(\zeta)$. The harmonic conjugate of $\log|z|$ is $\arg(z)$. Thus the function of $h(z)$ can be expressed as

$$h(z) = \int_\Gamma \sigma(\zeta) \arg(z - \zeta) ds \tag{24}$$

Note that the function $h(z)$ is only unique up to an addition constant. The addition of a constant to $h(z)$ results in a rotation of the conformal mapping defined in Eq. (21).

The practicality of this method depends on the efficient solution of the integral equation in Eq. (23). In order to solve this equation numerically, divide Γ into n intervals, Γ_j, $j = 1,2,...,n$ and assume $\sigma(\zeta)$ has a constant value, σ_j, on Γ_j. Let z_j be a fixed point of Γ_j. Now Eq. (23) can be approximated by the linear system of equations

$$\sum_{j=1}^{n} \sigma_j \int_{\Gamma_j} \log|z_i - \zeta| ds = -\log|z_i - z_o|, \quad i=1,2,....,n \tag{25}$$

There are two alternatives in computing the coefficients in this system. If the Γ_j are assumed to be straight lines, then the integrals can be calculated analytically. Otherwise, each integral must be computed numerically. Once these coefficients have been computed, the system can be solved to yield a step function which approximates the function $\sigma(\zeta)$. The values of σ_j are now used to estimate the functions $q(z)$ and $h(z)$:

$$q(z) = \sum_{j=1}^{n} \sigma_j \int_{\Gamma_j} \log|z - \zeta| ds$$

$$h(z) = \sum_{j=1}^{n} \sigma_j \int_{\Gamma_j} \arg(z - \zeta)ds \tag{26}$$

Again the above integrals would, in general, be computed numerically. These values of q(z) and h(z) would be substituted in Eq. (21) to yield the image in the unit disk of any given point z in the region D.

This integral equation method is a very efficient and accurate method. However, it has one deficiency in regard to grid generation and the numerical solution of partial differential equations. The transformation which is constructed maps the physical region D onto the canonical region, which in this case is the unit disk. The unit disk could be the computational region, or it could be mapped onto a rectangular region by an auxiliary transformation. In any case, what is needed is the mapping from the unit disk onto the physical region. Therefore an interpolation scheme would be needed to approximate the inverse of the computed mapping.

It is sometimes more efficient to generate the final grid by solving the Laplace system numerically with Dirichlet boundary conditions from the conformal transformations, especially if a fast Poisson solver can be applied.

4. Elementary Complex Transformations

An extensive list of complex mappings is compiled in Ref. [44]. However, these mappings are only for regions with special boundary curves. If a strictly conformal transformation is not necessary, then these mappings may be used to create what are called nearly conformal mappings. For example, suppose an airfoil shape can be modeled as the image of a circle under the Joukowski transformation

$$z = \zeta + 1/\zeta \qquad (27)$$

Under the inverse transformation, a given airfoil will map to a curve which is nearly circular. The region about the nearly circular curve can be mapped onto the region about a circular region by a simple algebraic transformation. One scheme for accomplishing this final mapping would be to divide each complex number on a given ray from the center by the modulus of the complex number on the curve. The composite mapping in this case would be a nearly-conformal mapping of the exterior of the airfoil onto the exterior of a circle. The inverse mapping, which could be explicitly defined, would define a nearly-orthogonal O-type grid about the airfoil.

Analytic functions are not only of value in mapping regions about airfoils, but are also helpful in the more general problem of generating grids in the neighborhood of boundary points with slope discontinuities. With most algebraic methods of grid generation, these slope discontinuities will propagate into the physical region resulting in non-smooth grid lines and the associated increase in truncation error in the numerical solution of partial differential equations. The general idea can be conveyed with the following example. Suppose we have a region where the boundary has an interior angle of θ at the point z_o. Under the mapping

$$\zeta = (z - z_o)^{\pi/\theta} \qquad (28)$$

the corner is eliminated. While this simple mapping may be useful in transforming the interior of a contour, the mapping of the exterior region would not be one-to-one. The elimination of corners for regions surrounding a contour can be effected by applying the Karman-Trefftz mapping defined by

$$\frac{\zeta - 1}{\zeta + 1} = \left(\frac{z - z_o}{z + \bar{z}_o}\right)^{\alpha} \qquad (29)$$

where $\bar{z}_o$ is the conjugate of z_o. The exponent α depends on the exterior angle and the region should be translated, if necessary, so that $\bar{z}_o$ is an interior point of the contour.

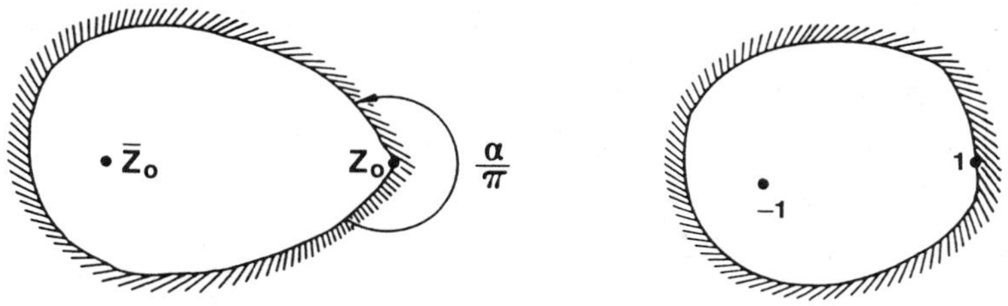

This transformation may be applied successively to eliminate any number of corners on the boundary of the physical region.

Elementary complex functions can therefore serve to precondition a region. Corners which are to map to sides of a computational rectangle can be eliminated. Conversely, right-angle corners can be formed at points of the physical reigon which are to map to vertices of the computational region thereby eliminating problems of extreme nonorthogonality.

The trend in treating more complicated regions is to break the mapping up into a sequence of more simple mappings. Contours, such as airfoils, are generally mapped to near-circles by one or more simple transformations, and then the near-circle is mapped to a circle by a series transformation, e.g., the Theodorsen procedure. It is necessary for convergence that the near circle be sufficiently near to being a circle. A series for the differential form is generally superior to the usual Theodorsen form for general bodies. This series appears in terms of arc length and surface angle, rather than the polar coordinates of the Theodorsen form which can lead to infinite derivatives and multiple values. The ordering of the points can break down in the Theodorsen form for closely spaced points also. The differential form is applicable, however, as long as there are no corners, even for twisted contours. In this and other series transformations, the differential form is usually more tolerant of odd shapes.

Multiple-body configurations can be treated by a sequence of transformations which map each body to a circle in succession, while maintaining previously established circles. Another procedure, invovles iteratively mapping each body to a cirlce with no special consideration of the others. This process generally requires only a few iterations to converge. Some recent applications are noted in Ref. [1] and [5].

XI. ADAPTIVE GRIDS

In an adaptive grid, the physics of the problem at hand must ultimately direct the grid points to distribute themselves so that a functional relationship on these points can represent the physical solution with sufficient accuracy. The idea is to have the grid points move as the physical solution develops, concentrating in regions of large variation in the solution as they emerge. The mathematics controls the points by sensing the gradients in the evolving physical solution, evaluating the accuracy of the discrete representation of the solution, communicating the needs of the physics to the points, and finally by providing mutual communication among the points as they respond to the physics. The basic techniques involved then are as follows:

(1) a means of distributing points over the field in an orderly fashion, so that neighbors may be easily identified and data can be stored and handled efficiently.

(2) a means of communication between points so that a smooth distribution is maintained as points shift their position.

(3) a means of representing continuous functions by discrete values on a collection of points with sufficient accuracy, and a means for evaluation of the error in this representation.

(4) a means for communicating the need for a redistribution of points in the light of the error evaluation, and a means of controlling this redistribution.

Several considerations are involved here, some of which are conflicting. The points must concentrate, and yet no region can be allowed to become devoid of points. The distribution also must retain a sufficient degree of smoothness, and the grid must not become too skewed, else the truncation error will be increased as noted in Chapter V. This means that points must not move independently, but rather each point must somehow be coupled at least to its neighbors. Also, the grid points must not move too far or too fast, else oscillations may occur. Finally the solution error, or other driving measure, must be sensed, and there must be a mechanism for translating this into motion of the grid. The need for a mutual influence among the points calls to mind either some elliptic system, thinking continuously, or some sort of attraction (repulsion) between points, thinking discretely. Both approaches have been taken with some success, and both are discussed below. It should be noted that the use of an adaptive grid may not necessarily increase the computer time, even though more computations are necessary, since convergence properties of the solution may be improved, and certainly fewer points will be required.

With the time derivatives at fixed values of the physical coordinates transformed to time derivatives taken at fixed values of the curvilinear coordinates, no interpolation is required when the adaptive grid moves. Thus, as given by Eq. (III-116),

$$\left(\frac{\partial u}{\partial t}\right)_{\xi^1,\xi^2,\xi^3} = \left(\frac{\partial u}{\partial t}\right)_{x_1,x_2,x_3} + \sum_{i=1}^{3} \frac{\partial u}{\partial x_i}\frac{\partial x_i}{\partial t} \tag{1}$$

where x_i and ξ^i are the cartesian and curvilinear coordinates, respectively. The computation thus can be done on a fixed grid in the transformed space, without need of interpolation, even though the grid points are in motion in physical space. The influence of the motion of the grid points is registered through the grid speeds, $(x_i)_t$, appearing in the transformed time derivative. This is the appropriate approach when the grid evolves with the solution at each time step. Some methods, however, change the grid only at selected time steps, and here interpolation must be used to transfer the values from the old grid to the new since the grid movement is not continuous.

In the following discussion, the problem of grid adaption will be formulated as a variational problem, the ideas being developed first in one dimension and then extended to multiple dimensions.

1. One-Dimensional Adaption

A. Equidistribution

A number of studies of numerical solutions of boundary-value problems in ordinary differential equations have shown that the error can be reduced by distributing the grid points so that some positive weight function, $w(x)$, is equally distributed over the field, i.e.,

$$\int_{x_i}^{x_{i+1}} w(x)dx = \text{constant} \tag{2}$$

or, in discrete form,

$$\Delta x_i w_i = \text{constant} \tag{3}$$

where Δx_i is the grid interval, i.e., $\Delta x_i = x_{i+1} - x_i$. (The subscript here indicates position on the line in this one-dimensional case.) With this condition, the grid interval will, of course, be small where the weight function is large, and vice versa. Thus if the weight function is some measure of the error, or the solution variation, the grid points will be closely spaced in regions of large error, or solution variation, and widely spaced where the solution is smooth. (It may be more appropriate in some cases to replace the equal sign in Eq. (2) and (3) with "less than or equal", and thus to "sub-equidistribute" the weight function.)

This approach has also been applied to redistribute the grid points (or to add points) at each time step, or at certain intervals, in numerical solutions of initial/boundary-value problems in one-dimensional partial differential equations. A number of references to the use of equidistribution are cited in Ref. [45]. It can be shown that the point distribution is asymptotically optimal if some error measure is distributed evenly, and that this optimum error is rather stable under perturbations of the point distribution. Thus it is not necessary to locate the grid points with excessive accuracy.

B. Equidistribution by transformation

The nonuniform point distribution can be considered to be a transformation, $x(\xi)$, from a uniform grid in ξ-space, with the coordinate ξ serving to identify the grid points. The grid points are conveniently defined by successive integer values of ξ, making $\Delta\xi = 1$ by construction and the maximum value of ξ, i.e., N, equal to the total number points on the line. Then $\Delta x = x_\xi \Delta\xi = x_\xi$, so that x_ξ represents the variation in x between grid points. Hence the equidistribution statement, Eq. (3), can be represented as

$$x_\xi w = \text{constant} \tag{4}$$

With the weight function w taken as a function of ξ this is just the Euler equation for the minimization of the integral

$$I_1 = \int_0^1 w(\xi)x_\xi^2 d\xi \tag{5}$$

(From the calculus of variations, the function $x(\xi)$ for which the integral $\int F(x,x_\xi)d\xi$ is an extremum is given by the solution of the differential equation $d/d\xi(\partial F/\partial x_\xi) - \partial F/\partial x = 0$. This equation is called the Euler's variational equation.) The integral (5) can be taken to represent the energy of a system of springs, with spring constants $w(\xi)$, spanning each grid interval, considering all the points to have been expanded from a common point so that x_ξ is the extension of the spring at ξ. The grid point distribution resulting from the equidistribution thus represents the equilibrium state of such a spring system, i.e., the state of minimum energy. Since x_ξ represents the distance between grid points, this variational problem can also be interpreted as the minimization of the cumulative spacing between the grid points in the least-squares sense, subject to the weight function $w(\xi)$.

If the weight function is taken to be a function of x, instead of ξ, then the integral for which Eq. (4) is the Euler equation is

$$I_2 = \int_0^1 [w(x)x_\xi]^2 d\xi \tag{6}$$

The variational problem in this case is the least-squares minimization over the grid of the cumulative grid point spacing weighted by the weight function.

Integration over ξ, as in both these cases, constitutes a summation over the grid points, with x_ξ representing the spacing between grid points. In the first case above, i.e., Eq. (5), the weight function $w(\xi)$, being a function of ξ, is associated with the grid points themselves, not with their locations. In the second case, Eq. (6), however, the weight function $w(x)$ is associated with the locations of the grid points, rather than directly with the points. Since there is a relation $x(\xi)$ representing the locations of the grid points, any weight function can obviously be transformed from one argument to the other. However, in deriving the Euler equations for a variational problem it is only the direct dependence that is considered in the partial derivatives $\partial F/\partial x$ or $\partial F/\partial \xi$, i.e., whether the weight function is determined by the identity of the grid point or by the location of the grid point, $w(x)$, although implicit differentiation is used in the total derivatives $d/d\xi(\partial F/\partial x_\xi)$ and $d/dx(\partial F/\partial \xi_x)$.

The constant in Eq. (4) can be evaluated by normalizing x to the interval $(0,L)$. If ξ is normalized to $(1,N)$ we have from Eq. (4),

$$x_\xi = \frac{C}{w}$$

and hence

$$C = \frac{L}{\int_1^N \frac{d\xi}{w}} \tag{7}$$

so that

$$x_{\xi} = \frac{L}{w\int_1^N \frac{d\xi}{w}} \tag{8}$$

Since $\xi_x = 1/x_{\xi}$, the transformation is then determined by

$$\xi(x) = 1 + \frac{1}{L}\left(\int_o^x w d\bar{x}\right)\left(\int_1^N \frac{d\xi}{w}\right) \tag{9}$$

Thus

$$\Delta\xi = \frac{1}{L}\left(\int_{x_i}^{x_{i+1}} w dx\right)\left(\int_1^N \frac{d\xi}{w}\right) \tag{10}$$

so that Eq. (2) is realized by taking equal increments in ξ, i.e., ξ varying by equal increments between grid points as was stated initially. From Eq. (8) the grid point spacing is given by

$$\Delta x_i = \frac{L}{w\int_1^N \frac{d\xi}{w}} \tag{11}$$

An alternative viewpoint results from integrating over x, instead of over ξ, i.e., summing over the grid intervals rather than over the grid points. Since ξ identifies the grid points, ξ_x represents the change in ξ, i.e., the number of grid points per unit distance, and hence is the grid point density. Eq. (4) is now the Euler equation for minimization of the integral

$$I_3 = \int_o^1 \frac{\xi_x^2}{w(x)} dx \tag{12}$$

Here the integral in question is $\int F(\xi_x, \xi))dx$, so that the Euler equation is given by $d/dx(\partial F/\partial \xi_x) - \partial F/\partial \xi = 0$.

Since ξ_x can be considered to represent the point density, this variational problem represents a minimization over the field of the density of grid points in the least-squares sense, subject to the weight function, and thus produces the smoothest point distribution attainable. Here the weight function w(x) is associated with the grid point locations, not directly with the points. If the weight function is associated with the points themselves, rather than the locations, then $w = w(\xi)$ and the integral for which Eq. (4) is the Euler equation is

$$I_4 = \int_0^1 [\frac{\xi_x}{w(\xi)}]^2 dx \tag{13}$$

This variational problem is the least-squares minimization over the field of the cumulative point density weighted by the weight function.

The constant in Eq. (4) is evaluated in this form by writing,

$$\xi_x = \frac{w}{C}$$

so that with the normalization as defined above,

$$C = \frac{1}{N-1} \int_0^L w dx \tag{14}$$

The transformation then is given by

$$\xi(x) = 1 + (N-1) \frac{\int_0^x w d\bar{x}}{\int_0^L w dx} \tag{15}$$

Thus

$$\Delta\xi = (N-1)\frac{\int_{x_i}^{x_{i+1}} wdx}{\int_o^L wdx} \tag{16}$$

so that again Eq. (2) is realized by taking equal increments in ξ. The point spacing is now given by

$$\Delta x_i = \frac{\int_o^L wdx}{(N-1)w} \tag{17}$$

The grid and solution may be determined separately, perhaps even in an iterative fashion. However, the transformation allows the grid and solution to be dynamically coupled so that both evolve together. With the spring analogy approach, Eq. (8) supplies the following differential equation for the grid:

$$x_\xi w = \frac{L}{\int_1^N \frac{d\xi}{w}} \tag{18}$$

which supplies an additional differential equation to be solved simultaneously with the differential equation system of the physical problem at hand, with the grid point location x as an additional dependent variable, and ξ being taken as the independent variable. Similarly, with the smoothness approach, the differential equation for the grid is

$$x_\xi w = \frac{1}{N-1}\int_o^L wdx \tag{19}$$

Eq. (18) and (19) really differ only by the way the constant is evaluated, i.e., whether by integration over ξ or over x. This is a real difference in implementation, though, since integration over ξ is dependent on the grid, but integration over x is not. Thus, with the spring analogy approach, the weight function is associated with the grid points, i.e., with ξ, and the grid adjusts to achieve a uniform value of $w\Delta x$. The uniform value reached, however, is dependent on the grid since the right-hand side of Eq. (18) is dependent on the point distribution. In contrast, in the smoothness approach, where the weight function is associated with the the grid points, i.e., with x, the grid adjusts to achieve a specified uniform value of $w\Delta x$, since the right-hand side of Eq. (19) is an integral in physical space, independent of the grid. In the first approach, the points move to change the spacing x_ξ between points, while in the second the points move to change the point density ξ_x. (Note that Eq. (4) can also be written as $\xi_x/w =$ constant.) Either approach is viable, unless it is intended that the uniform value of $w\Delta x$ be fixed beforehand, as would be the case if the weight function is taken to be representative of truncation error and a certain bound is to be imposed on this error. The smoothness approach, i.e., integration over x, has been the most widely used because it is natural in most physical problems to associate the weight function with some physical property which varies in space.

Implementation of the two forms proceeds as follows: The form based on the grid point density is implemented using Eq. (15). With the solution $u(\xi)$ known on the current grid points at a given time step, the weight function is evaluated at each point and then the integral in the denominator of Eq. (15) is evaluated by numerical quadrature, i.e., by summing the product $w\Delta x$ over the grid points using coefficients in the summation appropriate to whatever type of numerical quadrature is intended. The integral in the numerator is similarly evaluated out to values of the upper limit x that produce the successive integral values of ξ which define the grid points. Thus we have x_i defined by

$$\int_o^{x_i} w(x)dx = \frac{i-1}{N-1} \int_o^L w(x)dx \qquad (i=2,3.....,N-1) \qquad (20)$$

These values of x_i then are the new grid point locations, and the solution proceeds to the next time step.

The spring analogy form, however, requires iteration. Here we have, from Eq. (8), the point locations x_i defined by

$$x_i = L \frac{\int_1^i \frac{d\xi}{w(\xi)}}{\int_1^N \frac{d\xi}{w(\xi)}} \qquad (i=2,3.....,N-1) \qquad (21)$$

With the solution known at a given time step, the weight function is evaluated at each grid point, and the integral in the denominator is evaluated numerically as before. Then the integral in the numerator is evaluated with the upper limit set at the successive integral values of ξ as indicated, and this defines a changed point distribution, x_i. The complication here is that the integral in the denominator, i.e., the constant in Eq. (4), depends on the point distribution, amounting to a sum of $1/w$ over the points since $\Delta\xi = 1$ by construction regardless of the distribution.

(By contrast, the corresponding integral in Eq. (20), i.e., the constant in Eq. (4), does not depend on the point distribution, being simply an integral of a function in physical space.) Therefore, this integral must be re-evaluated using the changed point distribution.

The integral in the numerator is then also re-evaluated for each point, thus changing the point distribution again. This process must be continued until convergence before the final new point distribution is obtained. The solution then proceeds to the next time step. The necessity for iteration with the spring analogy form clearly makes this form more difficult to implement than the grid point density form. Since no particular advantages of the former have been noted, preference naturally falls to the latter.

A number of examples of both the point density form and the spring analogy form, as well as other applications of the use of one-dimensional equidistribution are cited in the survey of adaptive grids given as Ref. [45].

C. Weight functions

As noted above, the effect of the weight function w is to reduce the point spacing x_ξ where w is large, and therefore the weight function should be set as some measure of the solution error, or as some measure of the solution variation. The simplest choice is just the solution gradient, i.e.,

$$w = u_x \tag{22}$$

In this case, Eq. (4) becomes

$$x_\xi u_x = \text{constant}$$

which then reduces to

$$u_\xi = \text{constant}$$

With the solution gradient as the weight function the point distribution adjusts so that the same change in the solution occurs over each grid interval, as illustrated below:

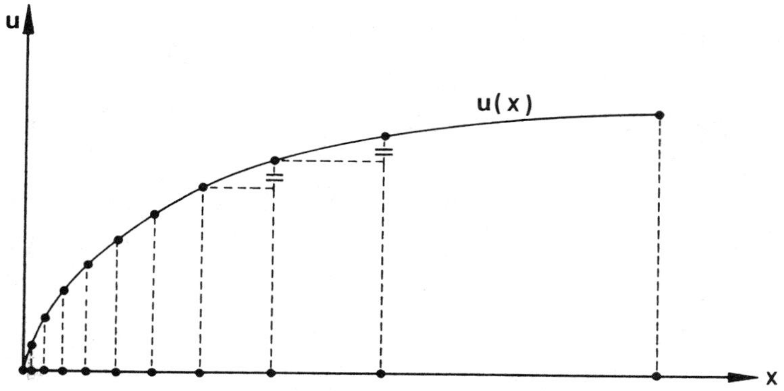

This choice for the weight function has the disadvantage of making the spacing infinitely large where the solution is flat, however.

A closely-related choice, also based on the solution gradient, is the form

$$w = \sqrt{1 + u_x^2} \tag{23}$$

An increment of arc length, ds, on the solution curve u(x) is given by

$$ds^2 = dx^2 + du^2 = (1+u_x^2)dx^2$$

so that this form of the weight function may be written

$$w = s_x$$

and then Eq. (4) becomes

$$x_\xi s_x = \text{constant}$$

which reduces to

$$s_\xi = \text{constant}$$

Thus, with the weight function defined by Eq. (23), the grid point distribution is such that the same increment in arc length on the solution curve occurs over each grid interval. For the curve shown above this gives the following point distribution:

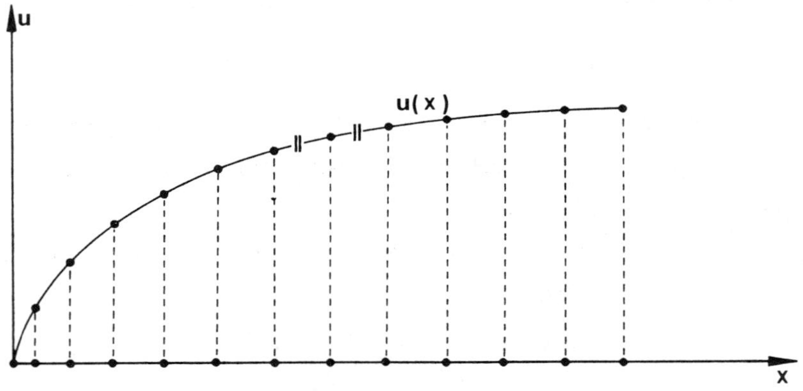

Unlike the previous choice, this weight function gives uniform spacing when the solution is flat. The concentration of points in the high-gradient region, however, is not as great. This concentration can be increased, while still maintaining uniform spacing where the solution is flat, by altering the weight function to

$$w = \sqrt{1 + \alpha^2 u_x^2} \tag{24}$$

where α is a parameter to be specified. Considering u to be plotted against x/α, we have for an increment of arc length on this solution curve

$$d\bar{s}^2 = [d(x/\alpha)]^2 + du^2 = [1 + u_{(x/\alpha)}][d(x/\alpha)]^2$$

so that this weight function is equivalent to

$$w = \bar{s}_{(x/\alpha)}$$

and Eq. (4) becomes

$$(x/\alpha)_\xi \, \bar{s}_{(x/\alpha)} = \frac{\text{constant}}{\alpha}$$

which reduces to

$$\bar{s}_\xi = \text{constant}$$

Thus we have equal increments of arc length on the solution curve with u plotted against x/α in this case. Now division of the abscissa by α for a flat curve would simply reduce the spacing by the same factor. However, since the slope steepens as the curve is compressed to the left by this change of scale, the effect on the spacing where the curve is not flat will be a greater reduction in spacing.

In fact, since the 1 in the weight function given by Eq. (24) tends to produce equal spacing, while the $\alpha^2 u_x^2$ tends to produce concentration in the high-gradient regions, with infinite spacing in flat regions, this weight function involves a weighted average between the tendency toward equal spacing and that toward concentration entirely in the high-gradient regions. The larger the value of α, the stronger will be the concentration in the high-gradient regions and the wider the spacing in the flat regions.

Now a disadvantage of all the above forms of the weight function is that regions near solution extrema, i.e., where $u_x = 0$ locally, are treated similar to flat regions, as is illustrated below for the form given by Eq. (22):

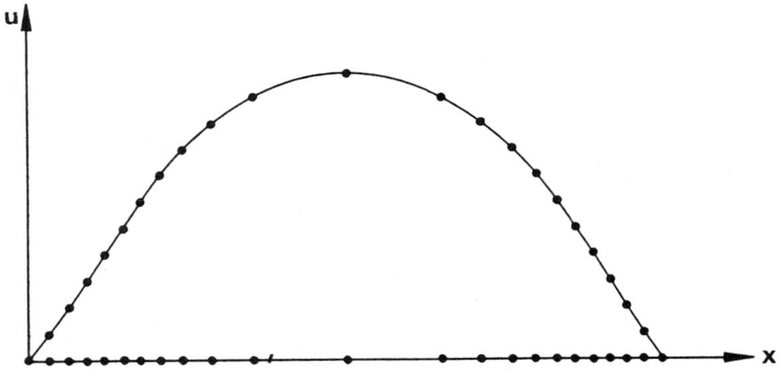

Although the distributions produced by the solution arc length forms, Eq. (23) and (24), would have closer spacings near the extrema, the effect is still the same, i.e., to concentrate points only near gradients, not extrema.

Concentration near solution extrema can be achieved by incorporating some effect of the second derivative u_{xx} into the weight function. A logical approach is to include this effect through consideration of the curvature of the solution curve:

$$K = \frac{u_{xx}}{(1+u_x^2)^{3/2}}$$

If the weight function is taken as

$$w = 1 + \alpha^2 |K| \tag{25}$$

then points will be concentrated in regions of high curvature of the solution curve, e.g., near extrema, with a tendency toward equal spacing in regions of zero curvature, i.e., where the solution curve is straight (not necessarily flat). This weight function, however, has the serious disadvantage of treating high-gradient regions with little curvature essentially the same as regions where the curve is flat. Thus in the curve shown above, nearly all the points would be concentrated near the maximum in the curve, with very wide spacing in the high-gradient regions on both sides.

A combination of the weight functions given by Eq. (24) and (25) provides the desired tendency toward concentration both in regions of high gradient and near extrema. The effect of the inclusion of the curvature is illustrated below (cf. Ref. [37]) with the function following):

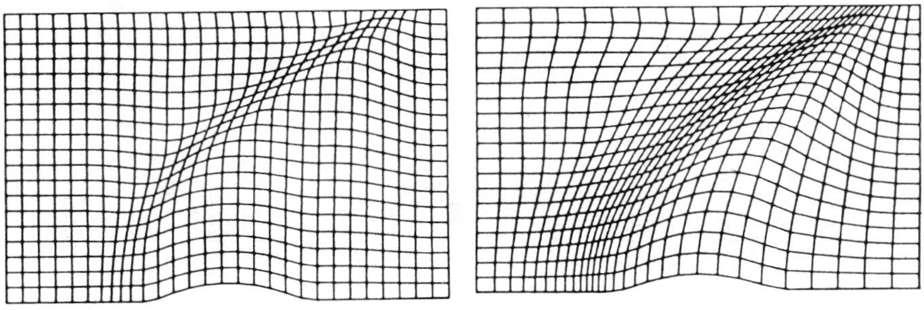

$$w = (1 + \beta^2|K|)\sqrt{1 + \alpha^2 u_x^2} \tag{26}$$

where α and β are parameters to be specified. Clearly, concentration near high gradients is emphasized by large values of α, while concentration near extrema (or other regions of large curvature) is emphasized by large β.

Another approach to the inclusion of the second derivative is simply to take the weight function as

$$w = 1 + \alpha|u_x| + \beta|u_{xx}| \tag{27}$$

where α and β are non-negative parameters to be specified.

With this form, (cf. Ref. [46] we have by Eq. (15), with $0 \leq \xi \leq 1$ and $0 \leq x \leq 1$,

$$\xi = \frac{x + \alpha\int_0^x |u_x|\,d\bar{x} + \beta\int_0^x |f(u_{xx})|\,d\bar{x}}{1 + \alpha\int_0^1 |u_x|\,dx + \beta\int_0^1 |f(u_{xx})|\,dx} \tag{28}$$

so that

$$\Delta\xi = \frac{(1 + \alpha|u_x| + \beta|f(u_{xx})|)\Delta x}{1 + \alpha\int_0^1 |u_x|\,dx + \beta\int_0^1 |f(u_{xx})|\,dx} \tag{29}$$

Then with R_1 defined as

$$R_1 = \frac{\alpha\int_0^1 |u_x|\,dx}{1 + \alpha\int_0^1 |u_x|\,dx + \beta\int_0^1 |f(u_{xx})|\,dx} \tag{30}$$

we have

$$\frac{|u_x|\Delta x}{\int_0^1 |u_x| dx} = \frac{\Delta\xi}{R_1} - \frac{(1 + \beta|f(u_{xx})|)\Delta x}{\alpha\int_0^1 |u_x| dx} \leq \frac{\Delta\xi}{R_1} \tag{31}$$

Since $\Delta\xi = 1/N$, where N+1 is the number of points on the coordinate line, the maximum percentage change in the solution over a grid interval,

$$r = \frac{|u_x|\Delta x}{\int_0^1 |u_x| dx} \tag{32}$$

is related to the ratio R_1, which measures the relative emphasis put on concentration of points according to the solution gradient by

$$r < \frac{1}{NR_1} \tag{33}$$

A guide for the choice of α to limit the maximum percentage solution change over an interval to a value r can then be obtained using an equality in Eq. (33) with R_1 from Eq. (30) and neglecting the effect of the β term:

$$\alpha = \frac{1}{(rN - 1)\int_0^1 |u_x| dx} \tag{34}$$

The smallest possible value of r is 1/N.

With the second derivative term included, the value of β can be continually updated to keep the same relative emphasis on concentration according to this term, as measured by the ratio R_2:

$$R_2 = \frac{\beta\int_0^1 |f(u_{xx})| dx}{1 + \alpha\int_0^1 |u_x| dx + \beta\int_0^1 |f(u_{xx})| dx} \tag{35}$$

The transformation can then be written as

$$\xi = (1 - R_2)\,\frac{x + \alpha\int_o^x |u_x|\,d\overline{x}}{1 + \alpha\int_o^1 |u_x|\,dx} + R_2\,\frac{\int_o^x |f(u_{xx})|\,d\overline{x}}{\int_o^1 |f(u_{xx})|\,dx} \tag{36}$$

where R_2 is considered to be constant. In this form, the transformation appears as the weighted average of one based on the solution gradient and one related to the second derivative.

The replacement of Eq. (24) with the form given by Eq. (27), with $\beta = 0$, still leaves a reasonable form for the weight function, but the clear association with the geometric properties of the solution curve are lost. In this case the weight function corresponding to Eq. (23) would, after substitution in Eq. (4), leads to the condition

$$x_\xi + u_\xi = \text{constant}$$

which corresponds to an equal distribution of the distance between points on the solution curve along a right-angle path formed by Δx and Δu from one point to the next. While this distance has some indirect relation to arc length on the solution curve (the chord length being the hypotenuse of the right triangle formed by this Δx and Δu), the direct association with arc length would seem to be preferable. Following the same reasoning, the use of solution curve curvature, rather than simply the second derivative, is also preferable. Therefore, the form given by Eq. (26) is probably more appropriate than that of Eq. (27). A number of other variations have been used, of course, as is noted in Ref. [45].

Since the numerical evaluation of higher derivatives can be subject to considerable computational noise, the use of formal truncation error expressions as the weight function is usually not practical, hence the emphasis above on solution gradients and curvature. Some problems may arise even with solution curvature, i.e., with second derivatives, in rough transits. It is common in any case to limit the grid point movement at each time step and/or to smooth the new point distribution.

For systems of equations involving more than one physical variable, one approach is to use the most rapidly-varying or dominant physical variable in the definition of the weight function. Another is to use some average of the variations of the several variables. It is also possible to use entirely different grids for different physical variables, with values transfered among the grids by interpolation. Examples of each of these approaches are cited in Ref. [45] and [5].

2. Multiple-Dimensional Adaption

A. Adaption along fixed lines

In multiple dimensions, adaption should in general occur in all directions in a mutually dependent manner. However, when the solution varies predominately in a single direction, one-dimensional adaption of the forms discussed above can be applied with the grid points constrained to move along one family of fixed curvilinear coordinate lines, and applications of this approach are noted in Ref. [45].

The fixed family of lines is established by first generating a full multi-dimensional grid by any of the grid generation techniques discussed in the earlier chapters, with the curvilinear coordinate lines of one family therein then being taken as the fixed lines. The points generated for this initial grid, together with some interpolation procedure, e.g., cubic splines, serve to define the fixed lines along which the points will move during the adaption. The one-dimensional adaption discussed above is then applied with x replaced by arc length along these lines.

Examples (cf. Ref. [46]) of application of the point density form discussed above in this manner are shown in the following figures. The first figure shows an adaptive grid for a combustion problem, where the adaption is along fixed radial lines. The flame front is clearly visable here because of the strong concentration of points therein:

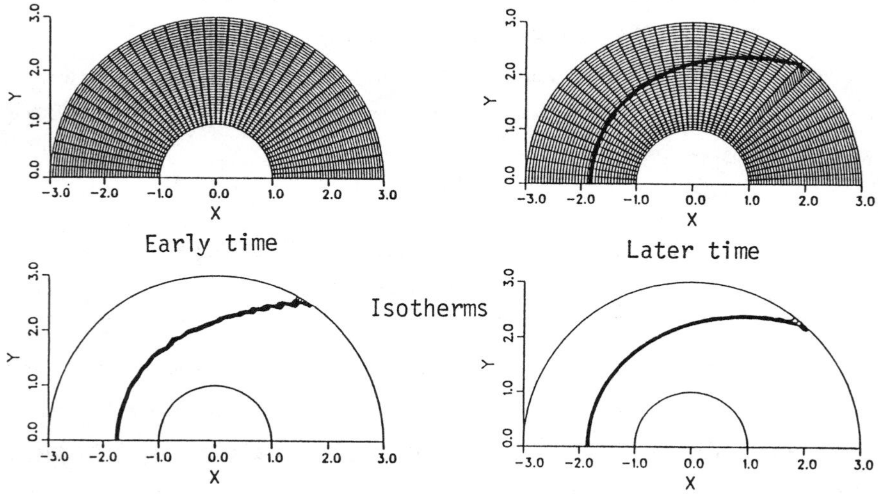

The oscillations evident with the fixed grid are removed by the grid adaption. An extension of this problem appears next with a flowing gas. This gives an example of the use of separate adaptive grids for different physical variables of the problem, one for the combustion and one for the fluid mechanics, with values transfered between the two grids by interpolation.

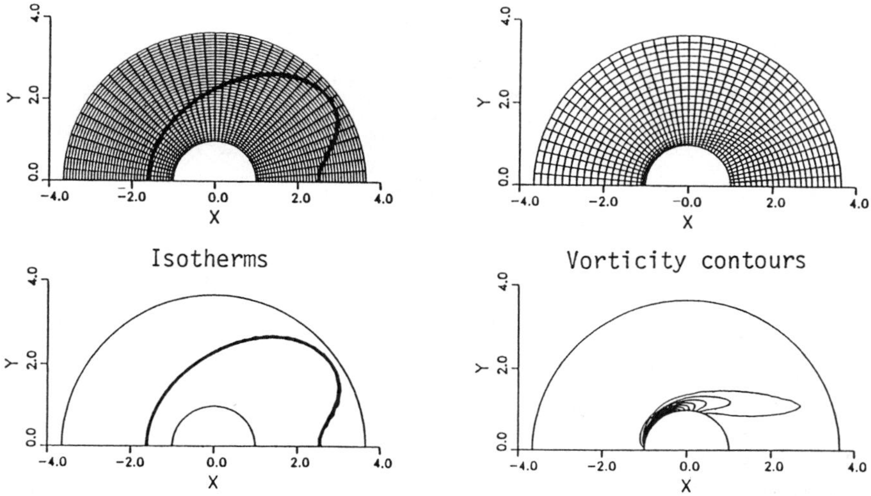

Adaption through the spring analogy is illustrated next with adaption along fixed lines between the body and outer boundary in a hypersonic flow problem (cf. Gnoffo in Ref. [45]). Here the concentration of points makes the shock location evident in the grid:

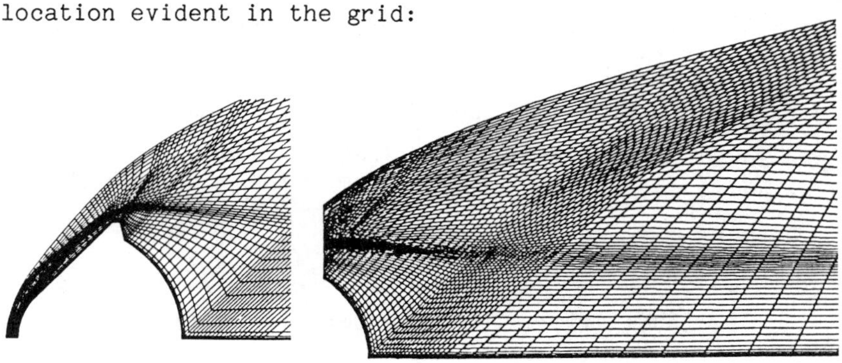

Another obvious application of of adaption along fixed lines is adaption of boundary points along a fixed boundary in two dimensions (cf. Nakamura in Ref. [45]). An example of such adaption along a boundary as a shock forms appears below:

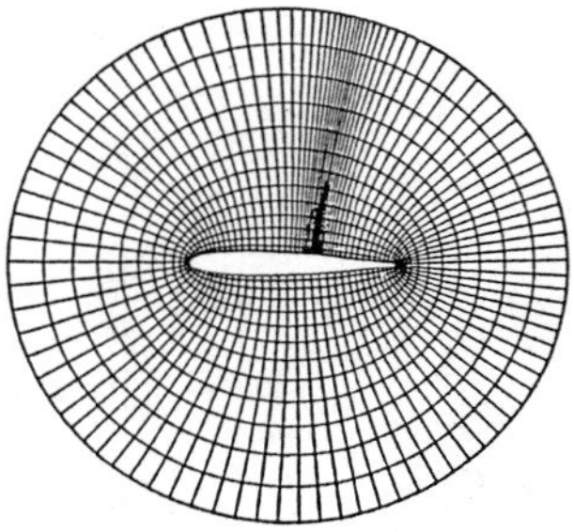

B. Uncoupled adaption

One step beyond this one-dimensional adaption along fixed lines is the application of successive one-dimensional adaptions separately in each of the curvilinear coordinate directions. This proceeds in the same manner as for the adaption on the fixed lines, simply using the latest grid to re-define the coordinate lines to serve as the "fixed" lines in the next direction of adaption, cf. Ref. [56] and [57]. In the latter a torsion spring analogy is used, as well as the tension springs discussed above, incorporating resistance to movement away from orthogonality. This is done in effect by adding the term $v(\xi)(x-x_o)^2$ to the integral of Eq. (5), where $v(\xi)$ is a second weight funciton and x_o is the arc length location of the intersection of the normal from the adjacent grid line with the line on which the adaption is occurring.

C. Coupled adaption

The final grid in the one-dimensional adaption discussed above will, of course, be the result of the grid point movement along the one family of fixed lines, and therefore the smoothness of the original grid may not be preserved as the grid adapts. Some restrictions on the point movement have generally been necessary in order to prevent excessive grid distortion.

In multiple dimensions, in general it is desirable to couple the adaption in the different directions in order to maintain sufficient smoothness in the grid. One approach to such coupling is to generate the entire grid anew at each stage of the adaption from some basic grid generation system, be it algebraic or based on partial differential equations. The structure of the grid generation system serves to maintain smoothness in the grid as the adaption proceeds. In this approach, which is analogous to the one-dimensional equidistribution discussed above, the new point locations are determined directly from the grid generation system, and then the grid point speeds, $\dot{x}_i$, for use in the transformed time derivatives, Eq. (1), are calculated from the change in the point locations by difference expressions. Another approach is to determine the grid point speeds directly through some process and then to calculate the new point locations by integrating these point speeds.

D. Weight functions

The one-dimensional weight function, Eq. (23), based on arc length on the solution curve can be generalized to higher dimensions as follows: Consider a hyperspace of dimensionality one greater than that of the physical space, with the solution, u, being the extra coordinate. Let the unit vector in the solution direction be $\underset{\sim}{e}$, this being orthogonal to the physical space. Then the position vector in this hyperspace is given by

$$\underset{\sim}{R} = \underset{\sim}{i}x + \underset{\sim}{j}y + \underset{\sim}{k}z + \underset{\sim}{e}u = \underset{\sim}{r} + \underset{\sim}{e}u \qquad (37)$$

where $\underset{\sim}{r}$ is the position vector in physical space. Now, following Eq. (III-5), the covariant metric element, denoted G_{ij}, in the hyperspace will be

$$G_{ij} = \underset{\sim}{R}_{\xi^i} \cdot \underset{\sim}{R}_{\xi^j} = (\underset{\sim}{r}_{\xi^i} + \underset{\sim}{e}u_{\xi^i}) \cdot (\underset{\sim}{r}_{\xi^j} + \underset{\sim}{e}u_{\xi^j}) \tag{38}$$

$$= g_{ij} + u_{\xi^i} u_{\xi^j}$$

where g_{ij} is the metric element in physical space. Now

$$u_{\xi^i} = \underset{\sim}{\nabla} u \cdot \underset{\sim}{r}_{\xi^i} \tag{39}$$

so that

$$u_{\xi^i} u_{\xi^j} = (\underset{\sim}{\nabla} u \cdot \underset{\sim}{r}_{\xi^i})(\underset{\sim}{\nabla} u \cdot \underset{\sim}{r}_{\xi^j})$$

and then

$$G_{ij} = g_{ij} + (\underset{\sim}{\nabla} u \cdot \underset{\sim}{r}_{\xi^i})(\underset{\sim}{\nabla} u \cdot \underset{\sim}{r}_{\xi^j}) \tag{40}$$

It can be shown that

$$\det|G_{ij}| = (1 + |\underset{\sim}{\nabla} u|^2)\det|g_{ij}| \tag{41}$$

(This has been verified for one and two dimensions.)

In one dimension this reduces to the expression for arc length on the solution curve, i.e.,

$$\sqrt{G} = \sqrt{1 + u_x^2}\, x_\xi$$

In two dimensions Eq. (41) gives an expression for area on the solution surface:

$$\sqrt{G} = \sqrt{1 + |\nabla u|^2}\,\sqrt{g} \tag{42}$$

Thus the extension of the one-dimensional weight function based on arc length on the solution curve to two dimensions is that based on area on the solution surface:

$$w = \sqrt{1 + |\nabla u|^2} \tag{43}$$

The extension of this form to three dimensions would also seem logical, but has not been verified.

3. Variational Approach

Considering the grid from a continuous viewpoint, it occurs that something should be minimized by the grid rearrangement, and thus a variational approach is logical. This is the natural extension of the equidistribution concept discussed above to multiple dimensions. The development in this section is a generalization of that in Ref. [47]. (cf. Ref. [1] for earlier related work.)

A. Variational formulation

The variational formulation for multiple dimensions can be constructed in analogy with the one-dimensional equidistribution discussed in Section 1. Thus in general a weighted integral measure of the accumulation of some grid property Q, either over the grid points, i.e.,

$$I = \int wQ\,d\underline{\xi} \tag{44}$$

or over the physical field, i.e.,

$$I = \int wQd\underline{x} \tag{45}$$

where w is the weight function, will be minimized. The resulting Euler equations then will constitute the grid generation system. In formulating the variational problem there are basically three decision points.

First, if the integration is taken over $\underline{\xi}$ then the integral represents a summation over the grid points, while integration over $\underline{x}$ represents a summation over cell volumes in physical space. With integration over $\underline{\xi}$ it is thus the accumulation of some property over the grid points that is minimized, while with integration over $\underline{x}$ the accumulation over the physical cell volumes is minimized.

The second question concerns the weight function. If the weight function is directly dependent on $\underline{\xi}$, then the weight is associated with the grid points, while with weight functions dependent directly on $\underline{x}$ the weight is associated with location in physical space. As noted in Section 1 it is this direct dependence of the weight function that figures in the partial derivatives $\partial F/\partial x$ and $\partial F/\partial \xi$ in the Euler equations, the fact that a change of variable could be effected by the transformation $\underline{x}(\underline{\xi})$ notwithstanding. In most applications the weight function will be based on some solution gradient and hence will be naturally taken as a function of position in physical space, $\underline{x}$.

Finally, there is the choice of what property is to be accumulated to be minimized. This choice depends, of course, on what is expected from the grid. Among the grid properties that might be considered are the following in computational space (integration over grid points, i.e., $d\underline{\xi}$):

(1). square of cell volume:

$$Q = [\mathbf{r}_1 \cdot (\mathbf{r}_2 \times \mathbf{r}_3)]^2 = \det|g_{ij}| = g$$

(2). inverse cell volume:

$$Q = 1/\sqrt{g}$$

(3). sum squares of cell edge lengths (average of of squares of diagonal lengths):

$$Q = \sum_i (\mathbf{r}_i \cdot \mathbf{r}_i) = \sum_i g_{ii}$$

(4). cell area squared/volume ratio:

$$Q = \frac{\sum_i (\mathbf{r}_j \times \mathbf{r}_k)^2}{\sqrt{g}} = \frac{1}{\sqrt{g}} \sum_i (g_{jj} g_{kk} - g_{jk}^2) \qquad (i,j,k)\text{cyclic}$$

(5). cell skewness based on edge tangents:

$$Q = \sum_i (\mathbf{r}_j \cdot \mathbf{r}_k)^2$$

$$= \sum_i g_{jk}^2 \qquad (i,j,k)\text{cyclic}$$

(6). cell skewness based on face normals:

$$Q = \sum_i [(\mathbf{r}_i \times \mathbf{r}_j) \cdot (\mathbf{r}_k \times \mathbf{r}_i)]^2$$

$$= \sum_i (g_{ij} g_{ik} - g_{ii} g_{jk})^2 \qquad (i,j,k)\text{cyclic}$$

In two dimensions the two orthogonaly properties, (5) and (6), are equivalent

These six properties correspond in order to the use of the following properties in physical space, where the integration is over the physical field ($d\underline{x}$):

(1). inverse point density:

$$Q = \sqrt{g} = \frac{1}{\sqrt{\det|g^{ij}|}}$$

(2). square of point density:

$$Q = \frac{1}{g}$$

(3). $Q = \sqrt{g} \sum_i (g^{jj}g^{kk} - g^{jk^2})$ (i,j,k)cyclic

(4). $Q = \sum_i g^{ii} = \sum_i |\nabla\xi^i|^2$

(5). $Q = g^{3/2} \sum_i (g^{ij}g^{ik} - g^{ii}g^{jk})^2$ (i,j,k)cyclic

(6). $Q = g^{3/2} \sum_i (g^{jk})^2$

$= g^{3/2} \sum_i (\nabla\xi^j \cdot \nabla\xi^k)^2$ (i,j,k) cyclic

Similar representations of other grid properties can also be considered, of course. The one-dimensional forms of properties (1) and (3) in the computational space reduce to x_ξ^2, while those of properties (2) and (4) become $1/x_\xi$. Therefore, in analogy with the one-dimensional equidistribution in Section 1, a weight function with properties (1) and (3) that is a function of $\underline{x}$ should actually be squared in the integral (cf. Eq. (6)), i.e.,

$$I = \int w^2(\underline{x})Q d\underline{\xi} \qquad (1) \text{ and } (3) \qquad (46a)$$

while $w(\underline{x})$ with properties (2) and (4) appears as (cf. Eq. (12))

$$I = \int \frac{Q}{w(\underline{x})} d\underline{\xi} \qquad (2) \text{ and } (4) \qquad (46b)$$

Similarly, weight functions that are functions of $\underline{\xi}$ should appear as (cf. Eq. (5) and (13))

$$I = \int w(\underline{\xi})Q d\underline{\xi} \qquad (1) \text{ and } (3) \qquad (47a)$$

$$I = \int \frac{Q}{w^2(\underline{\xi})} d\underline{\xi} \qquad (2) \text{ and } (4) \qquad (47b)$$

The construction for integration in the physical space is analogous, but noting that (1) and (3) correspond to $1/\xi_x$, while (2) and (4) correspond to ξ_x^2, in one dimension (cf. (5), (6), (13) and (12), respectively):

$$I = \int w(\underline{\xi})Q d\underline{x} \quad \text{or} \quad I = \int w^2(\underline{x})Q d\underline{x} \qquad (1) \text{ and } (3) \qquad (48a)$$

$$I = \int \frac{Q d\underline{x}}{w^2(\underline{\xi})} \quad \text{or} \quad I = \int \frac{Q d\underline{x}}{w(\underline{x})} \qquad (2) \text{ and } (4) \qquad (48b)$$

The grid for which the weighted accumulation of the property Q is minimized is obtained, by the calculus of variations, as the solution of the Euler variational equations for the integral I. If the integration is over $\underline{\xi}$ these equations are

$$\sum_{j=1}^{3} \frac{\partial}{\partial \xi_j} \frac{\partial F}{\partial (x_i)_{\xi j}} - \frac{\partial F}{\partial x_i} = 0 \qquad (i=1,2,3) \qquad (49)$$

where F is the integrand of the integral I. With integration over $\underline{x}$ the variational equations are

$$\sum_{j=1}^{3} \frac{\partial}{\partial x_j} \frac{\partial F}{(\xi^i)_{x_j}} - \frac{\partial F}{\partial \xi^i} = 0 \qquad (i=1,2,3) \qquad (50)$$

These partial differential equations then constitute the generation system for the grid. Note that the equations resulting from Eq. (50) must be transformed using the relations in Chapter III so that the curvilinear coordinates become the independent variables. The equations given by Eq. (49), however, will already be in this form.

A grid generation system which involves competitive emphasis on various grid properties can be constructed by casting the integral to be minimized as a weighted average of several of the above integrals, each of which represents an accumulation of a different grid property. Since the various grid properties do not all have the same dimensions, it is necessary to scale the various integrals involved, as is done below for the Brackbill-Saltzman construction.

There clearly is no unique construction of the variational formulation for adaptive grids, and this is an area that is not yet fully developed. The constructions given later in this chapter are logical and illustrative of the procedure, but should not be considered definitive.

B. Euler equations

The derivation of the Euler equations, hence the grid generation system, is straightforward but may be algebraically involved. The following developments simplify the derivation somewhat. Consider first the integral over the grid points

$$I = \int F[\underline{g}, w(\underline{x})] d\underline{\xi} \qquad (51)$$

where $\underline{g}$ is the covariant metric tensor, with elements g_{ij} defined by Eq. (III-5), and $w(\underline{x})$ is a weight function dependent on $\underline{x}$. The Euler equations then are given by Eq. (49). As shown in Appendix B, the Euler equations produce the following generation system (with $\partial F/\partial w$ written as F'):

$$\sum_{j=1}^{3} \sum_{k=1}^{3} [A_{jk} \underset{\sim}{r}_{\xi^j \xi^k} + A'_{jk} (\underset{\sim}{\nabla} w \cdot \underset{\sim}{r}_{\xi^j}) \underset{\sim}{r}_{\xi^k}$$

$$+ \sum_{m=1}^{3} \sum_{n=1}^{3} (\frac{\partial A_{jk}}{\partial g_{mn}}) (\underset{\sim}{r}_{\xi^m} \cdot \underset{\sim}{r}_{\xi^j \xi^n}$$

$$+ \underset{\sim}{r}_{\xi^n} \cdot \underset{\sim}{r}_{\xi^j \xi^m}) \underset{\sim}{r}_{\xi^k}] - F' \underset{\sim}{\nabla} w = 0 \qquad (52)$$

where

$$A_{jk} = \frac{\partial F}{\partial g_{jk}} + \frac{\partial F}{\partial g_{kj}} \qquad (53)$$

Here the gradient of the weight function in the last term is expressed using Eq. (III-42), with $\underset{\sim}{a}^i$ given by Eq. (III-33). It should be noted that if the weight function in the integral (51) had been defined as a function of $\underline{\xi}$ instead of $\underline{x}$, a result different from Eq.(52) would have been obtained for the generation system (cf. Eq. (9) of Appendix B). The two-dimensional form of Eq. (52) is given as Eq. (10) of Appendix B.

With the variational problem formulated in the physical space, and the weight functions dependent on $\underline{x}$, we have the integral

$$I = \int F[G, w(\underline{x})]d\underline{x} \tag{54}$$

where G is the contravariant metric tensor, i.e., with elements g^{ij} from Eq.(III-37). Then from the Euler equations given by Eq. (50), cf. Appendix B, the generation system is (with $\partial F/\partial w$ written as F'),

$$\sum_{k=1}^{3} \{[A_{ik}\nabla^2\xi^k + (A'_{ik}) \cdot \underset{\sim}{\nabla}\xi^k] + \sum_{m=1}^{3}\sum_{n=1}^{3} \frac{\partial A_{ik}}{\partial g^{mn}} \cdot [\underset{\sim}{\nabla}(\underset{\sim}{\nabla}\xi^m \cdot \underset{\sim}{\nabla}\xi^n)] \cdot \underset{\sim}{\nabla}\xi^k\} = 0 \qquad (i=1,2,3) \tag{55}$$

with

$$A_{ik} = \frac{\partial F}{\partial g^{ik}} + \frac{\partial F}{\partial g^{ki}}$$

C. Brackbill-Saltzman construction

As noted in Chapter V there is a need for smoothness in the grid in order to reduce certain terms in the truncation error of a solution done on the grid. The quantity $\nabla\xi^i \cdot \nabla\xi^i$ is the extension to multiple dimensions of the ξ_x^2 used above with the smoothness form in Eq. (12). Therefore to maximize the smoothness of the grid it is logical to minimize the integral of this quantity over the physical field:

$$I_s = \iiint \sum_{i=1}^{3} \nabla\xi^i \cdot \nabla\xi^i \, d\underline{x} = \iiint \sum_{i=1}^{3} g^{ii} d\underline{x} \qquad (57)$$

This amounts to a minimization of the linear point density in the least-squares sense. The property used here is that given as (4) on p.396, which corresponds to the ratio of the squares of the cell face areas to the cell volume when the accumulation is over the grid points, as given by property (4) on p. 395. The corresponding integral over the grid points is

$$I_s = \iiint \frac{1}{\sqrt{g}} \sum_{i=1}^{3} (g_{jj}g_{kk} - g_{jk}{}^2) d\underline{\xi} \qquad (i,j,k)\text{cyclic} \qquad (58)$$

Substitution of F from Eq. (57) into Eq. (19) of Appendix B then yields the elliptic grid generation system

$$\nabla^2\xi^i = 0 \qquad (i = 1,2,3) \qquad (59)$$

Thus the smoothest grid is that for which the curvilinear coordinates satisfy Laplace's equation.

Emphasis on orthogonality and/or on concentration of grid lines can also be incorporated into the grid generation system by basing the system on the Euler equations for additional variational principles. Orthogonality can be emphasized by minimizing the integral I_o defined with property (6) on p. 396 as

$$I_o = \iiint \sum_{i=1}^{3} g^{3/2}(\nabla\xi^j \cdot \nabla\xi^k)^2 \, d\underline{x}$$

$$= \iiint g^{3/2} \sum_{i=1}^{3} (g^{jk})^2 d\underline{x} \qquad (i,j,k) \text{ cyclic} \qquad (60)$$

since each of these dot products vanishes for an orthogonal grid. (Recall that $\nabla\xi^i$ is normal to the coordinate surface on which ξ^i is constant, cf. Chapter III.) The inclusion of the $g^{3/2}$, the cube of the Jacobian of the transformation, as a weight function in I_o is somewhat arbitrary, and causes orthogonality to be emphasized more strongly in the larger cells. With the accumulation over the grid points, this corresponds to the use of the square of the dot product of the cell face normals in the variational statement (property (6) on p. 395). The corresponding integral over the grid points is

$$I_o = \iiint \sum_{i=1}^{3} (g_{ij}g_{ik} - g_{ii}g_{jk})^2 d\underline{\xi} \qquad (i,j,k) \text{ cyclic} \qquad (61)$$

Finally, concentration can be emphasized by minimizing the integral I_w defined by

$$I_w = \iiint w^2(\underline{x})\sqrt{g} \, d\underline{x} \qquad (62)$$

where $w(\underline{x})$ is a specified weight function. This causes the cells to be small where the weight function is large, and uses property (1) on p. 396, i.e., the inverse point density. With the accumulation over the grid points this corresponds to the use of the square of the cell volume (property (1) on p. 395), and the integral over the grid points is

$$I_w = \iiint w^2(x) g d\underline{\xi} \quad (63)$$

The grid generation system is obtained by minimizing a weighted sum I of these three integrals:

$$I = I_s + \lambda_o \left(\frac{N}{L}\right)^7 I_o + \lambda_w \left(\frac{N}{L}\right)^5 \frac{1}{W} I_w \quad (64)$$

where N is a characteristic number of points, L is a characteristic length, and W is the average weight function over the field:

$$W = \frac{1}{V} \iiint w^2 d\underline{x} \quad (65)$$

with V being the volume of the field. This scaling in the weighted sum is obtained as follows: From the above expressions for I_s, I_o, and I_w we have

$$I_s \sim \frac{N^2}{L^2} L^3, \qquad I_o \sim \frac{L^5}{N^5} L^3, \qquad I_w \sim W \frac{L^3}{N^3} L^3$$

Therefore, the three terms in Eq. (64) should stand in the ratios given. In two dimensions the factors on I_o and I_w both become $(N/L)^4$, since the Jacobian is then proportional to $(L/N)^2$, rather than to $(L/N)^3$. The characteristic length and number of points might logically be taken as the cube roots of the volume and the total number of points in the field, respectively, in three dimensions, the square root being used in two dimensions.

Emphasis is varied among the competing features of smoothness, orthogonality, and adaptivity by the choice of the coefficients λ_o and λ_w. For example, a large λ_o will result in a grid that is nearly orthogonal, at the cost of smoothness and concentration, with an analogous effect of λ_w. The Euler equations for this variational problem, which will be the weighted sums of those for the individual integrals, form the system of partial differential equations from which the coordinate system is generated. These equations will be quasilinear, second-order partial differential equations, with coefficients which are quadratic functions of the first derivatives, and are derived in general as described in the preceeding section and Appendix B.

Clearly the integral I_s, Eq. (57), is the multi-dimensional generalization of the one-dimensional smoothness integral I_3, Eq. (12), without the weight function, and the integral I_w, in Eq. (63), is the extension of the one-dimensional spring analogy integral I_1, Eq. (5), to multiple dimensions, with the spring extension x_ξ generalizing to the volume, i.e., (the Jacobian $\sqrt{g}$ in three dimensions, area in two). This variational approach thus is a generalization of the one-dimensional equidistribution discussed above to multiple dimensions. All of the discussion of weight functions given above in regard to equidistribution therefore has relevance here to the weight function of the integral I_w, Eq. (63). (The role of the constant in the equidistribution weight function, e.g., the 1 in Eq. (23), etc., which tends to produce a linear transformation, is taken by the smoothness integral I_s of Eq. (57), which tends to produce an equally-spaced grid in multiple dimensions.)

For the three integrals given by Eq. (58), (61), and (63) we have, respectively, with (i,j,k) cyclic,

$$F_s = \sqrt{G} \sum_{i=1}^{3} g^{ii} = \frac{1}{\sqrt{G}} \sum_{i=1}^{3} (g_{jj}\, g_{kk} - g_{jk}^2) \tag{66}$$

$$F_o = \sum_{k=1}^{3} (g\, g^{ij})^2 = \sum_{k=1}^{3} (g_{ki} g_{kj} - g_{kk} g_{ij})^2 \tag{67}$$

$$F_w = g w(\underline{x}) \tag{68}$$

Here, of course, from Eq. (III-14), $g = \det|g_{ij}|$

In two dimensions, $g_{13} = g_{23} = 0$ and $g_{33} = 1$, so that these functionals reduce to

$$F_s = \frac{g_{11} + g_{22}}{\sqrt{g_{11} g_{22} - g_{12}^2}} \tag{69}$$

$$F_o = g_{12}^2 \tag{70}$$

$$F_w = (g_{11}\, g_{22} - g_{12}^2) w^2(x_1, x_2) \tag{71}$$

(Here an additive constant in F_s has been dropped since only derivatives of F contribute to the Euler equations.) Then using Eq. (1) of (Appendix B) the two-dimensional generation system based on concentration alone is

$$\begin{aligned} &2w^2[g_{22}\underline{r}_{\xi\xi} + g_{11}\underline{r}_{\eta\eta} - 2g_{12}\underline{r}_{\xi\eta} - (\underline{r}_\eta \cdot \underline{r}_{\xi\xi})\underline{r}_\eta \\ &- (\underline{r}_\xi \cdot \underline{r}_{\eta\eta})\underline{r}_\xi + (\underline{r}_\eta \cdot \underline{r}_{\xi\eta})\underline{r}_\xi + (\underline{r}_\xi \cdot \underline{r}_{\xi\eta})\underline{r}_\eta] \\ &+ 4w\{g_{22}(\nabla w \cdot \underline{r}_\xi)\underline{r}_\xi + g_{11}(\nabla w \cdot \underline{r}_\eta)\underline{r}_\eta \\ &- g_{12}[(\nabla w \cdot \underline{r}_\xi)\underline{r}_\eta + (\nabla w \cdot \underline{r}_\eta)\underline{r}_\xi]\} - 2gw\nabla w = 0 \end{aligned} \tag{72}$$

and the generation system based only on orthgonality is

$$2[2g_{12}\mathbf{r}_{\xi\eta} + (\mathbf{r}_\xi \cdot \mathbf{r}_{\eta\eta})\mathbf{r}_\xi + (\mathbf{r}_\eta \cdot \mathbf{r}_{\xi\xi})\mathbf{r}_\eta + (\mathbf{r}_\xi \cdot \mathbf{r}_{\xi\eta})\mathbf{r}_\eta + (\mathbf{r}_\eta \cdot \mathbf{r}_{\xi\eta})\mathbf{r}_\xi] = 0 \qquad (73)$$

The generation system based on smoothness (from Eq. (66)) is more complicated, but may be constructed from the relations given in Appendix B. The complete generation system then is obtained as the linear combination of the concentration system, Eq. (72), the orthgonality system, Eq. (73), and the smoothness system.

In Ref. [47] this combination is written in the form

$$b_1x_{\xi\xi} + b_2x_{\xi\eta} + b_3x_{\eta\eta} + a_1y_{\xi\xi} + a_2y_{\xi\eta} + a_3y_{\eta\eta} + \lambda_w' g w w_x = 0 \qquad (74a)$$

$$a_1x_{\xi\xi} + a_2x_{\xi\eta} + a_3x_{\eta\eta} + c_1y_{\xi\xi} + c_2y_{\xi\eta} + c_3y_{\eta\eta} + \lambda_w' g w w_y = 0 \qquad (74b)$$

where

$$a_i = a_{si} + w^2\lambda_w' a_{wi} + \lambda_o' a_{oi}$$

$$b_i = b_{si} + w^2\lambda_w' b_{wi} + \lambda_o' b_{oi}$$

$$c_i = c_{si} + w^2\lambda_w' c_{wi} + \lambda_o' c_{oi}$$

with

$a_{s1}=-A\alpha$, $b_{s1}=B\alpha$, $c_{s1}=C\alpha$,

$a_{s2}=2A\beta$, $b_{s2}=-2B\beta$, $c_{s2}=-2C\beta$,

$a_{s3}=-A\gamma$, $b_{s3}=B\gamma$, $c_{s3}=C\gamma$.

$a_{o1}=x_\eta y_\eta$, $\quad b_{o1}=x_\eta^2$, $\quad c_{o1}=y_\eta^2$,

$a_{o2}=x_\xi y_\eta+x_\eta y_\xi$, $\quad b_{o2}=2(2x_\xi x_\eta+y_\xi y_\eta)$, $\quad c_{o2}=2(x_\xi x_\eta+2y_\xi y_\eta)$,

$a_{o3}=x_\xi y_\xi$, $\quad b_{o3}=x_\xi^2$, $\quad c_{o3}=y_\xi^2$.

$a_{w1}=-x_\eta y_\eta$, $\quad b_{w1}=y_\eta^2$, $\quad c_{w2}=x_\eta^2$,

$a_{w2}=x_\xi y_\eta+x_\eta y_\xi$, $\quad b_{w2}=-2y_\xi y_\eta$, $\quad c_{w2}=-2x_\xi x_\eta$,

$a_{w3}=-x_\xi y_\xi$, $\quad b_{w3}=y_\xi^2$, $\quad c_{w3}=x_\xi^2$.

$A=x_\xi y_\xi+x_\eta y_\eta$, $\quad B=y_\xi^2+y_\eta^2$, $\quad C=x_\xi^2+x_\eta^2$,

$\alpha=(x_\eta^2+y_\eta^2)/J^3$, $\quad \beta=(x_\xi x_\eta+y_\xi y_\eta)/J^3$, $\quad \gamma=(x_\xi^2+y_\xi^2)/J^3$.

$J=x_\xi y_\eta-x_\eta y_\xi$

Here the coefficients subscripted s, o, and w, arise from the smoothness, orthogonality, and concentration integrals, respectively. The coeffcients λ_w' and λ_o' are, taking account of the scaling discussed above in connection with Eq. (64),

$$\lambda_w' = \lambda_w \left(\frac{N}{L}\right)^4 \frac{1}{W} \tag{75a}$$

$$\lambda_o' = \lambda_o \left(\frac{N}{L}\right)^4 \tag{75b}$$

In one dimension, with $y_\xi = x_\eta = 0$, we have

$A = 0 \qquad \alpha = \dfrac{1}{x_\xi^3 y_\eta}$

$B = y_\eta^2 \qquad \beta = 0$

$C = x_\xi^2$

$$J = x_\xi y_\eta \qquad \gamma = \frac{1}{x_\xi y_\eta^3}$$

Also, for the smoothness integral:

$$a_{s1} = a_{s2} = a_{s3} = 0$$

$$b_{s1} = \frac{y_\eta}{x_\xi^3}, \quad b_{s2} = 0, \qquad b_{s3} = \frac{1}{x_\xi y_\eta}$$

$$c_{s1} = \frac{1}{x_\xi y_\eta}, \quad c_{s2} = 0, \qquad c_{s3} = \frac{x_\xi}{y_\eta^3}$$

For the concentration integral:

$$a_{w1} = 0 \qquad b_{w1} = y_\eta^2 \qquad c_{w1} = 0$$

$$a_{w2} = x_\xi y_\eta \qquad b_{w2} = 0 \qquad c_{w2} = 0$$

$$a_{w3} = 0 \qquad b_{w3} = 0 \qquad c_{w3} = x_\xi^2$$

and for the orthogonality integral:

$$a_{o1} = 0 \qquad b_{o1} = 0 \qquad c_{o1} = y_\eta^2$$

$$a_{o2} = x_\xi y_\eta \qquad b_{o2} = 0 \qquad c_{o2} = 0$$

$$a_{o3} = 0 \qquad b_{o3} = x_\xi^2 \qquad c_{o3} = 0$$

Then

$a_1 = 0$

$$b_1 = \frac{y_\eta}{x_\xi^3} + w^2\lambda_w' y_\eta^2$$

$a_2 = (w^2\lambda_w' + \lambda_o')x_\xi y_\eta$

$b_2 = 0$

$a_3 = 0$

$$b_3 = \frac{1}{x_\xi y_\eta} + \lambda_o' x_\xi^2$$

$$c_1 = \frac{1}{x_\xi y_\eta} + \lambda_o' y_\eta^2$$

$c_2 = 0$

$$c_3 = \frac{x_\xi}{y_\eta^3} + w^2\lambda_w' x_\xi^2$$

Now also

$$y_{\xi\xi} = x_{\eta\eta} = x_{\xi\eta} = y_{\xi\eta} = 0$$

and, taking the ξ-direction to be the one of interest, we also have $y_{\eta\eta} = 0$.

The generation system in one dimension then reduces to

$$\left(\frac{y_\eta}{x_\xi^3} + w^2\lambda_w' y_\eta^2\right)x_{\xi\xi} + \lambda_w'(x_\xi y_\eta)^2 w w_x = 0$$

The y_η^2 can be made a part of λ_w': $\overline{\lambda}_w = \lambda_w' y_\eta^2$ so that the one-dimensional generation system finally is

$$\left(\frac{1}{x_\xi^3} + w^2\overline{\lambda}_w\right)x_{\xi\xi} + \overline{\lambda}_w x_\xi^2\, w w_x = 0 \tag{76}$$

with, for the scaling,

$$\lambda_w' = \lambda_w\left(\frac{N}{L}\right)^3 \frac{1}{W} \tag{77}$$

This then is the differential equation that can be applied on a boundary curve, interpreting x as arc length and ξ as the curvilinear coordinate that varies along the particular boundary.

In three dimensions the calculation of the required partial derivatives of F in Eq. (52) for the concentration integral, i.e., F_w given by Eq. (68), may be expedited by noting that since $g = \det|g_{ij}|$,

$$\frac{\partial F}{\partial g_{ij}} = c_{ij} \qquad \text{and} \qquad \frac{\partial^2 F}{\partial g_{ij}\partial g_{kl}} = \frac{\partial c_{ij}}{\partial g_{kl}}$$

where c_{ij} is the signed cofactor of g_{ij}. These second derivatives vanish if k=i or l=j, and are equal to $\pm\, g_{mn}$ otherwise, where (i,k,m) and (j,l,n) are cyclic, the sign being negative when the progression from i to k is opposite to that from j to l.

D. Applications

The dynamically-adaptive grid is applied by constructing the partial differential equations which constitute the grid generation system from the Euler equations as discussed above. These equations are solved numerically by replacing all derivatives with difference expressions (typically second-order, central differences) in the same manner as discussed in Chapter IV. As noted in Chapter III, the time derivatives in the equations of the physical problem to be solved on the grid are transformed according to Eq. (III-116), with the result that the grid point speeds appear in the difference equations of the physical problem. The grid is re-generated at each time step, and these grid point speeds are determined from difference representations between time steps. Although the difference equations for the grid and those for the physical solution could be iterated together at each time step, the more common procedure is to solve each separately at each time step.

Grid points on boundaries may, of course, be held fixed, but it is more appropriate in most cases to allow the points to move along the boundary to adapt as in the field. This can be accomplished either by using Neumann boundary conditions in the grid generation systems, i.e., making the system orthogonal at the boundary (cf. Chapter VI), or by applying the one-dimensional form of the grid generation equations, Eq. (76), in terms of arc length, along the boundary.

Some rather spectacular two-dimensional results of the grid adapting to a reflected shock are shown below for supersonic internal flow over a step. The formation and multiple reflections of the shock are made evident by the grid adaption into the shock as it develops. Here the magnitude of the pressure gradient was used in the weight function, and both smoothing and bounding was applied to the weight function to control grid distortion.

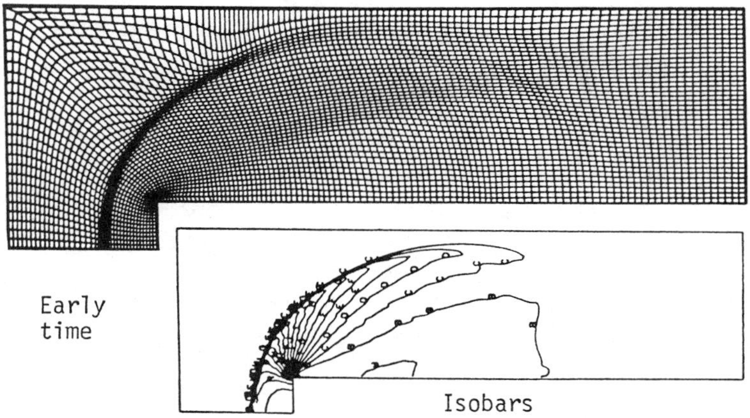

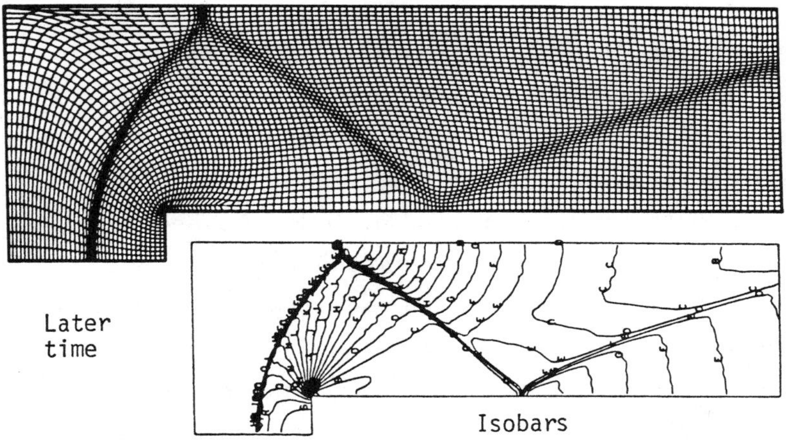

E. Extensions

In two dimensions, the departure of the grid from conformality can be controlled by basing F on the Cauchy-Riemann conditions:

$$F_c = \iint [(\xi_x - \eta_y)^2 + (\xi_y + \eta_x)^2]dxdy \tag{79}$$

and some applications are noted in Ref. [45].

Finally, another very useful addition to this composite variational system is a control on the grid point movement, which can be incorporated by taking F as

$$F_M = \sum_{i=1}^{3} (u_i - \dot{x}_i)^2 \tag{80}$$

where u_i is the fluid velocity and $\dot{x}_i$ is the grid speed. With $\dot{x}_i$ represented by a difference form, and with the fluid velocity evaluated at the previous time step, this F can be considered to be a function of x_i. Again some applications are noted in Ref. [45]. The following example shows the effectivenss of such control of the grid point movement:

• Without movement term

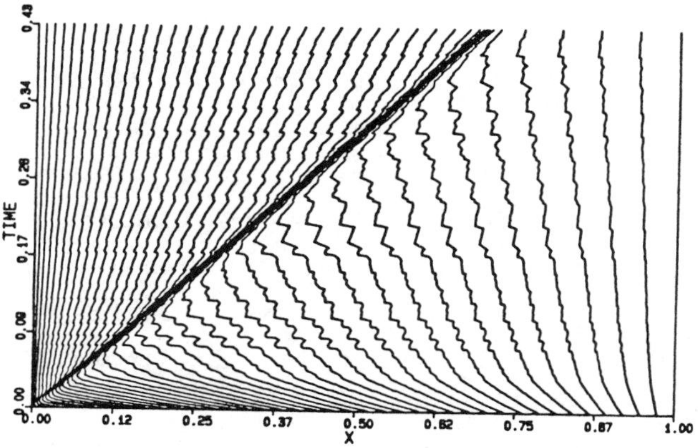

• With movement term - oscillations eliminated

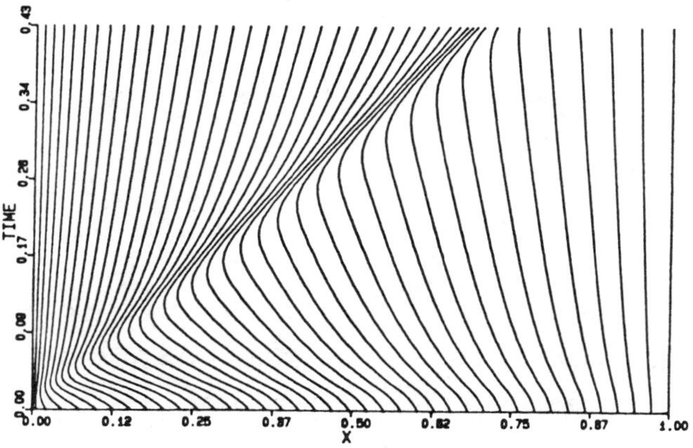

4. Other Approaches

Several other approaches are discussed in Ref. [45], three of which follow here.

A. Attraction-Repulsion

Another approach to adaptive grids is to let the grid points all move as if under the mutual influence of forces between all points. Here instead of generating new grid point locations through the solution of partial differential equations, the grid points move directly under the influence of mutual attraction or repulsion between points. This is accomplished by assigning to each point an attraction proportional to the difference between the magnitude of some measure of error (or solution variation) and the average magnitude of this measure over all the points. This causes points with values of this measure that exceed the average to attract other points, and thus to reduce the local spacing, while points with a measure less than the average will repel other points and hence increase the spacing.

$w > w_{avg.}$ $w < w_{avg.}$

This attraction is attenuated by an inverse power of the point separation distance in the transformed field. The collective attraction of all other points is then made to induce a velocity for each grid point. Since each point is influenced by all other points, this is effectively a type of elliptic generation system. Details of implementation are given in Ref. [48] and other references cited therein.

Smoothing through the addition of diffusion-like terms in the calculation of the grid evolution from the grid speeds has also been used. Reflections in boundaries in the transformed field are used to provide smooth grid motion near and on the boundaries. Since the transformed field is rectangular, this reflection is not complicated by the shape of the physical boundaries. A means of including terms that will induce rotational motion into the grid has been devised to cause the grid lines to align with lines of high gradients such as shocks.

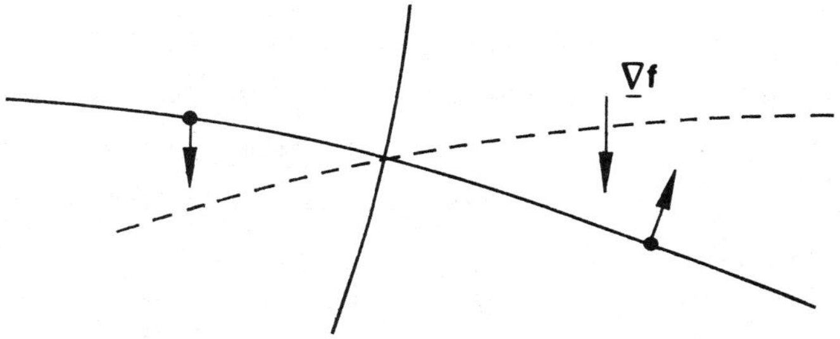

This procedure does not exercise any control over either the smoothness or orthogonality of the grid, so that distortion is possible. Collapse of points into each other is, however, impeded because attraction will become repulsion as the points approach each other, since the measure which drives the motion will drop below the average as the spacing decreases. Collapse is further impeded by the fact that the grid velocity decreases with the spacing. It has been found necessary to apply some limits and some damping of the grid speeds to prevent grid oscillation and distortion. In practice, the computed grid speeds are scaled so that the maximum over the field is a set value, but with the maximum scaling also limited. Provision is also made for exponential damping of the grid speeds according to the ratio of the maximum Jacobian to a specified value.

Since this procedure has all grid points moving to cause some measure to approach uniformity over the field, it can be considered an iterative approach to the equidistribution of this measure over the field. This occurs because the grid ceases to move when the measure is uniform, i.e., when the local value is equal to the average value everywhere. Therefore, the grid can be considered to move so as to minimize the variation in the measure over the field.

B. Reaction analogy

A different, but somewhat related, approach was noted in Ref. [45] and [5] based on a chemical reaction analogy. Here each grid interval is taken to represent a species concentration, and the reaction rate constants are made dependent on the difference between a local error measure for one grid interval compared with another. Each grid interval then is coupled with every other grid interval through reaction rate equations, so that each interval grows at the expense of others, and vice versa. A system of ordinary differential equations is solved for the intervals. This approach, as given, is somewhat inefficient, since there is no provision for limiting the effect to the nearer points. With each point affected equally by all other points, the number of ordinary differential equations to be solved is equal to the square of the total number of points.

The rate constants also contain factors designed to limit the range of variation of the grid intervals. The two-dimensional form given involves essentially applying the one-dimensional form separately along each family of curvilinear coordinate lines, with spacing in one cartesian coordinate being adjusted along one family of curvilinear lines, and the other cartesian coordinate being adjusted along the other family.

C. Moving finite elements

The moving finite element method of Miller (Ref. [49] - [50]) is a dynamically-adaptive finite element grid method in which the grid point locations are made additional dependent variables in a Galerkin formulation. The solution is expanded in piecewise linear functions, in terms of its values at the grid points and those of the grid point locations on each element. The residual is then required to be orthogonal to all the basis functions for both the solution and the grid. The grid point locations are thus obtained as part of the finite element solution. An internodal viscosity is introduced to penalize the relative motion between the grid points. This does not penalize the absolute motion of the points. An internodal repulsive force was also introduced to maintain a minimum point separation. Both of these effects are strong but of short range. A small long range attractive force is also introduced to keep the nodes more equally spaced in the absence of solution gradients. Small time steps are used in the initial development of the solution. The results show that the oscillations typically associated with shocks with fixed grids are removed with the adaptive grid, and that dispersion and dissipation are essentially eliminated. An order-of-magnitude increase in stability was also realized over conventional methods.

5. Correlations

The ultimate answer to numerical solution of partial differential equations may well be dynamically-adaptive grids, rather than more elaborate difference representations and solution methods. It has been noted by several authors that when the grid is right, most numerical solution methods work well. Oscillations associated with cell Reynolds number and with shocks in fluid mechanics computations have been shown to be eliminated with adaptive grids. Even the numerical viscosity introduced by upwind differencing is reduced as the grid adapts to regions of large solution variation. The results have clearly indicated that accurate numerical solutions can be obtained when the grid points are properly located.

It is also clear that there is considerable commonality among the various approaches to adaptive grids. All are essentially variational methods for the extremization of some solution property. The explicit use of varational principles allows effective control to be exercised over the conflicting requirements of smoothness, orthogonality, and concentration, and this is probably the most promising approach in multiple dimensions.

The adaptive grid is most effective when it is dynamically coupled with the physical solution, so that the solution and the grid are solved for together in a single continuous problem. The most fruitful directions for future effort thus are probably in the development and direct application of variational principles and in intimate coupling of the grid with the physical solution.

Exercises

1. Show that Eq. (4) is the Euler equation for the minimization of the integrals (5), (6), (12), and (13). Hint: For (6) note that in the term $d/d\xi\,(\partial F/\partial x_\xi)$, w must be differentiated with respect to ξ implicity, i.e., $w_\xi = w_x x_\xi$. A similar situation occurs with (13). Note, however, that implicit differentiation is not to be used in the term $\partial F/\partial x$ for (5) or in $\partial F/\partial \xi$ for (12).

2. Show that Eq. (4) is also the Euler equation for the integrals

$$\int_0^1 \frac{d\xi}{w(x)x_\xi}, \qquad \int_0^1 \frac{d\xi}{w^2(\xi)x_\xi}$$

$$\int_0^1 \frac{w(\xi)dx}{\xi_x}, \qquad \int_0^1 \frac{w^2(x)dx}{\xi_x}$$

3. With the weight function given by $w(x)=\sin(\pi x/L)$, find the grid point locations from Eq. (20). Note the concentration near $x=L/2$ where the weight function has its maximum value.

4. For $u(x)=(L/\pi)\sin(\pi x/L)$, obtain the point distribution from Eq. (20) using the weight functions from Eq. (22), (23), (25), (26) and (27). Use $\alpha=\beta=1$. Plot and compare.

5. Show that the average of the squares of the diagonal lengths is $\sum_i g_{ii}$.

6. Verify the correspondence between the six grid proper-

ties listed on p. 395 with the six listed on p. 396. Hint: Recall that $d\underline{x}=\sqrt{g}d\underline{\xi}$.

7. Verify that the one-dimensional forms of the first four properties on pp. 395-396 are as stated on p. 396 - 397. Hint: In one dimension take

$$g = \begin{bmatrix} x_\xi^2 & 0 & 0 \\ 0 & 1 & 0 \\ 0 & 0 & 1 \end{bmatrix}$$

8. Show that Eq. (59) is the Euler equation resulting from the integral given by Eq. (57).

9. Verify Eq. (72) and (73).

10. Show that with

$$F = \frac{\sum_{i=1}^{3} g^{ii}}{w^2(\underline{\xi})}$$

the generation system is

$$\nabla^2\xi^i = -(\sum_{k=1}^{3} |\nabla\xi^k|^2 \frac{1}{w} w_{\xi^i} + 2\frac{1}{w}\sum_{k=1}^{3} w_{\xi^k} \nabla\xi^k \cdot \nabla\xi^i$$

Hint: Use Eq. (19) of Appendix B.

11. Show that with $F = \sum_{i=1}^{3} g^{ii}$ the generation system consists of Laplace equations in the computational space.

12. Show that with

$$F = \frac{\sum_{i=1}^{3} g^{ii}}{w(\underline{x})}$$

the generation system is

$$\nabla^2\xi^i = \frac{1}{w}\nabla w \cdot \nabla\xi^i$$

13. Show that with $F = \sum_{i=1}^{3} g_{jk}^2$, (i,j,k) cyclic, the generation system is

$$2\sum_{i=1}^{3}[2g_{jk}\mathbf{r}_{\xi^j\xi^k} + (\mathbf{r}_{\xi^j} \cdot \mathbf{r}_{\xi^k\xi^k})\mathbf{r}_{\xi^j} + (\mathbf{r}_{\xi^j} \cdot \mathbf{r}_{\xi^j\xi^k})\mathbf{r}_{\xi^k}] = 0$$

Hint: Note that $\partial F/\partial g_{ij} = (1-g_{ij})g_{ji}$ and $\partial^2 F/\partial g_{ij}\partial g_{kl} = (1-g_{ij})\delta_{ij}\ \delta_{kj}\ \delta_{li}$.

APPENDIX A

DIFFERENTIAL-GEOMETRIC CONCEPTS ON SPACE CURVES AND SURFACES

1. Theory of Curves

In this appendix we consider only those parts of the theory of curves in space which are needed in the theory of surface geometry for the purpose of coordinate generation. Let C be a curve in space whose parametric equation is given as

$$\underline{r} = \underline{r}(\tau) \tag{1}$$

where τ is a parameter which takes values in a certain interval $a \leq \tau \leq b$.

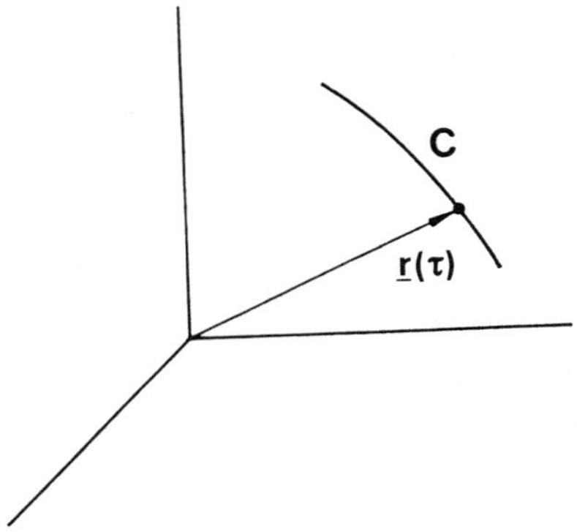

It is assumed that the real vector function $\underline{r}(\tau)$ is $p \geq 1$ times continuously differentiable for all values of τ in the specified interval, and at least one component of the first derivative

$$\underline{r}' = \frac{d\underline{r}}{d\tau} \tag{2}$$

is different from zero. Note that the parameter τ can be replaced by some other parameter, say s, provided that $ds/d\tau \neq 0$.

A. Tangent vector

Let us consider the arc length s as a parameter. Then the coordinates of two neighboring points on the curve are $\underset{\sim}{r}(s)$ and $\underset{\sim}{r}(s+h)$. The vector $\underset{\sim}{t}(s)$ defined as

$$\underset{\sim}{t}(s) = \lim_{h\to 0} \frac{\underset{\sim}{r}(s+h) - \underset{\sim}{r}(s)}{h} = \frac{d\underset{\sim}{r}}{ds} \tag{3}$$

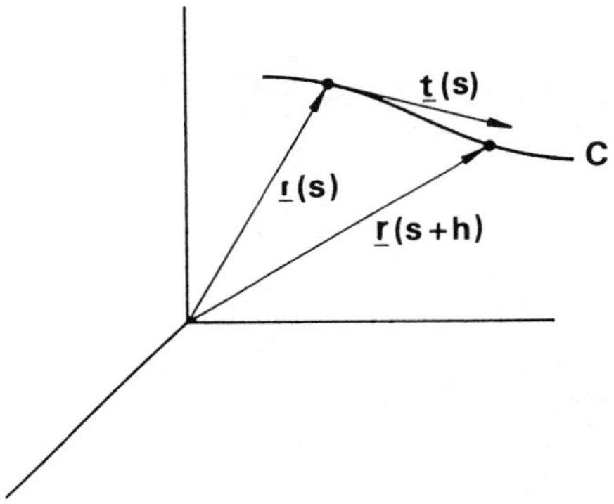

is the unit tangent vector at the point s on the curve. Since $|d\underset{\sim}{r}| = ds$, we immediately see that $|\underset{\sim}{t}(s)| = 1$.

If the curve C is referred to a general coordinate system ξ^i, then its parametric equations are given as

$$\xi^i = \xi^i(s), \qquad i = 1,2,3$$

In this case, using the chain rule of differentiation, we can write

$$\underset{\sim}{t}(s) = \sum_{i=1}^{3} \underset{\sim}{a}_i \frac{d\xi^i}{ds} \tag{4}$$

where $\underset{\sim}{a}_i$ are the covariant base vectors defined in Eq. (III-1).

B. Principal normal

Since $\underset{\sim}{t} \cdot \underset{\sim}{t} = 1$, a single differentiation with respect to s yields

$$\underset{\sim}{t} \cdot \frac{d\underset{\sim}{t}}{ds} = 0$$

so that the vector $d\underset{\sim}{t}/ds$ is orthogonal to $\underset{\sim}{t}$. The vector

$$\hat{\underset{\sim}{k}} = \frac{d\underset{\sim}{t}}{ds} \tag{5}$$

is called the <u>curvature vector.</u> The unit principal normal vector is then defined as

$$\underset{\sim}{p} = \hat{\underset{\sim}{k}}/|\hat{\underset{\sim}{k}}| \tag{6}$$

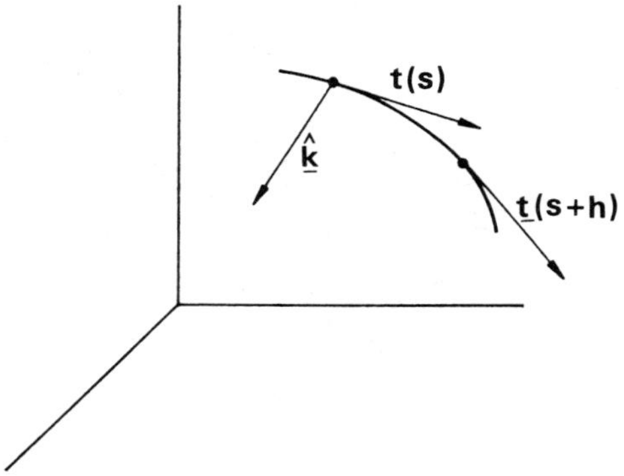

The magnitude $k(s) = |\hat{\underset{\sim}{k}}|$ and its reciprocal $\rho = 1/k(s)$ are, respectively, the curvature and the radius of curvature of the curve at the point under consideration. Both the curvature vector and the principal normal are directed toward the center of curvature of the curve at that point.

C. Normal and osculating planes

The totality of all vectors which are bound at a point of the curve and which are orthogonal to the unit tangent vector at that point lie in a plane. This plane is called the normal plane. The plane formed by the unit tangent and the principal normal vector is called the osculating plane.

D. Binormal vector

A unit vector $\underset{\sim}{b}(s)$ which is orthogonal to both $\underset{\sim}{t}$ and $\underset{\sim}{p}$ is called the binormal vector. Its orientation is fixed by taking $\underset{\sim}{t}$, $\underset{\sim}{p}$, $\underset{\sim}{b}$ to form a right-handed triad as shown below:

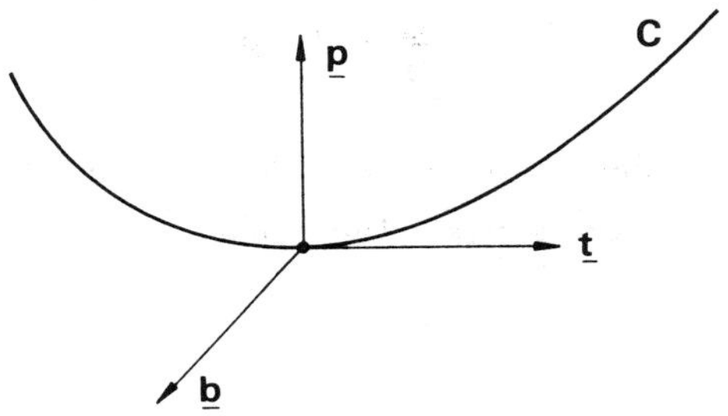

Thus

$$\underset{\sim}{b} = \underset{\sim}{t} \times \underset{\sim}{p} \tag{7}$$

Note that for plane curves the binormal $\underset{\sim}{b}$ is the constant unit vector normal to the plane, and the principal normal is the usual normal to the curve directed toward the center of curvature at that point.

The twisted curves in space have their binormals as functions of s. Because of twisting a new quantity called torsion appears, which is obtained as follows. Consider the obvious equations

$$\underset{\sim}{b} \cdot \underset{\sim}{b} = 1, \qquad \underset{\sim}{b} \cdot \underset{\sim}{t} = 0 \tag{8}$$

Differentiating each equation with respect to s, we obtain

$$\underset{\sim}{b} \cdot \frac{d\underset{\sim}{b}}{ds} = 0 \tag{9a}$$

$$\underset{\sim}{b} \cdot \frac{d\underset{\sim}{t}}{ds} + \frac{d\underset{\sim}{b}}{ds} \cdot \underset{\sim}{t} = 0 \tag{9b}$$

Thus

$$\frac{d\underset{\sim}{b}}{ds} \cdot \underset{\sim}{t} = -k\underset{\sim}{b} \cdot \underset{\sim}{p} = 0 \tag{9c}$$

From (9a,c) we find that $d\underset{\sim}{b}/ds$ is a vector which is orthogonal to both $\underset{\sim}{t}$ and $\underset{\sim}{b}$. Thus $d\underset{\sim}{b}/ds$ lies along the principal normal,

$$\frac{d\underset{\sim}{b}}{ds} = \pm \tau\underset{\sim}{p}$$

To decide about the sign we take the cross product of $\underset{\sim}{b}$ with $d\underset{\sim}{b}/ds$ and take it as a positive rotation about $\underset{\sim}{t}$:

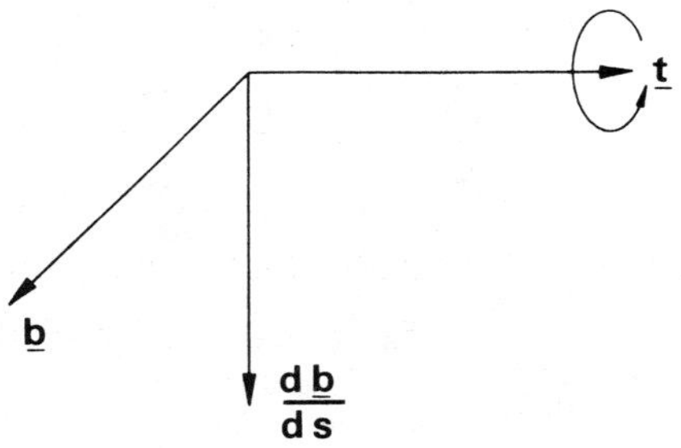

Thus

$$\underset{\sim}{b} \times \frac{d\underset{\sim}{b}}{ds} = \tau\underset{\sim}{t} \tag{10a}$$

and

$$\frac{d\underset{\sim}{b}}{ds} = -\tau \underset{\sim}{p} \tag{10b}$$

E. Serret-Frenet equations

A set of equations known as the Serret-Frenet equations, which are the intrinsic equations of a curve, are the following. Differentiating the equation

$$\underset{\sim}{p} = \underset{\sim}{b} \times \underset{\sim}{t}$$

with respect to s, we have

$$\frac{d\underset{\sim}{p}}{ds} = \tau \underset{\sim}{b} - k \underset{\sim}{t} \tag{11}$$

Equations (6), (10) and (11) are the Serret-Frenet equations, and are collected below:

$$\frac{d\underset{\sim}{t}}{ds} = k\underset{\sim}{p}\ , \qquad k = \text{curvature} \tag{12a}$$

$$\frac{d\underset{\sim}{b}}{ds} = -\tau\underset{\sim}{p}\ , \qquad \tau = \text{torsion} \tag{12b}$$

$$\frac{d\underset{\sim}{p}}{ds} = \tau \underset{\sim}{b} - k\underset{\sim}{t} \tag{12c}$$

For a plane curve, $\tau = 0$, so that

$$\underset{\sim}{b} = \text{constant}$$

$$\frac{d\underset{\sim}{t}}{ds} = k\underset{\sim}{p}, \qquad \frac{d\underset{\sim}{p}}{ds} = -\ k\underset{\sim}{t} \tag{13}$$

2. Geometry of Two-Dimensional Surfaces Embedded in E^3

Before taking up the main subject of surface theory, it is important to clarify the notations which are to be used in the ensuing development.

In an Euclidean E^3, a set of rectangular cartesian coordinates (x,y,z) can always be introduced. As before, in E^3 a general curvilinear coordinate system will be denoted by ξ^i (i = 1,2,3). With these curvilinear coordinates, a surface in E^3 will be denoted by ξ^ν = constant, where ν = 1,2,3. The following convention is adopted which maintains the right-handedness of the two remaining current coordinates: On the surface ξ^ν = constant, the current coordinates are ξ^α, ξ^β, where (ν,α,β) are cyclic.

A. First fundamental form

Let us consider the surface ξ^ν = constant. In this surface an element of length $ds^{(\nu)}$ is then given by

$$(ds^{(\nu)})^2 = d\underline{r} \cdot d\underline{r}$$

$$= \sum_{\alpha,\beta} \frac{\partial \underline{r}}{\partial \xi^\alpha} \cdot \frac{\partial \underline{r}}{\partial \xi^\beta} d\xi^\alpha d\xi^\beta$$

$$= \sum_{\alpha,\beta} \underline{a}_\alpha \cdot \underline{a}_\beta \, d\xi^\alpha d\xi^\beta \tag{14}$$

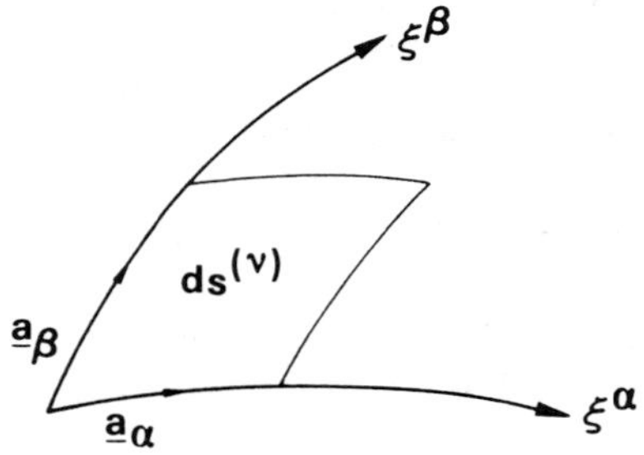

where the indices α and β will assume only the two values different from ν. Eq. (14) is called the first fundamental form of a surface.

B. Unit normal vector

The unit normal to the surface ξ^{ν} = constant is defined as

$$\underset{\sim}{n}^{(\nu)} = \frac{1}{\sqrt{gg^{\nu\nu}}} \frac{a_{\alpha} \times a_{\beta}}{|a_{\alpha} \times a_{\beta}|}, \tag{15}$$

where again (ν,α,β) are cyclic.

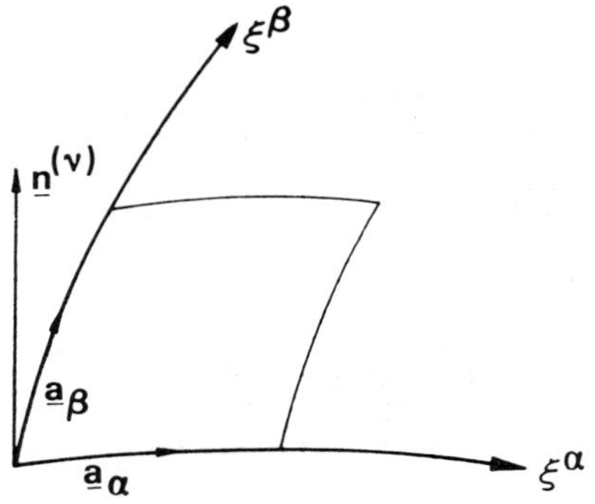

C. Second fundamental form

A plane containing the normal $\underset{\sim}{n}^{(\nu)}$ to the surface at a point P cuts the surface in different curves when rotated about the normal as an axis. Each curve so generated belongs both the surface and to the space E^3. A study of curvature properties of these curves reveals the curvature properties of the surfaces in which they lie. We decompose the curvature vector $\hat{\underset{\sim}{k}}$ at P of C, defined in Eq. (5), into a vector $\underset{\sim}{k}_n$ normal to the surface and a vector $\underset{\sim}{k}_g$ tangential to the surface as shown below:

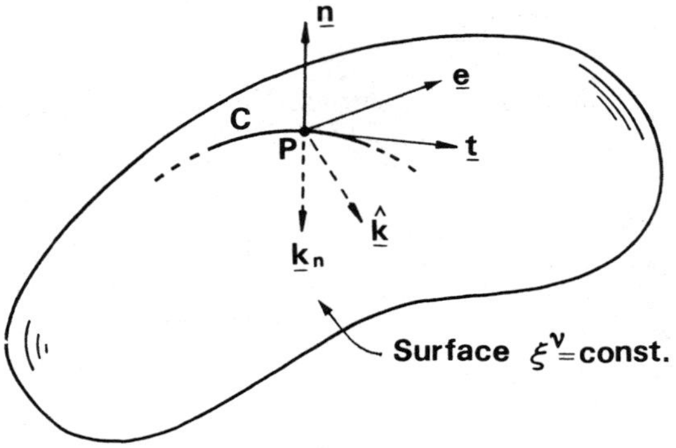

Thus

$$\hat{k}^{(\nu)} = k_n^{(\nu)} + k_g^{(\nu)} \tag{16}$$

The vector k_n is the normal curvature vector at the point P, and is given by

$$k_n^{(\nu)} = n^{(\nu)} k_n^{(\nu)} \tag{17}$$

where $k_n^{(\nu)}$ is its magnitude. To find an expression for $k_n^{(\nu)}$ we consider the equation

$$n^{(\nu)} \cdot t = 0$$

and differentiate it with respect to s (the arc length along the curve C) to have

$$k_n^{(\nu)} = - \frac{dn^{(\nu)} \cdot dr}{(ds)^2} \tag{18a}$$

Also, differentiating the equation

$$n^{(\nu)} \cdot a_\alpha = 0$$

with respect to ξ^β, we get

$$(\underline{n}^{(\nu)})_{\xi^\beta} \cdot \underline{a}_\alpha = -\underline{n}^{(\nu)} \cdot (\underline{a}_\alpha)_{\xi^\beta} \qquad (18b)$$

Further,

$$d\underline{n}^{(\nu)} = \sum_\alpha \frac{\partial \underline{n}^{(\nu)}}{\partial \xi^\alpha} d\xi^\alpha$$

$$d\underline{r} = \sum_\beta \frac{\partial \underline{r}}{\partial \xi^\beta} d\xi^\beta \qquad (18c)$$

Thus using Eq. (18b) and (18c) in (18a), we get

$$k_n^{(\nu)} = \sum_{\alpha,\beta} b\alpha\beta \frac{d\xi^\alpha d\xi^\beta}{(ds)^2} \qquad (19)$$

where

$$b_{\alpha\beta} = \underline{n}^{(\nu)} \cdot \underline{r}_{\xi^\alpha \xi^\beta} \qquad (20)$$

The two extreme values of $k_n^{(\nu)}$ are called the principal curvatures k_I and k_{II} and their sum is given by

$$k_I + k_{II} = \sum_{\alpha,\beta} g^{\alpha\beta} b_{\alpha\beta} \qquad (21)$$

The form

$$\sum_{\alpha,\beta} b_{\alpha\beta} d\xi^\alpha d\xi^\beta \qquad (22)$$

is called the second fundamental form.

3. Christoffel Symbols

Certain 3-index symbols, known as the Christoffel symbols, show up in a natural way when vectors or tensors are differentiated with respect to general coordinates introduced in a space. Here, by 'space' we mean a region in which arbitrary independent coordinates can be introduced; the number of independent coordinates determines the dimension of the sapce. A space is termed Eulclidean when rectangular cartesian coordinates can be introduced in it on a global scale. Examples are 2D or 3D regions in a plane or in a rectangular box, respectively. It must, however, be pointed out that in an Euclidean space, besides rectangular cartesian coordinates, any general coordinate system can be introduced without disturbing the basic nature of the space itself. Since this book is mainly concerned with the general coordinate systems in either 2D or 3D Euclidean spaces, or to 2D surfaces embedded in a 3D space, we shall restrict our attention to the Christoffel symbols for space and for surfaces only.

A. Space Christoffel symbols

From the definition of the base vectors $\underset{\sim}{a}_i$, we first note the following result. For any two indices i and k,

$$\frac{\partial \underset{\sim}{a}_i}{\partial \xi^k} = \frac{\partial}{\partial \xi^k}\left(\frac{\partial \underset{\sim}{r}}{\partial \xi^i}\right) = \frac{\partial}{\partial \xi^i}\left(\frac{\partial \underset{\sim}{r}}{\partial \xi^k}\right)$$

Thus

$$\frac{\partial \underset{\sim}{a}_i}{\partial \xi^k} = \frac{\partial \underset{\sim}{a}_k}{\partial \xi^i} \tag{23}$$

We now select any three indices, say i,j,k, and consider the following three equations,

$$\frac{\partial g_{ij}}{\partial \xi^k} = \frac{\partial}{\partial \xi^k}(\underset{\sim}{a}_i \cdot \underset{\sim}{a}_j)$$

$$\frac{\partial g_{jk}}{\partial \xi^i} = \frac{\partial}{\partial \xi^i}(\underset{\sim}{a}_j \cdot \underset{\sim}{a}_k)$$

$$\frac{\partial g_{ik}}{\partial \xi^j} = \frac{\partial}{\partial \xi^j}(\underset{\sim}{a}_i \cdot \underset{\sim}{a}_k)$$

Adding the second and third equations, and subtracting the first equation, while using Eq. (23), we get

$$\frac{\partial \underset{\sim}{a}_i}{\partial \xi^j} \cdot \underset{\sim}{a}_k = [ij,k] \tag{24}$$

where

$$[ij,k] = \frac{1}{2}\left(\frac{\partial g_{jk}}{\partial \xi^i} + \frac{\partial g_{ik}}{\partial \xi^j} - \frac{\partial g_{ij}}{\partial \xi^k}\right) \tag{25}$$

is called the Christoffel symbol of the <u>first kind</u>.

Eq. (24) implies that

$$\frac{\partial \underset{\sim}{a}_i}{\partial \xi^j} = \sum_k [ij,k]\, \underset{\sim}{a}^{(k)} \tag{26}$$

Taking the dot product on both sides of Eq. (26) by $\underset{\sim}{a}^l$, we obtain

$$\frac{\partial \underset{\sim}{a}_i}{\partial \xi^j} \cdot \underset{\sim}{a}^l = \Gamma^l_{ij} \tag{27}$$

where

$$\Gamma^i_{ij} = \sum_k g^{kl}\, [ij,k] \tag{28}$$

is called the Christoffel symbol of the <u>second kind</u>.

Eq.(27) implies that

$$\frac{\partial \underset{\sim}{a}_i}{\partial \xi^j} = \sum_l \Gamma^l_{ij} \underset{\sim}{a}_l \tag{29}$$

It must be noted that both kinds of Christoffel symbols are symmetric in the first two indices, viz.,

$$[ij,k] = [ji,k], \qquad \Gamma^l_{ij} = \Gamma^l_{ji}$$

It is also easy to show, based on the definition of Γ^l_{ij} that

$$\sum_i \Gamma^i_{ij} = \frac{1}{2g} \frac{\partial g}{\partial \xi^j} \tag{30}$$

The Christoffel symbols Γ^l_{ij} can be computed by using the following expanded formulae:

$$\Gamma^l_{ij} = \sum_k \sum_n g^{kl} \frac{\partial x_n}{\partial \xi^k} \frac{\partial^2 x_n}{\partial \xi^i \xi^j} \tag{31}$$

where the indices l,i,j range from 1 to 3 in 3D, or from 1 to 2 in 2D.

B. Christoffel symbols in a surface

The Christoffel symbols, (25) and (28) are applicable both to 2D and 3D Eulcidean spaces. In fact, if we take (25) and (28) as the definitions of some 3-index symbols without any consideration of an Euclidean space,then they are also applicable to an n-dimensional non-Euclidean space.

The Christoffel symbols for a 2D surface embedded in a 3D Euclidean space are defined exactly as for any other space. Since in a surface only two independent coordinates can be introduced, we again use the Greek indices to emphasize this point and write

$$[\alpha\beta,\delta] = \frac{1}{2}\left(\frac{\partial g_{\alpha\delta}}{\partial \xi^{\beta}} + \frac{\partial g_{\beta\delta}}{\partial \xi^{\alpha}} - \frac{\partial g_{\alpha\beta}}{\partial \xi^{\delta}}\right) \tag{32}$$

$$T^{\sigma}_{\alpha\beta} = \sum_{\delta} g^{\delta\sigma}\,[\alpha\beta,\delta] \tag{33}$$

as the Christoffel symbols of the first and second kind respectively, of a surface. Here the indices assume only two values.

An important point to note here is that for a 2D space the metric coefficients g_{ij} do not depend on one of the cartesian coordinate, say z. On the other hand for a 2D space formed by a surface in 3D Euclidean space the metric coefficients appearing in (32) and (33) depend on all three cartesian coordinates.

Gauss indirectly introduced the definition of the Christoffel symobls by arguing that in a surface the base vectors $\mathbf{r}_{\alpha}$, $\mathbf{r}_{\beta}$ and the unit normal $\mathbf{n}$ (Eq. (15)) form a triad of independent vectors. Thus any other vector in the surface can be presented as a linear combination of $\mathbf{r}_{\alpha}$, $\mathbf{r}_{\beta}$, $\mathbf{n}$. Following this argument, the second derivative of the position vector $\mathbf{r}$ can be expressed as

$$\mathbf{r}_{\alpha\beta} = \sum_{\delta} T^{\delta}_{\alpha\beta}\,\mathbf{r}_{\delta} + b_{\alpha\beta}\,\mathbf{n} \tag{34}$$

which are called the formulae of Gauss. Thus, for a surface ξ^3 = constant in which ξ^1, ξ^2 are the current coordinates, Eq. (34) is written as

$$\underset{\sim}{r}_{\alpha\beta} = \sum_{\delta} T^{\delta}_{\alpha\beta}\, \underset{\sim}{r}_{\delta} + b_{\alpha\beta}\, \underset{\sim}{n}^{(\nu)} \tag{35}$$

where Eq. (35) represents the second derivatives $\underset{\sim}{r}_{\xi^1\xi^1}$, $\underset{\sim}{r}_{\xi^1\xi^2}$, $\underset{\sim}{r}_{\xi^2\xi^2}$.

APPENDIX B

EULER EQUATIONS

1. Variational Principle in Transformed Space

Consider the integral

$$I = \int F[\underset{\sim}{g}, w(\underline{x})] d\underset{\sim}{\xi} \qquad (1)$$

where $\underset{\sim}{g}$ is the covariant metric tensor, with elements g_{ij} defined by Eq. (III-5), and $w(\underset{\sim}{x})$ is a weight function dependent on $\underset{\sim}{x}$.

A. Grid Generation System

The Euler equations then are given by

$$\sum_{j=1}^{3} \frac{\partial}{\partial \xi^j} \frac{\partial F}{\partial (x_i)_{\xi^j}} - \frac{\partial F}{\partial x_i} = 0 \qquad (i = 1,2,3) \qquad (2)$$

as has been noted. Since

$$(x_i)_{\xi^j} = (a_j)_i$$

and F depends on $(x_i)_{\xi^j}$ only through the elements of the metric tensor, $\underset{\sim}{g}$, we have

$$\frac{\partial F}{\partial (x_i)_{\xi^j}} = \frac{\partial F}{\partial (\underset{\sim}{a}_j)_i} = \frac{\partial F}{\partial \underset{\sim}{a}_j} \frac{\partial \underset{\sim}{a}_j}{\partial (\underset{\sim}{a}_j)_i} = \frac{\partial F}{\partial \underset{\sim}{a}_j} \underset{\sim}{e}_i \qquad (3)$$

where $\underset{\sim}{e}_i$ is the unit vector in the x_i-direction. Here the operation indicated by the notation, $\partial F/\partial \underset{\sim}{a}_j\, \underset{\sim}{e}_i$, is the simple replacement of $\underset{\sim}{a}_j$ by $\underset{\sim}{e}_i$ in F. Also, since F depends on

$\underset{\sim}{a}_j$ only through $\underset{\sim}{g}$, we have

$$\frac{\partial F}{\partial \underset{\sim}{a}_j} \underset{\sim}{e}_i = \sum_{k=1}^{3} \sum_{l=1}^{3} \frac{\partial F}{\partial g_{kl}} \frac{\partial g_{kl}}{\partial \underset{\sim}{a}_j} \underset{\sim}{e}_i$$

$$= \sum_{k=1}^{3} \sum_{l=1}^{3} \frac{\partial F}{\partial g_{kl}} \frac{\partial(\underset{\sim}{a}_k \cdot \underset{\sim}{a}_l)}{\partial \underset{\sim}{a}_j} \underset{\sim}{e}_i$$

or

$$\frac{\partial F}{\partial \underset{\sim}{a}_j} \underset{\sim}{e}_i = \sum_{k=1}^{3} \sum_{l=1}^{3} \frac{\partial F}{\partial g_{kl}} [\delta_{kj}(\underset{\sim}{a}_l \cdot \underset{\sim}{e}_i) + \delta_{lj}(\underset{\sim}{a}_k \cdot \underset{\sim}{e}_i)]$$

$$= \sum_{l=1}^{3} \frac{\partial F}{\partial g_{jl}} (\underset{\sim}{a}_l \cdot \underset{\sim}{e}_i) + \sum_{k=1}^{3} \frac{\partial F}{\partial g_{kj}} (\underset{\sim}{a}_k \cdot \underset{\sim}{e}_i)$$

Therefore,

$$\frac{\partial F}{\partial (x_i)_{\xi^j}} = \sum_{k=1}^{3} \left(\frac{\partial F}{\partial g_{jk}} + \frac{\partial F}{\partial g_{kj}}\right)(\underset{\sim}{a}_k \cdot \underset{\sim}{e}_i) \tag{4}$$

Since F depends on $\underset{\sim}{x}$ only through the weight function we have

$$\frac{\partial F}{\partial x_i} = \frac{\partial F}{\partial w} \frac{\partial w}{\partial x_i} = \frac{\partial F}{\partial w} (\underset{\sim}{\nabla} w)_i = \frac{\partial F}{\partial w} \underset{\sim}{e}_i \cdot \underset{\sim}{\nabla} w \tag{5}$$

Then the Euler Equations can be written as

$$\left\{ \sum_{j=1}^{3} \sum_{k=1}^{3} \left[\left(\frac{\partial F}{\partial g_{jk}} + \frac{\partial F}{\partial g_{kj}}\right)\underset{\sim}{a}_k\right]_{\xi^j} - \frac{\partial F}{\partial w} \underset{\sim}{\nabla} w\right\} \cdot \underset{\sim}{e}_i = 0 \qquad (i = 1,2,3)$$

or as the vector equation

$$\sum_{j=1}^{3} \sum_{k=1}^{3} [(\frac{\partial F}{\partial g_{jk}} + \frac{\partial F}{\partial g_{kj}})\underset{\sim}{a}_k]_{\xi^j} - \frac{\partial F}{\partial w} \underset{\sim}{\nabla} w = 0 \tag{6}$$

(Note that the symmetric elements of the metric tensor, $g_{jk} = g_{kj}$, are to be left as distinct elements in F until after the differentiation has been performed.)

Expanding the ξ^j-derivative, we then have

$$\sum_{j=1}^{3} \sum_{k=1}^{3} [(\frac{\partial F}{\partial g_{jk}} + \frac{\partial F}{\partial g_{kj}})\underset{\sim}{r}_{\xi^j\xi^k} + (\frac{\partial F}{\partial g_{jk}} + \frac{\partial F}{\partial g_{kj}})_{\xi^j} \underset{\sim}{r}_{\xi^k}]$$

$$- \frac{\partial F}{\partial w} \underset{\sim}{\nabla} w = 0$$

But also

$$(\frac{\partial F}{\partial g_{kj}})_{\xi^j} = \sum_{m=1}^{3} \sum_{n=1}^{3} [\frac{\partial}{\partial g_{mn}} (\frac{\partial F}{\partial g_{kj}})](g_{mn})_{\xi^j}$$

$$+ [\frac{\partial}{\partial g_{kj}}(\frac{\partial F}{\partial w})] \underset{\sim}{\nabla} w \cdot \underset{\sim}{r}_{\xi^j}$$

so that

$$(\frac{\partial F}{\partial g_{kj}})_{\xi^j} = \sum_{m=1}^{3} \sum_{n=1}^{3} \frac{\partial^2 F}{\partial g_{mn}\partial g_{kj}} (\underset{\sim}{r}_{\xi^m} \cdot \underset{\sim}{r}_{\xi^j\xi^n} + \underset{\sim}{r}_{\xi^n} \cdot \underset{\sim}{r}_{\xi^j\xi^m})$$

$$+ [\frac{\partial}{\partial g_{kj}}(\frac{\partial F}{\partial w})] (\underset{\sim}{\nabla} w \cdot \underset{\sim}{r}_{\xi^j})$$

Thus we have the grid generation system, with $\partial F/\partial w$ written as F',

$$\sum_{j=1}^{3} \sum_{k=1}^{3} [A_{jk} \, \underset{\sim}{r}_{\xi^j \xi^k} + A'_{jk} (\underset{\sim}{\nabla} w \cdot \underset{\sim}{r}_{\xi^j}) \underset{\sim}{r}_{\xi^k}$$

$$+ \sum_{m=1}^{3} \sum_{n=1}^{3} \frac{\partial A_{jk}}{\partial g_{mn}} (\underset{\sim}{r}_{\xi^m} \cdot \underset{\sim}{r}_{\xi^j \xi^n}$$

$$+ \underset{\sim}{r}_{\xi^n} \cdot \underset{\sim}{r}_{\xi^j \xi^m}) \underset{\sim}{r}_{\xi^k}] - F' \, \underset{\sim}{\nabla} w = 0 \qquad (7)$$

where

$$A_{jk} = \frac{\partial F}{\partial g_{jk}} + \frac{\partial F}{\partial g_{kj}} \qquad (8)$$

This is a quasi-linear, second-order partial differential equation for the cartesian coordinates $\underset{\sim}{r}$.

If the weight function depends directly on $\underset{\sim}{\xi}$, instead of on $\underset{\sim}{x}$ in Eq. (1), then $\partial F/\partial x_i = 0$ in Eq. (2). Also in this case, the $\underset{\sim}{\nabla} w \cdot \underset{\sim}{r}_{\xi^j}$ that appears on p. 439 and in the development that leads to Eq. (7) is replaced by simply w_{ξ^j}. Then Eq. (7) is replaced by

$$\sum_{j=1}^{3} \sum_{k=1}^{3} [A_{jk} \, \underset{\sim}{r}_{\xi^j \xi^k} + A'_{jk} \, w_{\xi^j} \, \underset{\sim}{r}_{\xi^k}$$

$$+ \sum_{m=1}^{3} \sum_{n=1}^{3} \frac{\partial A_{jk}}{\partial g_{mn}} (\underset{\sim}{r}_{\xi^m} \cdot \underset{\sim}{r}_{\xi^j \xi^n} + \underset{\sim}{r}_{\xi^n} \cdot \underset{\sim}{r}_{\xi^j \xi^m}) \underset{\sim}{r}_{\xi^k} \qquad (9)$$

for a weight function $w(\underset{\sim}{\xi})$ in Eq. (1).

B. Two-Dimensional Examples

In two dimensions, the generation system (7) becomes (with $\xi^1 = \xi$ and $\xi^2 = \eta$)

$$
\begin{aligned}
&\frac{\partial F}{\partial g_{11}} \underset{\sim}{r}_{\xi\xi} + \frac{\partial F}{\partial g_{22}} \underset{\sim}{r}_{\eta\eta} + 2 \frac{\partial F}{\partial g_{12}} \underset{\sim}{r}_{\xi\eta} \\
&\quad + \frac{\partial F'}{\partial g_{11}} (\underset{\sim}{\nabla} w \cdot \underset{\sim}{r}_{\xi}) \underset{\sim}{r}_{\xi} + \frac{\partial F'}{\partial g_{22}} (\underset{\sim}{\nabla} w \cdot \underset{\sim}{r}_{\eta}) \underset{\sim}{r}_{\eta} \\
&\quad + \frac{\partial F'}{\partial g_{12}} [(\underset{\sim}{\nabla} w \cdot \underset{\sim}{r}_{\xi}) \underset{\sim}{r}_{\eta} + (\underset{\sim}{\nabla} w \cdot \underset{\sim}{r}_{\eta}) \underset{\sim}{r}_{\xi}] \\
&\quad + \underset{\sim}{r}_{\xi} \{ \underset{\sim}{r}_{\xi} \cdot [2 \frac{\partial^2 F}{\partial g_{11}^2} \underset{\sim}{r}_{\xi\xi} + 4 \frac{\partial^2 F}{\partial g_{11} \partial g_{12}} \underset{\sim}{r}_{\xi\eta} \\
&\quad + (\frac{\partial^2 F}{\partial g_{12}^2} + \frac{\partial^2 F}{\partial g_{12} \partial g_{21}}) \underset{\sim}{r}_{\eta\eta}] \\
&\quad + \underset{\sim}{r}_{\eta} \cdot [2 \frac{\partial^2 F}{\partial g_{11} \partial g_{12}} \underset{\sim}{r}_{\xi\xi} + 2 \frac{\partial^2 F}{\partial g_{22} \partial g_{12}} \underset{\sim}{r}_{\eta\eta} \\
&\quad + (2 \frac{\partial^2 F}{\partial g_{11} \partial g_{22}} + \frac{\partial^2 F}{\partial g_{12}^2} + \frac{\partial^2 F}{\partial g_{12} \partial g_{21}}) \underset{\sim}{r}_{\xi\eta}]\} \\
&\quad + \underset{\sim}{r}_{\eta} \{[(\frac{\partial^2 F}{\partial g_{12}^2} + \frac{\partial^2 F}{\partial g_{12} \partial g_{21}}) \underset{\sim}{r}_{\xi\xi} \\
&\quad + 2 \frac{\partial^2 F}{\partial g_{22}^2} \underset{\sim}{r}_{\eta\eta} + 4 \frac{\partial^2 F}{\partial g_{22} \partial g_{12}} \underset{\sim}{r}_{\xi\eta}] \cdot \underset{\sim}{r}_{\eta} \\
&\quad + \underset{\sim}{r}_{\xi} \cdot [2 \frac{\partial^2 F}{\partial g_{11} \partial g_{12}} \underset{\sim}{r}_{\xi\xi} + 2 \frac{\partial^2 F}{\partial g_{22} \partial g_{12}} \underset{\sim}{r}_{\eta\eta} \\
&\quad + (2 \frac{\partial^2 F}{\partial g_{11} \partial g_{22}} + \frac{\partial^2 F}{\partial g_{12}^2} + \frac{\partial^2 F}{\partial g_{12} \partial g_{21}}) \underset{\sim}{r}_{\xi\eta}]\} - \frac{1}{2} F' \underset{\sim}{\nabla} w = 0
\end{aligned}
\tag{10}
$$

If the weight function depends on ξ, rather than on $\mathbf{x}$, the terms $\nabla w \cdot \mathbf{r}_\xi$ and $\nabla w \cdot \mathbf{r}_\eta$ in Eq. (10) become w_ξ and w_η, respectively, and the last term, $- 1/2\, F'\nabla w$, vanishes.

As an example, consider F_w from Eq. (XI-71). Then we have

$$\frac{\partial F}{\partial g_{11}} = w^2 g_{22}, \quad \frac{\partial F}{\partial g_{22}} = w^2 g_{11}, \quad \frac{\partial F}{\partial g_{12}} = -w^2 g_{21}, \quad \frac{\partial F}{\partial g_{21}} = -w^2 g_{12}$$

$$\frac{\partial A_{11}}{\partial g_{11}} = 0, \quad \frac{\partial A_{22}}{\partial g_{22}} = 2w^2, \quad \frac{\partial A_{11}}{\partial g_{12}} = 0$$

$$\frac{\partial A_{22}}{\partial g_{11}} = 2w^2, \quad \frac{\partial A_{22}}{\partial g_{22}} = 0, \quad \frac{\partial A_{22}}{\partial g_{12}} = 0$$

$$\frac{\partial A_{12}}{\partial g_{12}} = -w^2, \qquad \frac{\partial A_{12}}{\partial g_{11}} = \frac{\partial A_{12}}{\partial g_{22}} = 0$$

$$\frac{\partial F'}{\partial g_{11}} = 2wg_{22}, \qquad \frac{\partial F'}{\partial g_{22}} = 2wg_{11}, \qquad \frac{\partial F'}{\partial g_{12}} = \frac{\partial F'}{\partial g_{21}} = -2wg_{12}$$

Then the generation system based on concentration by Eq. (7) is

$$\begin{aligned} &2w^2[g_{22}\mathbf{r}_{\xi\xi} + g_{11}\mathbf{r}_{\eta\eta} - 2g_{12}\mathbf{r}_{\xi\eta} - (\mathbf{r}_\eta \cdot \mathbf{r}_{\xi\xi})\mathbf{r}_\eta \\ &\quad - (\mathbf{r}_\xi \cdot \mathbf{r}_{\eta\eta})\mathbf{r}_\xi + (\mathbf{r}_\eta \cdot \mathbf{r}_{\xi\eta})\mathbf{r}_\xi + (\mathbf{r}_\xi \cdot \mathbf{r}_{\xi\eta})\mathbf{r}_\eta] \\ &\quad + 4w\{g_{22}(\nabla w \cdot \mathbf{r}_\xi)\mathbf{r}_\xi + g_{11}(\nabla w \cdot \mathbf{r}_\eta)\mathbf{r}_\eta \\ &\quad - g_{12}[(\nabla w \cdot \mathbf{r}_\xi)\mathbf{r}_\eta + (\nabla w \cdot \mathbf{r}_\eta)\mathbf{r}_\xi]\} - 2wg\,\nabla w = 0 \end{aligned} \tag{11}$$

With F taken to be a measure of orthogonality, i.e., F_o from Eq. (XI-70), we have,

$$\frac{\partial F}{\partial g_{11}} = \frac{\partial F}{\partial g_{22}} = 0, \quad \frac{\partial F}{\partial g_{12}} = g_{21}, \quad \frac{\partial F}{\partial g_{21}} = g_{12}, \quad \frac{\partial A_{12}}{\partial g_{12}} = \frac{\partial A_{12}}{\partial g_{21}} = 1$$

The generation system based only on orthogonality then is

$$2[2g_{12}\mathbf{r}_{\xi\eta} + (\mathbf{r}_\xi \cdot \mathbf{r}_{\eta\eta})\mathbf{r}_\xi + (\mathbf{r}_\eta \cdot \mathbf{r}_{\xi\xi})\mathbf{r}_\eta$$

$$+ (\mathbf{r}_\xi \cdot \mathbf{r}_{\xi\eta})\mathbf{r}_\eta + (\mathbf{r}_\eta \cdot \mathbf{r}_{\xi\eta})\mathbf{r}_\xi] = 0 \qquad (12)$$

Finally, for the smoothness integral, Eq, (XI-69), the derivatives needed are

$$\frac{\partial F}{\partial g_{11}} = \frac{(g_{11} - g_{22})g_{22} - 2g_{12}^2}{2(g_{11}g_{22} - g_{12}^2)^{3/2}}$$

$$\frac{\partial F}{\partial g_{22}} = \frac{(g_{22} - g_{11})g_{11} - 2g_{12}^2}{2(g_{11}g_{22} - g_{12}^2)^{3/2}}$$

$$\frac{\partial F}{\partial g_{12}} = \frac{(g_{11} + g_{22})g_{21}}{2(g_{11}g_{22} - g_{12}^2)^{3/2}}$$

$$\frac{\partial F}{\partial g_{21}} = \frac{(g_{11} + g_{22})g_{12}}{2(g_{11}g_{22} - g_{12}^2)^{3/2}}$$

$$\frac{\partial A_{11}}{\partial g_{11}} = \frac{-g_{11}g_{22} + 4g_{12}^2 + 3g_{22}^2}{2(g_{11}g_{22} - g_{12}^2)^{5/2}} \; g_{22}$$

$$\frac{\partial A_{11}}{\partial g_{22}} = \frac{-g_{22}^2 g_{11} + 8g_{22}g_{12}^2 - g_{11}^2 g_{22} + 2g_{12}^2 g_{11}}{2(g_{11}g_{22} - g_{12}^2)^{5/2}}$$

$$\frac{\partial A_{11}}{\partial g_{12}} = - \frac{2g_{11}g_{22} + g_{12}^2}{(g_{11}g_{22} - g_{12}^2)^{5/2}} \; g_{12} = \frac{\partial A_{11}}{\partial g_{21}}$$

$$\frac{\partial A_{22}}{\partial g_{11}} = \frac{-g_{11}^2 g_{22} + 8g_{12}^2 g_{11} - g_{22}^2 g_{11} + 2g_{12}^2 g_{22}}{2(g_{11}g_{22} - g_{12}^2)^{5/2}}$$

$$\frac{\partial A_{22}}{\partial g_{22}} = \frac{-g_{11}g_{22} + 4g_{12}^2 + 3g_{11}^2}{2(g_{11}g_{22} - g_{12}^2)^{5/2}} \; g_{11}$$

$$\frac{\partial A_{22}}{\partial g_{12}} = - \frac{2g_{11}g_{22} + g_{12}^2}{(g_{11}g_{22} - g_{12}^2)^{5/2}} \; g_{12} = \frac{\partial A_{22}}{\partial g_{21}}$$

$$\frac{\partial A_{12}}{\partial g_{11}} = - \frac{g_{11}g_{22} + 2g_{12}^2 + 3g_{22}^2}{2(g_{11}g_{22} - g_{12}^2)^{5/2}} \; g_{12}$$

$$\frac{\partial A_{12}}{\partial g_{22}} = - \frac{g_{11}g_{22} + 2g_{12}^2 + 3g_{11}^2}{2(g_{11}g_{22} - g_{12}^2)^{5/2}} \; g_{12}$$

$$\frac{\partial A_{12}}{\partial g_{12}} = \frac{g_{11}^2 g_{22} + 2g_{11}g_{12}^2 + g_{11}g_{22}^2 + 2g_{22}g_{12}^2}{2(g_{11}g_{22} - g_{12}^2)^{5/2}} = \frac{\partial A_{12}}{\partial g_{21}}$$

The complete generation system is then obtained as the linear combination of the concentration system, Eq. (11), the orthogonality system, Eq. (12), and the smoothness system which is formed by substituting the above relations into the general equations (7). The three-dimensional case follows in an analogous fashion.

2. Variational Principle in Physical Space

With the variational problem formulated in the physical space, consider the integral

$$I = \int F[\overline{g}, w(\underset{\sim}{\xi})]d\underset{\sim}{x} \qquad (13)$$

where $\overline{g}$ is the contravariant metric tensor, i.e., with elements g^{ij} from Eq. (III-37), and the weight function is a function of $\underset{\sim}{\xi}$.

A. Grid Generation System

Then for the Euler equations, we have

$$\sum_{j=1}^{3} \frac{\partial}{\partial x_j} \frac{\partial F}{\partial (\xi^i)_{x_j}} - \frac{\partial F}{\partial \xi^i} = 0 \quad (i = 1,2,3) \qquad (14)$$

Now,

$$(\xi^i)_{x_j} = (\underset{\sim}{\nabla}\xi^i)_j = (\underset{\sim}{a}^i)_j$$

and F depends on $(\xi^i)_{x_j}$ only through $\overline{g}$. Then

$$\frac{\partial F}{\partial(\xi^i)_{x_j}} = \frac{\partial F}{\partial(\underset{\sim}{a}^i)_j} = \frac{\partial F}{\partial \underset{\sim}{a}^i} \frac{\partial \underset{\sim}{a}^i}{\partial(\underset{\sim}{a}^i)_j} = \frac{\partial F}{\partial \underset{\sim}{a}^i} \underset{\sim}{e}_j$$

Also, since F depends on $\underset{\sim}{a}^i$ only through g^{ik} (k = 1,2,3) we have

$$\frac{\partial F}{\partial \underset{\sim}{a}^i} \underset{\sim}{e}_j = \sum_{k=1}^{3} \sum_{l=1}^{3} \frac{\partial F}{\partial g^{lk}} \frac{\partial g^{lk}}{\partial \underset{\sim}{a}^i} \underset{\sim}{e}_j$$

$$= \sum_{k=1}^{3} \sum_{l=1}^{3} \frac{\partial F}{\partial g^{lk}} \frac{\partial(\underset{\sim}{a}^l \cdot \underset{\sim}{a}^k)}{\partial \underset{\sim}{a}^i} \underset{\sim}{e}_j$$

$$= \sum_{k=1}^{3} \sum_{l=1}^{3} \frac{\partial F}{\partial g^{lk}} (\delta_{li} \underset{\sim}{a}^k \cdot \underset{\sim}{e}_j + \delta_{ki} \underset{\sim}{a}^l \cdot \underset{\sim}{e}_j)$$

Therefore,

$$\frac{\partial F}{\partial(\xi^i)_{x_j}} = \sum_{k=1}^{3} \left(\frac{\partial F}{\partial g^{ki}} + \frac{\partial F}{\partial g^{ik}}\right) \underset{\sim}{a}^k \cdot \underset{\sim}{e}_j$$

Also, since F depends on $\underset{\sim}{\xi}$ only through the weight function, we have

$$\frac{\partial F}{\partial \xi^i} = \frac{\partial F}{\partial w} \frac{\partial w}{\partial \xi^i}$$

Then the Euler equations can be written

$$\sum_{j=1}^{3} \sum_{k=1}^{3} \frac{\partial}{\partial x_j} \left[\left(\frac{\partial F}{\partial g^{ik}} + \frac{\partial F}{\partial g^{ki}}\right) \underset{\sim}{a}^k \cdot \underset{\sim}{e}_j\right] - \frac{\partial F}{\partial w} \frac{\partial w}{\partial \xi^i} = 0$$

$$(i = 1,2,3)$$

or

$$\sum_{j=1}^{3} \sum_{k=1}^{3} [(\frac{\partial F}{\partial g^{ik}} + \frac{\partial F}{\partial g^{ki}}) \underset{\sim}{e}_j \cdot (\underset{\sim}{a}^k)_{x_j}$$

$$+ (\underset{\sim}{a}^k \cdot \underset{\sim}{e}_j) \frac{\partial}{\partial x_j} (\frac{\partial F}{\partial g^{ik}} + \frac{\partial F}{\partial g^{ki}})] - \frac{\partial F}{\partial w} \frac{\partial w}{\partial \xi^i} = 0 \qquad (i = 1,2,3)$$

Now

$$\underset{\sim}{e}_j \cdot (\underset{\sim}{a}^k)_{x_j} = \underset{\sim}{e}_j \cdot (\underset{\sim}{\nabla}\xi^k)_{x_j} = (\xi^k)_{x_j x_j}$$

and

$$\underset{\sim}{a}^k \cdot \underset{\sim}{e}_j = (\xi^k)_{x_j}$$

Then

$$\sum_{j=1}^{3} \sum_{k=1}^{3} [(\frac{\partial F}{\partial g^{ik}} + \frac{\partial F}{\partial g^{ki}}) \xi^k_{x_j x_j} + \frac{\partial}{\partial x_j} (\frac{\partial F}{\partial g^{ik}} + \frac{\partial F}{\partial g^{ki}}) \xi^k_{x_j}]$$

$$- \frac{\partial F}{\partial w} \frac{\partial w}{\partial \xi^i} = 0$$

or,

$$\sum_{k=1}^{3} (\frac{\partial F}{\partial g^{ik}} + \frac{\partial F}{\partial g^{ki}}) \nabla^2 \xi^k + \sum_{j=1}^{3} \sum_{k=1}^{3} \frac{\partial}{\partial x_j}(\frac{\partial F}{\partial g^{ik}} + \frac{\partial F}{\partial g^{ki}}) \xi^k_{x_j}$$

$$- \frac{\partial F}{\partial w} \frac{\partial w}{\partial \xi^i} = 0 \qquad (i = 1,2,3)$$

Now

$$\sum_{j=1}^{3} \frac{\partial}{\partial x_j} (\frac{\partial F}{\partial g^{ik}}) \xi^k_{x_j} = \underset{\sim}{\nabla}(\frac{\partial F}{\partial g^{ik}}) \cdot \underset{\sim}{\nabla}\xi^k$$

and

$$\underset{\sim}{\nabla}\left(\frac{\partial F}{\partial g^{ik}}\right) = \sum_{m=1}^{3}\sum_{n=1}^{3}\frac{\partial^2 F}{\partial g^{mn}\partial g^{ik}}\underset{\sim}{\nabla} g^{mn} + \left[\frac{\partial}{\partial g^{ik}}\left(\frac{\partial F}{\partial w}\right)\right]\sum_{l=1}^{3}\frac{\partial w}{\partial \xi^l}\underset{\sim}{\nabla}\xi^l$$

$$= \sum_{m=1}^{3}\sum_{n=1}^{3}\frac{\partial^2 F}{\partial g^{mn}\partial g^{ik}}[(\underset{\sim}{\nabla}\xi^m \cdot \underset{\sim}{\nabla})\underset{\sim}{\nabla}\xi^n + (\underset{\sim}{\nabla}\xi^n \cdot \underset{\sim}{\nabla})\underset{\sim}{\nabla}\xi^m]$$

Then the generation system is, with $\partial F/\partial w$ written as F',

$$\sum_{k=1}^{3}\{A_{ik}\nabla^2\xi^k + A'_{ik}\sum_{l=1}^{3}\frac{\partial w}{\partial \xi^l}\underset{\sim}{\nabla}\xi^l \cdot \underset{\sim}{\nabla}\xi^k$$

$$+ \sum_{m=1}^{3}\sum_{n=1}^{3}\frac{\partial A_{ik}}{\partial g^{mn}}[\underset{\sim}{\nabla}(\underset{\sim}{\nabla}\xi^m \cdot \underset{\sim}{\nabla}\xi^n)] \cdot \underset{\sim}{\nabla}\xi^k\}$$

$$- F'\frac{\partial w}{\partial \xi^i} = 0 \qquad (i = 1,2,3) \tag{15}$$

where

$$A_{ik} = \frac{\partial F}{\partial g^{ik}} + \frac{\partial F}{\partial g^{ki}} \tag{16}$$

This can also be written as

$$\sum_{k=1}^{3} A_{ik}\,\nabla^2\xi^k = S_i \qquad (i = 1,2,3) \tag{17}$$

$$S_i = -\sum_{k=1}^{3}\{\sum_{m=1}^{3}\sum_{n=1}^{3}\frac{\partial A_{ik}}{\partial g^{mn}}[\underset{\sim}{\nabla}(\underset{\sim}{\nabla}\xi^m \cdot \underset{\sim}{\nabla}\xi^n)] \cdot \underset{\sim}{\nabla}\xi^k$$

$$- A'_{ik}\sum_{l=1}^{3}\frac{\partial w}{\partial \xi^l}\underset{\sim}{\nabla}\xi^l \cdot \underset{\sim}{\nabla}\xi^k\} + F'\frac{\partial w}{\partial \xi^i} \tag{18}$$

Then

$$\nabla^2 \xi^i = \frac{1}{\det|\underset{\sim}{A}|} \sum_{k=1}^{3} C_{ik} S_k \tag{19}$$

where C_{ik} is the signed cofactor of A_{ki}.

If the weight function in the integral (13) is a function of $\underset{\sim}{x}$, rather than $\underset{\sim}{\xi}$, then $\partial F/\partial \xi^i = 0$ in the Euler equation (14), and Eq. (15) is replaced by

$$\sum_{k=1}^{3} \{[A_{ik} \nabla^2 \xi^k + A'_{ik} \underset{\sim}{\nabla} w \cdot \underset{\sim}{\nabla} \xi^k]$$

$$+ \sum_{m=1}^{3} \sum_{n=1}^{3} \frac{\partial A_{ik}}{\partial g^{mn}} [\underset{\sim}{\nabla}(\underset{\sim}{\nabla} \xi^m \cdot \underset{\sim}{\nabla} \xi^n)] \cdot \underset{\sim}{\nabla} \xi^k\} = 0 \tag{20}$$

In this case S_i of Eq. (18) are redefined as

$$S_i = - \sum_{k=1}^{3} \{ \sum_{m=1}^{3} \sum_{n=1}^{3} \frac{\partial A_{ik}}{\partial g^{mn}} [\underset{\sim}{\nabla}(\underset{\sim}{\nabla} \xi^m \cdot \underset{\sim}{\nabla} \xi^n)] \cdot \underset{\sim}{\nabla} \xi^k - A'_{ik} \underset{\sim}{\nabla} w \cdot \underset{\sim}{\nabla} \xi^k\} \tag{21}$$

APPENDIX C

CODE DEVELOPMENT AND COMPUTER EXERCISES

1. Code Development Exercises

1. Take the computational region to be a rectangle on which the curvilinear coordinates are defined to be equal to the indices of the field arrays, i.e., $\xi = I$ and $\eta = J$, with $x = X(I,J)$ and $y = Y(I,J)$.

   Make provision for reading in values of x and y on any segments of the boundary of the computational region. Generate x and y in the interior by interpolating linearly between the top and bottom boundaries. Plot the grid.

2. Modify the code to allow horizontal (in the computational plane) interpolation, as well, the choice being specified by input.

3. Now add the choice of interpolation from the four corners (tensor product interpolation).

4. Finally add the choice of transfinite interpolation.

5. Generalize the interpolation to cubic Hermite interpolation, with the grid being orthogonal at the boundary.

6. Generalize the interpolation to use the hyperbolic tangent distribution function, rather than being linear, with specified relative spacing on each end.

7. Modify the code to provide for reading in x and y on

any segment of any horizontal or vertical line in the computational region. Also provide for the interpolation to be done on any rectangular segment of the computational region (including a segment that is only a line.)

8. Add another field array TYPE (I,J) which is a flag to identify each point as one for which the x,y values are (1) fixed, e.g., specified points on the physical boundary, (2) out of the computation, e.g., points inside a slab, or (3) to be generated. Provide for the designation (1) and (2) to be made by input for any rectangular segment of the computational region, the default being to the designation (3).

9. Modify the dimensions of the field arrays so that an extra layer of points surrounds the computational region. Also add two more field arrays, ILINK (I,J) and JLINK (I,J). Provide for any segments of any horizontal or vertical lines to be designated as image points in TYPE by input, i.e., points for which the values of x and y are set equal to those at some other point. Also provide for the indices of these other points to be put in ILINK and JLINK by input.

10. Add an elliptic generator, based on Laplace equation, to the code. Use the algebraic generator (the interpolation) to provide the initial guess for point SOR iteration.

11. Add control functions to the elliptic generator. Let the control function be evaluated on the boundaries and interpolated into the field by transfinite interpolation.

2. Computer Exercises

1. Generate an algebraic grid between two concentric circles. Use linear interpolation between the circles.

2. Generate an algebraic grid between two ellipses, both of which are centered at the origin but which may have different eccentricities, using interpolation between the ellipses. Compare grids generated using linear and Hermite interpolation, the latter being orthogonal at the boundaries.

3. Generate a C-type algebraic grid for an ellipse inside an outer boundary formed by a semicircle replacing one side of a rectangle:

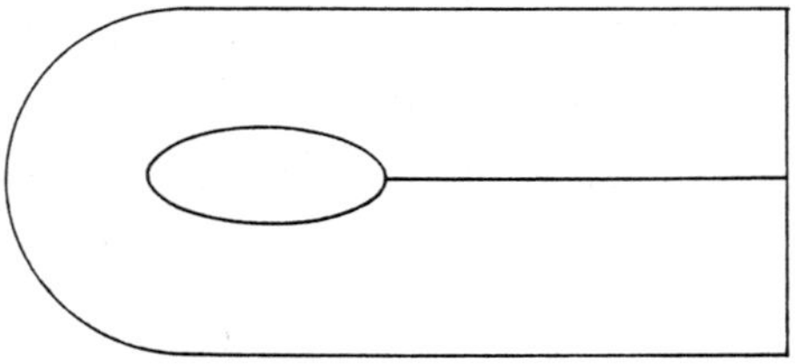

   Compare (1) vertical interpolation in the computational region boundary, (2) horizontal interpolation, (3) tensor product interpolation, and (4) transfinite interpolation, using linear interpolation in each case. Note that (2) and (3) are totally unreasonable.

4. Generate an algebraic grid for a circular simply-connected region by (1) unidirectional interpolation, (2) tensor product interpolation, and (3) transfinite interpolation. Note that here only (3) gives a reasonable grid. Compare linear and Hermite interpolation

for (3).

5. Repeat Exercise 4 with a triangular boundary.

6. Using the boundary configuration of Exercise 3, but with a hyperbolic tangent point distribution on the right-hand boundary of the physical region with smaller spacing at the centerline than at the top and bottom. Compare algebraic grids generated using (1) linear interpolation between the inner and outer boundaries, (2) nonlinear interpolation, based on the hyperbolic tangent, between the inner and outer boundaries, (3) transfinite interpolation with linear blending functions, and (4) transfinite interpolation using the boundary point distribution (in terms of relative arc length) as the blending functions. Note that only (2) and (4) preserve the boundary point distribution in the field.

7. Generate an algebraic grid for a square inside a rectangle using linear interpolation between the inner and outer boundaries. Note the propagation of the boundary slope discontinuities into the field. Generate a grid from an elliptic generation system for the same boundary point distribution and note the difference.

8. Generate an algebraic grid for a square inside a circle using linear interpolation between the inner and outer boundaries. Show that it is possible to position the points on the circle such that the grid overlaps the corners of the square. Generate a grid from an elliptic generation system for the same boundary point distribution and note the difference.

3. Listing of Routine for Computer Exercises

```
      SUBROUTINE INTERP
      PARAMETER(NI=20,NJ=20,N=5)
      COMMON/COORD/X(NI,NJ),Y(NI,NJ)
      COMMON/CONST/CHOICE,IMAX,JMAX,NA,DS1,DS2
      COMMON/ATTR/IAL(N),IAX(N),IAY(N),JAL(N),JAX(N),JAY(N)
      COMMON/COEF/AI(N),BI(N),CI(N),DI(N),AJ(N),BJ(N)
      COMMON/COEF/CJ(N),DJ(N)
      DIMENSION P(NI,NJ),Q(NI,NJ),XX(0:NI,0:NJ),YY(0:NI,0:NJ)
      DIMENSION X1(NI,NJ),X2(NI,NJ),Y1(NI,NJ),Y2(NI,NJ)
      INTEGER CHOICE
C
C     BOUNDARY INTERPOLATION
C
C       X      : X ARRAY OF XI-ETA COORDINATE
C       Y      : Y ARRAY OF XI-ETA COORDINATE
C       IMAX   : MAX. NUMBER OF GRID IN XI AXIS
C       JMAX   : MAX. NUMBER OF GRID IN ETA AXIS
C       NA     : MAX. NUMBER OF ATTRACTIONS
C       DS1    : SPECIFIED LENGTH OF INITIAL INTERVAL
C       DS2    : SPECIFIED LENGTH OF FINAL INTERVAL
C       ATTR   : ARRAY OF ATTRACTION TO LINES/POINTS
C       COEF   : ARRAY OF COEFFICIENT FOR ATTRACTION
C
C       CHOICE
C         1    : VERTICAL INTERPOLATION
C         2    : HORIZONTAL INTERPOLATION
C         3    : TENSOR PRODUCT INTERPOLATION
C         4    : TRANSFINITE INTERPOLATION
C         5    : HERMITE CUBIC INTERPOLATION
C         6    : HYPERBOLIC TANGENT INTERPOLATION
C         7    : ELLIPTIC GRID GENERATION ( SOR ITERATION )
C         8    : ATTRACTION TO COORDINATES
C
      IF(CHOICE.EQ.1) GO TO 100
      IF(CHOICE.EQ.2) GO TO 200
      IF(CHOICE.EQ.3) GO TO 300
      IF(CHOICE.EQ.4) GO TO 400
      IF(CHOICE.EQ.5) GO TO 500
      IF(CHOICE.EQ.6) GO TO 600
      IF(CHOICE.EQ.7) GO TO 700
      IF(CHOICE.EQ.8) GO TO 800
C
```

```
C    **** VERTICAL INTERPOLATION ****
C
 100  DO 110 I=1,IMAX
      DO 110 J=1,JMAX
      RJ1=FLOAT(JMAX-J)/FLOAT(JMAX-1)
      RJ2=FLOAT(J-1)/FLOAT(JMAX-1)
C    *** ( EQ. 8-1 )
      X(I,J)=RJ1*X(I,1)+RJ2*X(I,JMAX)
 110  Y(I,J)=RJ1*Y(I,1)+RJ2*Y(I,JMAX)
      RETURN
C
C    **** HORIZONTAL INTERPOLATION ****
C
 200  DO 210 I=1,JMAX
      DO 210 J=1,IMAX
      RI1=FLOAT(IMAX-I)/FLOAT(IMAX-1)
      RI2=FLOAT(I-1)/FLOAT(IMAX-1)
C    *** ( EQ. 8-1 )
      X(I,J)=RI1*X(1,J)+RI2*X(IMAX,J)
 210  Y(I,J)=RI1*Y(1,J)+RI2*Y(IMAX,J)
      RETURN
C
C    **** TENSOR PRODUCT INTERPOLATION ****
C
 300  DO 310 I=1,IMAX
      DO 310 J=1,JMAX
      RI1=FLOAT(IMAX-I)/FLOAT(IMAX-1)
      RI2=FLOAT(I-1)/FLOAT(IMAX-1)
      RJ1=FLOAT(JMAX-J)/FLOAT(JMAX-1)
      RJ2=FLOAT(J-1)/FLOAT(JMAX-1)
C    *** ( EQ. 8-69 )
      X(I,J)=RI1*RJ1*X(1,1)+RI1*RJ2*X(1,JMAX)
     * +RI2*RJ1*X(IMAX,1)+RI2*RJ2*X(IMAX,JMAX)
      Y(I,J)=RI1*RJ1*Y(1,1)+RI1*RJ2*Y(1,JMAX)
     * +RI2*RJ1*Y(IMAX,1)+RI2*RJ2*Y(IMAX,JMAX)
 310  CONTINUE
      RETURN
C
C    **** TRANSFINITE INTERPOLATION ****
C
 400  DO 410 I=1,IMAX
      DO 410 J=1,JMAX
      RI1=FLOAT(I-1)/FLOAT(IMAX-1)
      RI2=FLOAT(IMAX-I)/FLOAT(IMAX-1)
      X1(I,J)=RI1*X(IMAX,J)+RI2*X(1,J)
 410  Y1(I,J)=RI1*Y(IMAX,J)+RI2*Y(1,J)
      DO 420 I=1,IMAX
      DO 420 J=1,JMAX
```

```
      RJ1=FLOAT(J-1)/FLOAT(JMAX-1)
      RJ2=FLOAT(JMAX-J)/FLOAT(JMAX-1)
      X2(I,J)=RJ1*(X(I,JMAX)-X1(I,JMAX))+RJ2*(X(I,1)-X1(I,1))
420   Y2(I,J)=RJ1*(Y(I,JMAX)-Y1(I,JMAX))+RJ2*(Y(I,1)-Y1(I,1))
C   *** ( EQ. 8-73 )
      DO 430 I=1,IMAX
      DO 430 J=1,JMAX
      X(I,J)=X1(I,J)+X2(I,J)
430   Y(I,J)=Y1(I,J)+Y2(I,J)
      IF(CHOICE.NE.4) GO TO 740
      RETURN
C
C *** HERMITE CUBIC INTERPOLATION (ORTHOGONAL BOUNDARY) ***
C
500   DO 510 I=1,IMAX
      DO 510 J=1,JMAX
      XX(I,J)=X(I,J)
510   YY(I,J)=Y(I,J)
      DO 520 J=1,JMAX
      XX(0,J)=XX(IMAX-1,J)
      YY(0,J)=YY(IMAX-1,J)
      XX(IMAX+1,J)=XX(2,J)
520   YY(IMAX+1,J)=YY(2,J)
      DO 530 I=1,IMAX
      DO 530 J=1,JMAX
      RJJ=FLOAT(J-1)/FLOAT(JMAX-1)
C   *** ( EQ. 8-6 a and b, n=2 )
      PHI1=(1.+2.*RJJ)*(1.-RJJ)*(1.-RJJ)
      PHI2=(3.-2.*RJJ)*RJJ*RJJ
      PSI1=(1.-RJJ)*(1.-RJJ)*RJJ
      PSI2=(RJJ-1.)*RJJ*RJJ
C
C ** CAL. NORMAL DERIV. **
C
      XXI1=.5*(XX(I+1,1)-XX(I-1,1))
      XXI2=.5*(XX(I+1,JMAX)-XX(I-1,JMAX))
      YXI1=.5*(YY(I+1,1)-YY(I-1,1))
      YXI2=.5*(YY(I+1,JMAX)-YY(I-1,JMAX))
      UNIT1=SQRT(XXI1*XXI1+YXI1*YXI1)
      UNIT2=SQRT(XXI2*XXI2+YXI2*YXI2)
C   *** ( EQ. 3-108 )
      XN1=-YXI1/UNIT1*DS1
      XN2=-YXI2/UNIT2*DS2
      YN1=XXI1/UNIT1*DS1
      YN2=XXI2/UNIT2*DS2
C   *** ( EQ. 8-5 )
      XX(I,J)=PHI1*XX(I,1)+PHI2*XX(I,JMAX)+PSI1*XN1+PSI2*XN2
530   YY(I,J)=PHI1*YY(I,1)+PHI2*YY(I,JMAX)+PSI1*YN1+PSI2*YN2
```

```
      DO 540 I=1,IMAX
      DO 540 J=1,JMAX
      X(I,J)=XX(I,J)
 540  Y(I,J)=YY(I,J)
      RETURN
C
C   **** HYPERBOLIC TANGENT SPACING INTERPOLATION ****
C
 600  TOL=1.0E-10
C   *** ( EQ. 8-49, 50 and 51 )
      A=SQRT(DS2/DS1)
      B=1./(FLOAT(JMAX-1)*SQRT(DS1*DS2))
C   *** INITIAL GUESS BY SERIES EXPANSION
      DELTA=SQRT(6.*(B-1.))
      DO 610 IT=1,20
      RESID=SINH(DELTA)/(DELTA*B)-1.
      IF(ABS(RESID).LT.TOL) GO TO 630
 610  CALL AITKEN(DELTA,RESID,DELTO,RO,RSO)
      PRINT 620, RESID,DELTA,IT-1
 620  FORMAT(//, 5X, 'DELTA IS NOT CONVERGE ?', 5X, 2E15.5,
     * 5X, I3, //)
      GO TO 660
 630  CONTINUE
C   *** ( EQ. 8-52, 53 and 54 )
      DO 650 I=1,IMAX
      DO 650 J=2,JMAX-1
      RATIO=FLOAT(J-1)/FLOAT(JMAX-1)
      U=.5*(1.+TANH(DELTA*(RATIO-.5))/TANH(.5*DELTA))
      S=U/(A+(1.-A)*U)
      X(I,J)=X(I,1)+(X(I,JMAX)-X(I,1))*S
 650  Y(I,J)=Y(I,1)+(Y(I,JMAX)-Y(I,1))*S
 660  RETURN
C
C   **** ELLIPTIC GRID GENERATION ( SOR ITERATION ) ****
C   *** CAL. P AND Q ON THE BOUNDARY **
C
 700  DO 710 I=1,IMAX
      DO 710 J=1,JMAX
      XXI=.5*(X(I+1,J)-X(I-1,J))
      XXIXI=X(I+1,J)-2.*X(I,J)+X(I-1,J)
      XETA=.5*(X(I,J+1)-X(I,J-1))
      XETA2=X(I,J+1)-2.*X(I,J)+X(I,J-1)
      YXI=.5*(Y(I+1,J)-Y(I-1,J))
      YXIXI=Y(I+1,J)-2.*Y(I,J)+Y(I-1,J)
      YETA=.5*(Y(I,J+1)-Y(I,J-1))
      YETA2=Y(I,J+1)-2.*Y(I,J)+Y(I,J-1)
      IF(ABS(XETA2).LT.10E-3) XETA2=0.
      IF(ABS(YETA2).LT.10E-3) YETA2=0.
```

```
      RXI2=XXI*XXI+YXI*YXI
      RETA2=XETA*XETA+YETA*YETA
C   *** ( EQ. 8-70 )
      P(I,J)=(XXI*XXIXI+YXI*YXIXI)/RXI2
      Q(I,J)=(XETA*XETA2+YETA*YETA2)/RETA2
 710  CONTINUE
C
C ** INTERPOLATE P AND Q BETWEEN BOUNDARY **
C        P : VERTICAL, Q : HORIZONTAL
C
      DO 720 I=1,IMAX
      DO 720 J=1,JMAX
      RJ1=FLOAT(JMAX-J)/FLOAT(JMAX-1)
      RJ2=FLOAT(J-1)/FLOAT(JMAX-1)
      RI1=FLOAT(IMAX-I)/FLOAT(IMAX-1)
      RI2=FLOAT(I-1)/FLOAT(IMAX-1)
      P(I,J)=RJ1*P(I,1)+RJ2*P(I,JMAX)
      Q(I,J)=RI1*Q(1,J)+RI2*Q(IMAX,J)
 720  CONTINUE
C
C ** INITIAL GUESS WITH TRANSFINITE INTEPOLATION **
C
      GO TO 400
 740  CONTINUE
C
C   *** ITERATION ( SOR ) ***
C
      ITMAX=200
      TOL=10.E-5
      W=1.8
      DO 760 IT=1,ITMAX
      ERRX=0.
      ERRY=0.
      DO 750 J=2,JMAX-1
      DO 750 I=2,IMAX-1
      XXI=.5*(X(I+1,J)-X(I-1,J))
      YXI=.5*(Y(I+1,J)-Y(I-1,J))
      XXIXI=X(I+1,J)+X(I-1,J)
      YXIXI=Y(I+1,J)+Y(I-1,J)
      XETA=.5*(X(I,J+1)-X(I,J-1))
      YETA=.5*(Y(I,J+1)-Y(I,J-1))
      XXIETA=.25*(X(I+1,J+1)-X(I+1,J-1)-X(I-1,J+1)+X(I-1,J-1))
      YXIETA=.25*(Y(I+1,J+1)-Y(I+1,J-1)-Y(I-1,J+1)+Y(I-1,J-1))
      XETA2=X(I,J+1)+X(I,J-1)
      YETA2=Y(I,J+1)+Y(I,J-1)
C   *** ( EQ. 6-18 and 6-20 )
      G11=XXI*XXI+YXI*YXI
      G22=XETA*XETA+YETA*YETA
```

```
      G12=XXI*XETA+YXI*YETA
      XTEMP=.5*(G22*(P(I,J)*XXI+XXIXI)+G11*(Q(I,J)*XETA+XETA2)
     * -2.*G12*XXIETA)/(G11+G22)
      YTEMP=.5*(G22*(P(I,J)*YXI+YXIXI)+G11*(Q(I,J)*YETA+YETA2)
     * -2.*G12*YXIETA)/(G11+G22)
      XTEMP=W*XTEMP+(1.-W)*X(I,J)
      YTEMP=W*YTEMP+(1.-W)*Y(I,J)
      ERRX=AMAXO(ERRX,ABS(XTEMP-X(I,J)))
      ERRY=AMAXO(ERRY,ABS(YTEMP-Y(I,J)))
      X(I,J)=XTEMP
      Y(I,J)=YTEMP
  750 CONTINUE
      IF(ERRX.LT.TOL.AND.ERRY.LT.TOL) GO TO 780
  760 CONTINUE
      PRINT 770,ERRX,ERRY,IT-1
  770 FORMAT(//, 5X, 'X AND Y ARE NOT CONVERGE ?', 2E15.5,
     * 5X, I5, //)
  780 CONTINUE
      IF(CHOICE.EQ.8) GO TO 830
      RETURN
C
C    **** ATTRACTION TO COORDINATE LINE/POINT ****
C
  800 DO 810 I=1,IMAX
      DO 810 J=1,JMAX
      P(I,J)=0.
  810 Q(I,J)=0.
      DO 820 NS=1,NA
      DO 820 I=1,IMAX
      DO 820 J=1,JMAX
      XL=FLOAT(I-IAL(NS))
      XI=FLOAT(I-IAX(NS))
      XJ=FLOAT(J-IAY(NS))
      YL=FLOAT(J-JAL(NS))
      YI=FLOAT(I-JAX(NS))
      YJ=FLOAT(J-JAY(NS))
C    *** ( EQ. 6-30 )
      P(I,J)=P(I,J)-AI(NS)*(XL/ABS(XL))*EXP(-CI(NS)*ABS(XL))
     * -BI(NS)*(XI/ABS(XI))*EXP(-DI(NS)*SQRT(XI*XI+XJ*XJ))
      Q(I,J)=Q(I,J)-AJ(NS)*(YL/ABS(YL))*EXP(-CJ(NS)*ABS(YL))
     * -BJ(NS)*(YJ/ABS(YJ))*EXP(-DJ(NS)*SQRT(YI*YI+YJ*YJ))
  820 CONTINUE
      GO TO 400
  830 CONTINUE
      RETURN
      END
```

4. Examples for Computer Exercises

1.

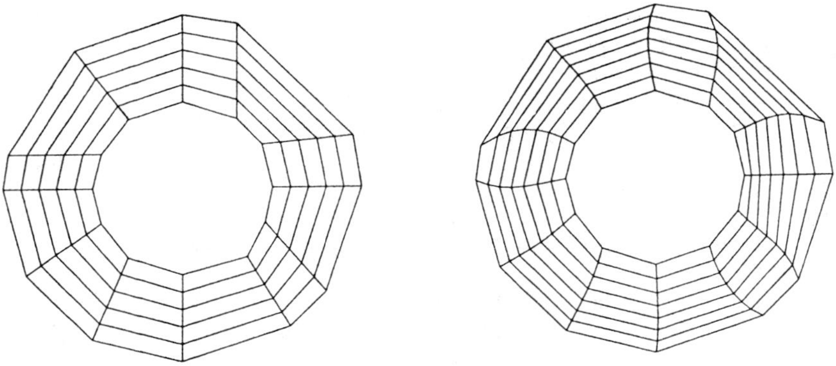

2.

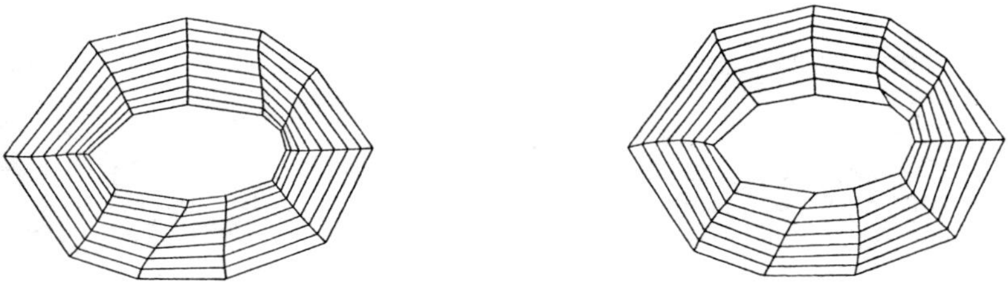

3.

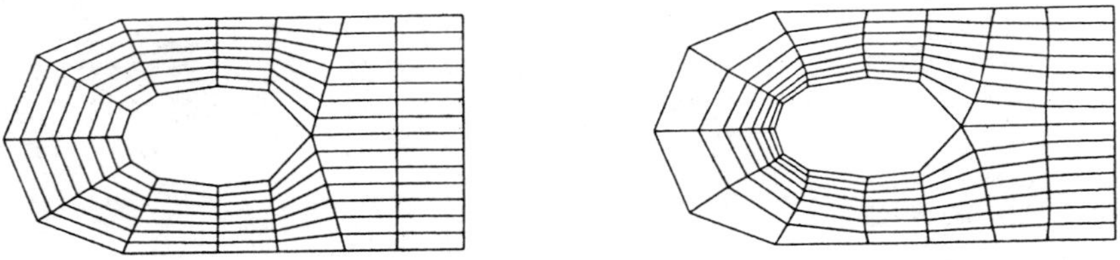

4.

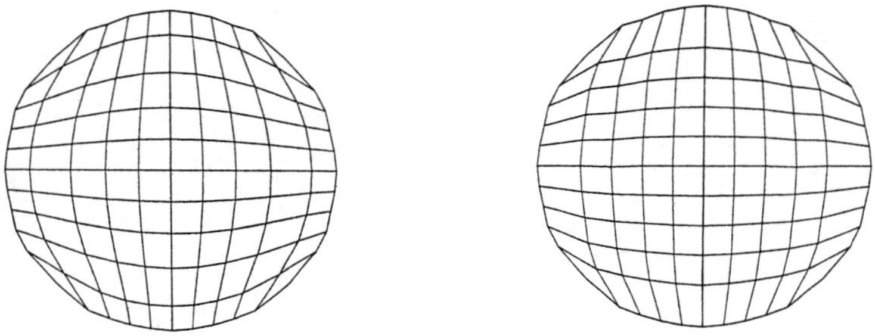

5.

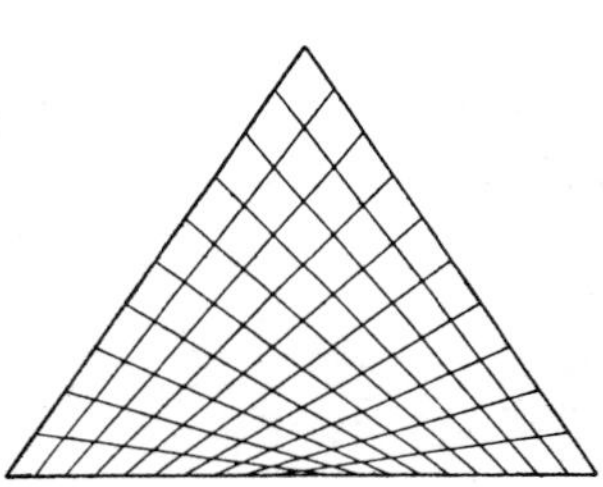

6.

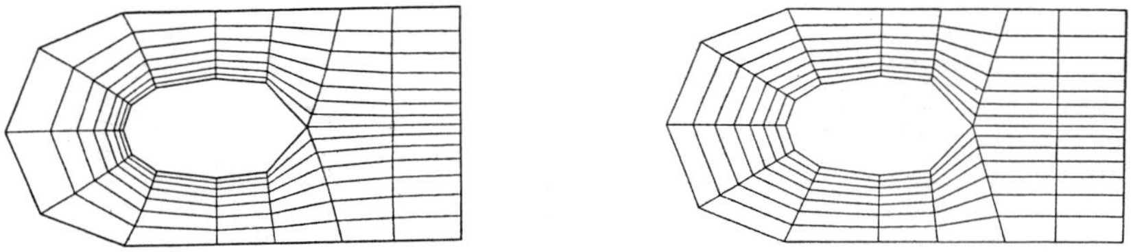

7.

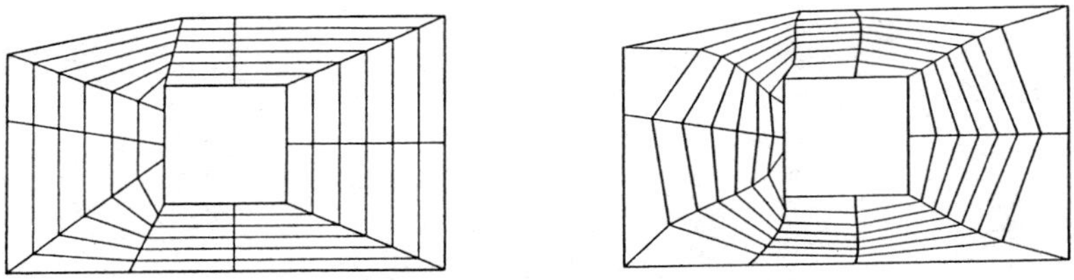

8.

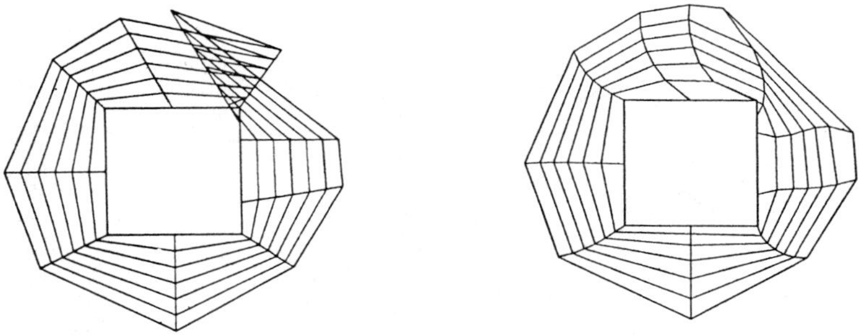

Attraction

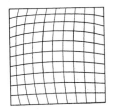

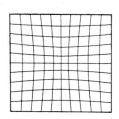

REFERENCES

1. Thompson, Joe F., Warsi, Z. U. A. and Mastin, C. W., "Boundary-Fitted Coordinate Systems for Numerical Solution of Partial Differential Equations - A Review", Journal of Computational Physics, 47, 1, 1982.

2. Thompson, Joe F. (Ed.) Numerical Grid Generation, North-Holland 1982. (Also published as Vol. 10 11 of Applied Mathematics and Computation, 1982).

3. Smith, Robert E., (Ed), Numerical Grid Generation Techniques, NASA Conference Publication 2166, NASA Langley Research Center, 1980.

4. Ghia, K. N. and Ghia, U., (Ed), Advances in Grid Generation, FED-Vol. 5, ASME Applied Mechanics, Bioengineering, and Fluids Engineering Conference, Houston, 1983.

5. Thompson, Joe F., "Grid Generation Techniques in Computational Fluid Dynamics", AIAA Journal, 22, 1505, 1984.

6. Halsey, Douglas, "Conformal Grid Generation for Multi-element Airfoils", Numerical Grid Generation, Ed. Joe F. Thompson, North-Holland, 585, 1982.

7. Ives, David C., "Conformal Grid Generation", Numerical Grid Generation, Ed. Joe F. Thompson, North-Holland, 107, 1982.

8. Smith, Robert E., "Three-Dimensional Algebraic Grid Generation", AIAA-83-1904, AIAA 6th Computational Fluid

Dynamics Conference, Danvers, Massachusetts, 1983.

9. Thompson, J. F. and Warsi, Z. U. A., "Three-Dimensional Grid Generation from Elliptic Systems", AIAA-83-1905, AIAA 6th Computational Fluid Dynamics Conference, Danvers, Massachussetts, 1983.

10. Coleman, Roderick M., "Generation of Boundary-Fitted Coordinate Systems Using Segmented Computational Regions", Numerical Grid Generation, Ed. Joe F. Thompson, North-Holland, 633, 1982.

11. Rubbert, P. E. and Lee, K. D., "Patched Coordinate Systems", Numerical Grid Generation, Ed. Joe F. Thompson, North-Holland, 235, 1982.

12. Thomas, P. D., "Numerical Generation of Composite Three-Dimensional by Quasilinear Elliptic Systems", Numerical Grid Generation, Ed. Joe F. Thompson, North-Holland, 667, 1982.

13. Miki, Kazuyoshi and Takagi, Toshiyuki, "A Domain Decomposition and Overlapping Method for the Generation of Three-Dimensional Boundary-Fitted Coordinate Systems", Journal of Computational Physics, 53, 319, 1984.

14. Thompson, Joe F., unpublished research, 1984.

15. Warsi, Z. U. A., "Tensors and Differential Geometry Applied to Analytic and Numerical Coordinate Generation", MSUU-EIRS-81-1, Mississippi State University, 1981.

16. Eiseman, P. R., "Geometric Methods in Computational Fluid Dynamics", ICASE 80-11, NASA Langley Research

Center, 1980.

17. Thompson, Joe F. and Mastin, C. Wayne, "Order of Difference Expressions on Curvilinear Coordinate Systems", Advances in Grid Generation, FED-Vol. 5, Ed. K. N. Ghia and U. Ghia, ASME Applied Mechanics, Bioengineering, and Fluids Engineering Conference, Houston, 1983.

18. Vinokur, Marcel, "On One-Dimensional Stretching Functions for Finite-Difference Calculations", Journal of Computational Physics, 50, 215, 1983.

19. Warsi, Z. U. A., "Basic Differential Models for Coordinate Generation", Numerical Grid Generation, Ed. Joe F. Thompson, North-Holland, 41, 1982.

20. Mastin, C. Wayne and Thompson, Joe F., "Elliptic Systems and Numerical Transformations", Journal of Mathematical Analysis and Applications, 62, 52, 1978.

21. Mastin, C. Wayne and Thompson, Joe F., "Transformation of Three-Dimensional Regions onto Rectangular Regions by Elliptic Systems", Numerische Mathematik, 29, 397, 1978.

22. Mastin, C. Wayne and Thompson, Joe F., "Discrete Quasiconformal Mappings", Journal of Applied Mathematics and Physics (ZAMP), 29, 1978.

23. Mastin, C. W. and Thompson, J. F., "Quasiconformal Mappings and Grid Generation", SIAM Journal On Scientific and Statistical Computing, 5, 305, 1984.

24. Sorenson, R. L., "A Computer Program to Generate Two-Dimensional Grids About Airfoils and Other Shapes by

the Use of Poisson's Equations", NASA Ames Research Center, NASA TM 81198, 1980.

25. Sorenson, Reese L., "Grid Generation by Elliptic Partial Differential Equations for a Tri-Element Augmentor-Wing Airfoil", Numerical Grid Generation, Ed. Joe F. Thompson, North-Holland, 653, 1982.

26. Sorenson, R. L. and Steger, J. L., "Grid Generation in Three Dimensions by Poisson Equations with Control of Cell Size and Skewness at Boundary Surfaces", Advances in Grid Generation, FED-Vol. 5, Ed. K. N. Ghia and U. Ghia, ASME Applied Mechanics, Bioengineering, and Fluids Engineering Conference, Houston, 1983.

27. Warsi, Z. U. A., "A Note on the Mathematical Formulation of the Problem of Numerical Coordinate Generation", Quarterly of Applied Mathematics, 41, 221, 1983.

28. Steger, J. L. and Chaussee, D. S., "Generation of Body Fitted Coordinates Using Hyperbolic Partial Differential Equations", SIAM J. Sci. Stat. Comput. 1, 431, 1980.

29. Steger, J. L. and Sorenson, R. L., "Use of Hyperbolic Partial Differential Equations to Generate Body Fitted Coordinates", Numerical Grid Generation Techniques, Ed. Robert E. Smith, NASA-CP-2166, 463, 1980.

30. Nakamura, S., "Marching Grid Generation Using Parabolic Partial Differential Equations", Numerical Grid Generation, Ed. Joe F. Thompson, North-Holland, 775, 1982.

31. Smith, Robert E., "Algebraic Grid Generation", Numer-

ical Grid Generation, Ed. Joe F. Thompson, North-Holland, 137, 1982.

32. Eiseman, P. R., "A Multi-Surface Method of Coordinate Generation", Journal of Computational Physics, 33, 118, 1979.

33. Eiseman, Peter R. and Smith, Robert, "Mesh Generation Using Algebraic Techniques", Numerical Grid Generation Techniques, Ed. Robert E. Smith, NASA CP-2166, 1980.

34. Eiseman, Peter R., "Automatic Algebraic Coordinate Generation", Numerical Grid Generation, Ed. Joe F. Thompson, North-Holland, 447, 1982.

35. Eiseman, Peter R., "Coordinate Generation with Precise Controls Over Mesh Properties", Journal of Computational Physics, 47, 331, 1982.

36. Eiseman, Peter R., "High Level Continuity for Coordinate Generation with Precise Controls", Journal of Computational Physics, 47, 352, 1982.

37. Eiseman, Peter R. "Grid Generation for Fluid Mechanics Computations", Annual Review of Fluid Mechanics, Vol. 17, 1985.

38. Eiseman, Peter R. unpublished result referred to in Ref. [33].

39. Roberts, A., "Automatic Topology Generation and Generalized B-Spline Mapping", Numerical Grid Generation, Ed. Joe F. Thompson, North-Holland, 465, 1982.

40. Gordon, William J. and Thiel, Linda C., "Transfinite Mappings and Their Application to Grid Generation",

Numerical Grid Generation, Ed. Joe F. Thompson, North-Holland, 171, 1982.

41. Gordon, W. J. "Blending Function Methods of Bivariate and Multivariate Interpolation", SIAM J. of Numerical Analysis, 8, 158, 1971.

42. Eiseman, Peter R., "Orthogonal Grid Generation", Numerical Grid Generation, Ed. Joe F. Thompson, North-Holland, 193, 1982.

43. Ives, D. C. and Siddons, W. D., "Orthogonal Grid Generation", AIAA-84-1248, AIAA/SAE/ASME 20th Joint Propulsion Conference, Cincinnati, 1984.

44. Kober, H., Dictionary of Conformal Representations, Dover, New York, 1952.

45. Thompson, Joe F., "A Survey of Dynamically-Adaptive Grids in the Numerical Solution of Partial Differential Equations", to appear in Journal of Numerical Mathematics, 1984. (also AIAA-84-1606, AIAA Fluid and Plasma Dynamics Conference, Snowmass, Colorado, 1984).

46. Dwyer, H. A., Smooke, Mitchell, D. and Kee, Robert J., "Adaptive Gridding for Finite Difference Solutions to Heat and Mass Transfer Problems", Numerical Grid Generation, Ed. Joe F. Thompson, North-Holland, 339, 1982.

47. Brackbill, J. U. and Saltzman, J. S., "Adaptive Zoning for Singular Problems in Two Dimensions", Journal of Computational Physcis, 46, 342, 1982.

48. Anderson, Dale, A., and Rai, M. M., "The Use of Solution Adaptive Grids in Solving Partial Differential

Equations", Numerical Grid Generation, ed. Joe F. Thompson, North-Holland, 317, 1982.

49. Miller, Keith and Miller, Robert N., "Moving Finite Elements. I", SIAM Journal of Numerical Analysis, 18, 1019, 1981.

50. Miller, Keith, "Moving Finite Elements. II", SIAM Journal of Numerical Analysis, 18, 1033, 1981.

51. Weatherill, N. C. and Forsey, C. R. "Grid Generation and Flow Calculations for Complex Aircraft Geometrics Using a Multi-Block Scheme", AIAA-84-1665, AIAA 17th Fluid Dynamics, Plasma Dynamics, and Lasers Conference, Snowmass, CO, 1984.

52. Rai, M. M. "A Conservative Treatment of Zonal Boundaries for Euler Equations Calculations", AIAA-84-0164, AIAA 22nd Aerospace Sciences Meeting, Reno, NV, 1984.

53. Hessenius, K. A. and Rai, M. M., "Applications of a Conservative Zonal Scheme to Transient and Geometrically Complex Problems", AIAA-84-1532, AIAA 17th Fluid Dynamics, Plasma Dynamics, and Lasers Conference, Snowmass, CO, 1984.

GRID ILLUSTRATION REFERENCES

The grids used for illustration are drawn from those in the works cited here:

Page

20 Anderson, O. L., Davis, R. T., Hankins, G. B., and Ewards, D. E., "Solution of Viscous Internal Flows on Curvilinear Grids Generated by the Schwarz-Christoffel Transformation." in Ref. [2].

21 Kumar, D., Hester, L. R., and Thompson, J. F., "Development of Partial Channel Flow for Arbitrary Input Velocity Distribution Using Boundary-Fitted Coordinate Systems", in Nonsteady Fluid Dynamics, ASME Winter Annual Meeting, San Francisco, 53, 1978.

22, 23 McWhorter, John C., "Solid Mechanics Applications of Boundary Fitted Coordinate Systems", in Ref. [2].

24, 25 Lee, K. D., Huang, Yu, N. J., and Rubbert, P. E., "Grid Generation for General Three-Dimensional Configurations", in Ref. [3].

27, 28 Coleman, Roderick M., "Generation of Boundary-Fitted Coordinate Systems Using Segmented Computational Regions", in Ref. [2].

29 Thompson, J. F., Thames, F. C., and Mastin, C. W., "'TOMCAT' - A Code for Numerical Generation of Boundary- Fitted Curvilinear Coordinate Systems on Fields Containing any Number of Arbitrary Two-Dimensional Bodies", Journal of Computational Physics,

245, 1977.

30 Reddy, R. N. and Thompson, Joe F., "Numerical Solution of Incompressible Navier-Stokes Equations in the Integro-Differential Formulation Using Boundary-Fitted Coordinate Systems", Proceedings of the AIAA 3rd Computational Dynamics Conference, Albuquerque, 1977.

33 Thompson, J. F., "A Boundary-Fitted Coordinate Code for General Two-Dimensional Regions with Obstacles and Boundary Intrusions", Technical Report E-83-8, U.S. Army Engineer Waterways Experiment Station, Vicksburg, Mississippi, 1983.

34 Thompson, J. F., Thames, F. C., and Mastin, C. W., "'TOMCAT' - A code for Numerical Generation of Boundary-Fitted Curvilinear Coordinate Systems on Fields Containing any Number of Arbitrary Two-Dimensional Bodies", Journal of Comptuational Physics, 245, 1977.

35, 37 Thompson, J. F., Thames, F. C., and Mastin, C. W., "'TOMCAT' - A Code for Numerical Generation of Boundary-Fitted Curvilinear Coordinate Systems on Fields Containing any Number of Arbitrary Two-Dimensional Bodies", Journal of Computational Physics, 245, 1977.

38 Halsey, Douglas, "Conformal Grid Generation for Multi-Element Airfoils", in Ref. [2].

39 Long, W. S., "Two-Body Coordinate System Generation Using Body-Fitted Coordinate System and Complex Variable Transformation", M.S. thesis, Mississippi State University, 1977.

40 Coleman, R. M. "NUMESH: A Computer Program to Generate Finite Difference Meshes for Arbitrary Double-Connected Two-Dimensional Regions," CMLD-77-05, David W. Taylor Naval Ship Research and Development Center, 1977.

45 Haussling, Henry J., "Solution of Nonlinear Water Wave Problems Using Boundary-Fitted Coordinate Systems," in Ref. [2].

48 Sorenson, Reese L., "Grid Generation by Elliptic Partial Differential Equations for a Tri-Element Augmentor-Wing Airfoil," in Ref. [2].

52 Thompson, Joe F., General Curvilinear Coordinate Systems," in Ref. [2].

53a Chen, Brian C-J, Sha, W. F., Doria, M. L., Schmidt, R. C., and Thompson, J. F., "BODYFIG-IFE: A Computer Code for the Three-Dimensional Steady-State/Transient Single-Phase Rod-Bundle Thermal-Hydraulic Analysis, NUREG/CR-1874, ANL-80-127, Argonne National Laboratory, 1980.

54 Dulikravich, Djordje S., "Fast Generation of Three-Dimensional Computational Boundary-Conforming Periodic Grids of C-Type," in Ref. [2].

54 Rai Man Mohan, "An Implicit, Conservative, Zoned-Boundary Scheme for Euler Equation Calculations", AIAA-85-0488, AIAA 23rd Aerospace Sciences Meeting, Reno, 1985.

55 Ives, David D., "Conformal Grid Generation," in Ref. [2].

55 Rai, Man Mohan, "A Relaxation Approach to Patched-Grid Calculations with the Euler quations", AIAA-85-0295, AIAA 23rd Aerospace Sciences Meeting, Reno, 1985.

56 Eiseman, P. R., "Alternating Direction Adaptive Grid Generation", AIAA-83-1937, AIAA 6th Computational Fluid Dynamics Conference, Danvers, Mass. 1983.

57 Nakahashi, Kazuhiro and Deiwert, George S., "A Practical Adaptive Grid Method for Complex Fluid-Flow Problems", NASA TM 85989, NASA Ames Research Center, 1984.

57a Jain, Sunil K., "Embedded-Grid Generation with Complete Continuity Across Interfaces for Multi-Element Airfoils." ASE 84-270, Ph-D Dissertation, Mississippi State University, 1984.

57b Sorenson, Reese L., "Grid Generation by Elliptic Partial Differential Equations for a Tri-Element Augmentor-Wing Airfoil," in Ref. [2].

57c Rubbert, P. E. and Lee, K. D., "Patched Coordinate Systems," in Ref. [2].

58 Halsey, Douglas, "Conformal Grid Generation for Multi-element Airfoils," in Ref. [2].

65a Jain, Sunil K., "Embedded-Grid Generation With Complete Contintuity Across Interfaces for Multi-Element Airfoils." ASE 84-270, Ph-D Dissertation, Mississippi State University, 1984.

65b Sorenson, Reese L., "Grid Generation by Elliptic Partial Differential Equations for a Tri-Element Augmentor-Wing Airfoil," in Ref. [2].

69a Miki, Kazuyoshi and Takagi, Toshiyuki, "A Domain Decom- position and Overlapping Method for the Generation of Three-Dimensional Boundary-Fitted Coordinate Systems", Journal of Computational Physics, 53, 319, 1984.

69b Steger, J. L., Dougherty, F. C., and Benek, J. A., "A Chimera Grid Scheme," Advances in Grid Generation, ASME Fluids Engineering Conference, Houston, June 1983.

190a Johnson, Billy H. and Thompson, Joe F., "A Discussion of Boundary-Fitted Coordinate Systems and Their Applicability to the Numerical Modeling of Hydraulic Problems," Miscellaneous Paper H-78-9, U. S. Army Engineer Waterways Experiment Station, Vicksburg, Mississippi, 1978.

190b Chae, Yeon Seok, "An Investigation of a Navier-Stokes Solution for Quasi-Three-Dimensional flow". ASE 84-282, M.S. thesis, Mississippi State University, 1984.

193 Thompson, J. F., Thames, F. C., and Mastin, C. W., "Automatic Numerical Generation of Body-Fitted Curvilinear Coordinate System for Fields Containing any Number of Arbitrary Two-Dimensional Bodies," Journal of Computatonal Physics, 15, 299, 1974.

194 Thompson, J. F., Thames, F. C., and Mastin, C. W., "TOMCAT" - A Code for Numerical Generation of Boundary-Fitted Curvilinear Coordinate Systems on Fields Containing any Number of Arbitrary Two-Dimensional Bodies," Journal of Computational Physics, 245, 1977.

196 Shanks, S. P. and Thompson, J. F., "Numerical Solution of the Navier-Stokes Equation for 2D Hydrofoil in or Below a Free Surface," Proceedings of the 2nd International Conference on Numerical Ship Hydrodynamics, Berkeley, 1977.

208 Thompson, Joe F., General Curvilinear Coordinate Systems," in Ref. [2].

232a Sorenson, Reese L., "Grid Generation by Elliptic Partial Differential Equations for a Tri-Element Augmentor-Wing Airfoil," in Ref. [2].

232b Sorenson, R. L., "A Computer Program to Generate Two-Dimensional Grids About Airfoils and Other Shapes by the Use of Poisson's Equations," NASA TM 81198, 1980.

350a Ives, David C., "Conformal Grid Generation," in Ref. [2].

350b Dulikravich, Djordje S., "Fast Generation of Three-Dimensional Computational Boundary-Conforming Periodic Grids of C-Type," in Ref. [2].

383 Dywer, H. A., Smooke, Mitchell, D. and Kee, Robert J., "Adaptive Gridding for Finite Difference Solutions to Heat and Mass Transfer Problems", Numerical Grid Generation, Ed. Joe F. Thompson, North-Holland, 339, 1982.

388, 389a Dwyer, H. A., Smooke, Mitchell, D. and Kee, Robert J., "Adaptive Gridding for Finite Difference Solutions to Heat and Mass Transfer Problems," in Ref. [2].

389b Gnoffo, Peter A., "A Vectorized Finite-Volume, Adaptive Grid Algorithm for Navier-Stokes

Calculations," in Ref. [2].

390 Nakamura, S., "Adaptive Grid Relocation Algorithm for Transonic Full Potential Calculators Using One-Dimensional or Two-Dimensional Diffusion Equations," Advances in Grid Generation, ASME Fluids Engineering Conference, Houston, 1983.

411, 412 Saltzman, Jeffery and Brackbill, Jeremiah, "Applications and Generalizations of Variational Methods for Generating Adaptive Meshes," in Ref. [2].

350a Ives, David C., "Conformal Grid Generation," in Ref. [2].

350b Dulikravich, Djordje S., "Fast Generation of Three- Dimensional Computational Boundary-Conforming Periodic Grids of C-Type," in Ref. [2].

413 Bell, J. B. and Shulin, G. R., "An Adaptive Grid Finite Difference Method for Conservation Laws", Journal of Computational Physics, 52, 569, 1983.

INDEX